Photochemical Conversion
and Storage of Solar Energy

PHOTOCHEMICAL CONVERSION AND STORAGE OF SOLAR ENERGY

Edited by

John S. Connolly
Solar Energy Research Institute
Golden, Colorado

1981   ACADEMIC PRESS

A Subsidiary of Harcourt Brace Jovanovich, Publishers

New York   London   Toronto   Sydney   San Francisco

CHEMISTRY

6871 - 4415

Proceedings of the Third International Conference on Photochemical
Conversion and Storage of Solar Energy Held in Boulder, Colorado on
August 3–8, 1980 Sponsored by the Solar Energy Research Institute
Golden, Colorado and the Division of Chemical Sciences Office of
Basic Energy Sciences U.S. Department of Energy, Washington, D.C.

ACADEMIC PRESS, INC.
111 Fifth Avenue, New York, New York 10003

United Kingdom Edition published by
ACADEMIC PRESS, INC. (LONDON) LTD.
24/28 Oval Road, London NW1 7DX

Library of Congress Cataloging in Publication Data
Main entry under title:

Photochemical conversion and storage of solar energy.

    "Preceedings of the third international conference,
Boulder, Colorado, August 3-8, 1980."
    Includes bibliographical references and index.
    1. Solar energy--Congresses. 2. Photochemistry--
Industrial applications--Congresses. 3. Energy storage
--Congresses. I. Connolly, John S.
TJ810.P48   621.47'1   81-12853
ISBN 0-12-185880-4   AACR2

PRINTED IN THE UNITED STATES OF AMERICA

81 82 83 84   9 8 7 6 5 4 3 2 1

Contents

Contributors   vii
Preface   ix

1. Simulating Photosynthetic Quantum Conversion   1
   Melvin Calvin

2. Current Status of Biomimetic Systems for Solar Energy
   Conversion   27
   Joseph J. Katz and James C. Hindman

3. Quantum Harvesting and Energy Transfer   79
   Christian K. Jørgensen

4. Photochemical Electron Transfer Reactions in Homogeneous
   Solutions   97
   Vincenzo Balzani and Franco Scandola

5. Photoinduced Water Splitting in Heterogeneous Solution   131
   Michael Grätzel

6. Photoinduced Generation of Hydrogen and Oxygen from
   Water   161
   Jean-Marie Lehn

7. Photogalvanic Cells and Effects   201
   Mary D. Archer and M. Isabel C. Ferreira

8. Chemical Aspects of Photovoltaic Cells   229
   Sigurd Wagner

9. Electrochemical Photovoltaic Cells   243
   Rüdiger Memming and John J. Kelly

10. Photoelectrosynthesis at Semiconductor Electrodes   **271**
     Arthur J. Nozik

11. Photochemical Energy Storage: An Analysis of Limits   **297**
     James R. Bolton, Alan F. Haught, and Robert T. Ross

General Discussion   **341**

Contributed Papers   **347**

Index   *437*

Contributors

Numbers in parentheses indicate the pages on which the authors' contributions begin.

Mary D. Archer (201), Department of Physical Chemistry, University of Cambridge, Lensfield Road, Cambridge CB2 1EP, United Kingdom

Vincenzo Balzani (97), Istituto Chimico "G. Ciamician," Università di Bologna, 40216 Bologna, Italy

James R. Bolton (297), Photochemistry Unit, Department of Chemistry, The University of Western Ontario, London, Ontario N6A 5B7, Canada

Melvin Calvin (1), Chemical Biodynamics Division, Lawrence Berkeley Laboratory, University of California, Berkeley, California 94720

M. Isabel C. Ferreira (201), Ciencias Exactas e Tecnologia, Universidade do Minho, Braga, Portugal

Michael Grätzel (131), Institut de Chimie Physique, Ecole Polytechnique Fédérale, CH-1015 Lausanne, Switzerland

Alan F. Haught (297), United Technologies Research Center, Silver Lane, East Hartford, Connecticut 06108

James C. Hindman (27), Chemistry Division, Argonne National Laboratory, 9700 South Cass Avenue, Argonne, Illinois 60439

Christian K. Jørgensen (79), Départment de Chimie minerale, analytique et appliquée, Université de Genéve, CH 1211 Geneva, 4 Switzerland

Joseph J. Katz (27), Chemistry Division, Argonne National Laboratory, 9700 South Cass Avenue, Argonne, Illinois 60439

John J. Kelly (243), Philips Research Laboratories, Eindhoven, The Netherlands

Jean-Marie Lehn (161), Institut Le Bel, Université Louis Pasteur, 6700 Strasbourg, France

Rüdiger Memming (243), Philips GmbH Forschungs-Laboratorium Hamburg, Vogt-Kölln-Strasse 30, Postfach 54 08 40, 2000 Hamburg 54, Federal Republic of Germany

Arthur J. Nozik (271), Photoconversion Research Branch, Solar Energy Research Institute, 1617 Cole Boulevard, Golden, Colorado 80401

Robert T. Ross (297), Department of Biochemistry, The Ohio State University, 484 West 12th Avenue, Columbus, Ohio 43210

Franco Scandola (97), Istituto Chimico, Università di Ferrara, Ferrara, Italy

Sigurd Wagner (229), Department of Electrical Engineering and Computer Science, Princeton University, Princeton, New Jersey 08544

Preface

The energy crises of the 1970s demonstrated perhaps more than any event since World War II the fragility of civilized society. Nevertheless, our continued dependence on and profligate use of fossil fuels have led to rampant inflation that has brought the entire world to the brink of disaster.

The items on the international agenda include not only energy, but also the interrelated problems of population, the global increase of atmospheric carbon dioxide, and the possibility that fresh water may prove to be the limiting resource of the planet. Yet the use of our most abundant energy source, the sun, has increased only marginally over the past decade.

To increase the world-wide utilization of solar energy will require the resolve of every nation—developing as well as developed—to employ existing solar technologies and to contribute, according to their abilities, to the fundamental and applied research needed to lay the basis for the technologies of the future.

Photoconversion mechanisms, i.e., direct quantum conversion processes, offer the best hope for discovering new molecular systems and, eventually, developing the devices needed to harness the sun's energy cleanly and efficiently. Photoconversion embraces the areas of photobiology, photochemistry, and photoelectrochemistry as well as the more mature field of photovoltaics.

Natural photosynthesis, of course, has the advantage of some three billion years of evolutionary development of elegant molecules sequestered in equally elegant macromolecular structures that carry out specific reactions at efficiencies that are optimal for their own survival. Purely synthetic systems, on the other hand, offer the prospect of much greater flexibility and higher solar efficiencies, together with the possibility of "tuning" the chemistry for production of specific fuels and chemicals. The processes of interest include not only photochemical water splitting, but also photofixation of molecular nitrogen and photoreduction of carbon dioxide.

The eleven chapters in this volume are concerned primarily with photochemical and photoelectrochemical mechanisms and represent the plenary lectures presented at the Third International Conference on Photochemical Conversion and Storage of Solar Energy held at the University of Colorado at Boulder on August 3–8, 1980. This conference series was initiated in 1976 by James R. Bolton at The University of Western Ontario, and the second conference was held in 1978 at Cambridge

University under the direction of Mary D. Archer. The Fourth International Conference will be held at The Hebrew University in August 1982 with Joseph Rabani as Chairman. The International Organizing Committee (IOC) is looking forward to future conferences in this series being held in the Far East as well as in Europe and in North America.

In addition to the invited papers, these proceedings include the discussions following each lecture, together with short abstracts of 136 contributed papers that were presented in poster sessions.[1] The volume thus contains a blend of reviews of the state of the art and original research contributions in the areas of photochemistry and photoelectrochemistry in the context of solar energy conversion.

The prospects for artificial photosynthesis are discussed from rather different viewpoints in Chapter 1 by Calvin and in Chapter 2 by Katz and Hindman. Some spectroscopic aspects of fluorescent concentrators for photovoltaic cells are reviewed in Chapter 3 by Jørgensen. In Chapter 4, Balzani and Scandola delineate the requirements for homogeneous photoredox chemistry in inorganic systems, while the use of inorganic components coupled with catalysts in heterogeneous assemblies for photochemical water splitting is treated independently by Grätzel and Lehn in Chapters 5 and 6, respectively. Archer and Ferreira review progress in and the prospects for photogalvanic cells in Chapter 7. Some chemical problems encountered in the fabrication and manufacture of photovoltaic devices are discussed in Chapter 8 by Wagner. Recent advances in photoelectrochemistry are treated from two points of view: electrochemical photovoltaic cells are described in Chapter 9 by Memming and Kelly, and photoelectrosynthetic reactions at the semiconductor–electrolyte interface are discussed in Chapter 10 by Nozik. Lastly, the crucial problem of the thermodynamic limits on photoconversion *and* storage of solar energy is treated in Chapter 11 by Bolton, Haught, and Ross. (It might be mentioned here that this paper was actually written at the Conference itself; the three authors, who initially had somewhat different perspectives on the subject, distilled their ideas into a unified framework under the pressure of a unique deadline!)

The poster sessions were organized, as closely as possible, along the lines of the topics of the plenary lectures. The distinctions between related sub-disciplines, however, are not always facile, as befits a field undergoing such rapid growth. It is not clear, for example, whether photoredox reactions involving micelles should be considered "homogeneous" or "heterogeneous." Are coordination compounds of transition metals classified as organometallic or inorganic compounds? Likewise, should electron-transfer reactions be categorized as "biomimetic systems" if they contain porphyrins and chlorophylls? This is not really important. What is important is that this exciting field is growing and progress is being made at a remarkable rate.

Moreover, it is becoming increasingly clear that an intellectually coherent discipline is beginning to emerge. This field, which might be called "molecular photo-

---

[1] Extended abstracts of the contributed program have been published in "Book of Abstracts, Third International Conference on Photochemical Conversion and Storage of Solar Energy," SERI/TP–623–797 (J. S. Connolly, ed.), Solar Energy Research Institute, Golden, Colorado 1980. (480 pp.).

conversion,'' is concerned primarily with two aspects: In homogeneous systems (loosely defined), the emphasis is on photoredox reactions as models for artificial photosynthesis; while in heterogeneous systems, semiconductors are being modified to exhibit catalytic properties, and photocatalysts are being developed such that they possess semiconductor properties.

The subtle distinctions (and occasional rivalries) among traditional scientific disciplines are thus beginning to dissolve. It is the hope of the IOC that the Third International Conference on Photochemical Conversion and Storage of Solar Energy will have contributed to the evolution of this new field.

The success of this symposium was due to the efforts of many people. The IOC is largely responsible for the broad international representation at this conference (298 delegates from 18 countries). The members of the local organizing committee are to be commended for their careful attention to the many details involving the logistical arrangements. I also want to thank the session chairmen for keeping the program on schedule and for stimulating the discussions. Special thanks are due to Kate Blattenbauer of the SERI Conferences Group and to Jeanna Finch, my administrative assistant, for her extraordinary efforts and good cheer before, during and after the conference.

I am deeply grateful to Kathrine Castañon Deems, who not only carefully typed the eleven plenary lectures and the discussions, but also served as my editorial assistant. She made many helpful suggestions with respect to formatting the volume and extended the capabilities of the word-processing equipment to accommodate the equations, tables and figures.

Finally, it is my great pleasure to acknowledge the financial support of the contributing sponsors, who underwrote the costs of the social events and, especially, the Solar Energy Research Institute and the Division of Chemical Sciences, Office of Basic Energy Sciences, U.S. Department of Energy, which made this conference possible.

John S. Connolly
Golden, Colorado
May 25, 1981

CHAPTER 1

SIMULATING PHOTOSYNTHETIC QUANTUM CONVERSION[1]

Melvin Calvin

Chemical Biodynamics Division
Lawrence Berkeley Laboratory
University of California
Berkeley, California
U.S.A.

I.   INTRODUCTION

I would like to describe an artificial photosynthetic system that we have been trying to construct which will achieve the essential function of the chloroplast, that is, the capture of solar quanta and their conversion into a stable chemical form (1-3). I know that the synthetic system will not perform in exactly the same structural fashion as the green plant. However, all we wish to reproduce is the function(s) of the natural system, i.e., the capture of quanta and their storage in a stable chemical form.

Recently I encountered an editorial entitled "The Photochemistry of the Future" written by the Italian photochemist, Giacomo Ciamician, published in Science in October 1912 (4):

"Modern civilization is the daughter of coal, for this offers to mankind the solar energy in its most concentrated form; that is in a form in which it has been accumulated in a long series of centuries. Modern man uses it with increasing eagerness and thoughtless prodigality for the conquest of the world and, like the mythical gold of the Rhine, coal is today the greatest source of energy and wealth. . . . . Is fossil solar energy the only one that may be used in modern life and civilization? That is the question."

---

[1]Supported, in part, by the Division of Chemical Sciences, Office of Basic Energy Sciences, U.S. Department of Energy, under Contract W-7405-eng-48.

1

The best quantum converting "machine" that we know today is the green plant, with its chloroplasts, and the possible uses that we can make of that machine are already well established. For example, petroleum, which is being exhausted, will have to be replaced by many different methods of utilizing solar energy, with the green plant being the most common (5). We are "going back" to the "green machine" today by converting the carbohydrate of the plant into a more useful liquid fuel by a fermentation process to make alcohol. The processes and technology for this first step are already available and in use in some areas of the world (specifically Brazil) (6) and are currently under development here in the United States (7,8).

The second stage of "going back" to the "green machine" involves the discovery of plants (and development of their agronomy) which carry the reduction of carbon dioxide all the way to hydrocarbons, instead of stopping at carbohydrates as most plants do (5). There are some plants already in use commercially that can go all the way to hydrocarbons, for example the Hevea rubber tree, and we have found others which will go all the way to hydrocarbons and which will grow in the semiarid areas which are available in many places in the world. Examples of plants that produce hydrocarbons are members of the Euphorbiaceae family, which has over 2000 different species of latex-producers (9).

The final method of "going back" to the "green machine" will be to produce synthetically the quantum capturing and converting function of the chloroplast (10-14) so that we will not require either plants or the land on which to grow them.

We have approached the problem of developing a truly synthetic system for the conversion of solar energy into stored chemical energy in a series of steps (10,11). We are using the green-plant chloroplast as a model to guide us in the construction of efficient and useful ways of capturing and storing solar quanta (15). In general, while a sensitizer can absorb a photon and reach an excited state, we must still find a way for that excited-state energy to be stored in a permanent form. The principal way to do this is by inducing an electron-transfer reaction, i.e., moving the electron from a donor system to some acceptor system. Basically the problem is to prevent the back reaction; the system has to be adjusted in such a way that the excited energetic state that we start with does not fall back to some ground state without storage of chemical energy.

The green plant has long since "learned" how to perform these energy-efficient capturing and converting reactions that take place in the lamellae of the chloroplasts which are constructed of a collection of membranes with two distinct types of surfaces. One membrane surface is characterized by a distribution of rather large particles and another surface is characterized by a distribution of much smaller protein particles. The membrane appears to have two sides between which the quantum conversion process occurs. The details of what is occurring in the plant chloroplast are not absolutely known at the present time, but a simplified representation of what is known about the photosynthetic electron-transfer scheme (the so-called Z-scheme) has been published (Fig. 1). There are two quanta involved which move an electron from a donor water molecule through two successive quantum acts to reach a level approaching that of molecular hydrogen. These two quantum acts (so-called I and II) may take place on two protein molecules on each side of the membrane. This is the kind of physical and chemical model that has guided us in the design of our artificial photosynthetic system.

　　In order to simulate the natural quantum conversion act in the green plant we constructed an insulating membrane, placed somewhere between the two quantum acts, which would allow the two steps to take place on opposite sides. This rather naive

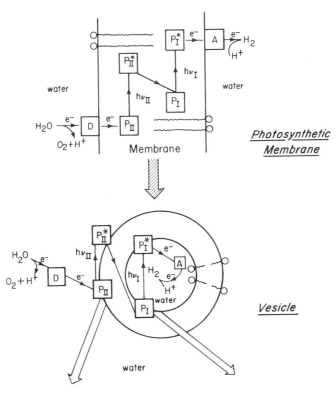

FIGURE 1. Photoinduced electron-transfer schemes for photosynthetic membranes and vesicles.

concept is a simplified method of reducing the natural photosynthetic quantum conversion act to a level where it might be
synthetically reproduced. The diagrammatic representation of
this photoelectron transfer scheme is developed in Figures 1
and 2. The bilipid membrane is represented by phospholipid
molecules on either side; on one side is a donor system and on
the other side an acceptor catalyst. There are two sensitizers, one on each side of the membrane (labeled photosystem
I and photosystem II). The electron is taken from a water
molecule on the left-hand side of the membrane by the first
quantum act. It falls back part way and then is taken by the
second quantum act up to the level of molecular hydrogen and
injected into the acceptor catalyst. One could try to construct such an artificial, flat membrane on a solid or liquid
surface; a number of research groups are doing this with some
degree of success (16-18). It seemed to us, however, that we
could achieve this kind of a system by constructing bilipid
vesicles.

These liposomes were in various forms, one of which was a
simple bilipid membrane sphere. By bilipid, I mean two layers
of phospholipids, tail to tail, in which the surface   tension

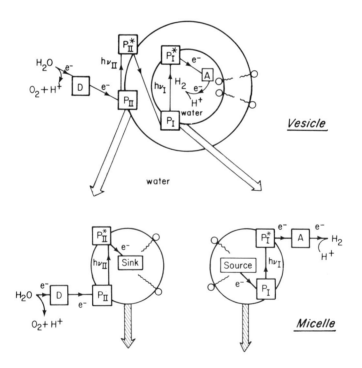

FIGURE 2. Photoinduced   electron-transfer   schemes   for
vesicles and micelles.

and surface properties of the lipids and the water are such that when the monolayer on the water system is broken up it tends to form spheres. This is a spontaneously formed spherical vesicle (liposome) with water on the inside and water on the outside. When this phenomenon occurs in the presence of whatever substance is placed on the inside of the liposome, for example, in the presence of A, then A will be captured on the inside. If a sensitizer is present, it will be captured either inside or outside, depending on the nature of that sensitizer. We can wash away the molecules in the water outside and place B (a separate water-soluble complex) back on the outside, thus obtaining an asymmetric system.

The basic property of the chloroplast is its ability to transfer an electron across a phase boundary, e.g., across a water-oil boundary, and this seems to be one of the essential physical requirements. Two phase boundaries can be generated separately by making an oil droplet with a monolipid layer which is a micelle and not a vesicle, as shown in Figure 2. Thus, the two phase-boundary reactions can be performed separately; one sensitizer can be placed on one of the oil droplets and a second sensitizer on the other oil droplet. We have, however, carried out a photoinduced electron transfer through the wall of an actual vesicle across both phase boundaries (19-21).

II. EXPERIMENTAL

A. Catalyst Construction

We need a catalyst (labeled D in Figs. 1 and 2) to obtain electrons from water, and an acceptor catalyst (A) that can transfer the electrons to protons in order to generate hydrogen. The sensitizers in the plant are clearly the chlorophyll molecules in one form or another, together with their associated pigments, but the catalyst that leads to hydrogen generation on one side and oxygen generation on the other side of the membrane is not so well known. Before we tried to generate these catalysts, we used a donor system which did not require water and oxygen generation and an acceptor system which did not immediately require the generation of hydrogen. In other words, when we began this series of experiments, we performed "simpler" chemistry before trying to synthesize the hydrogen and oxygen generating catalysts. The natural hydrogen generating catalyst is rather well known, an iron-sulfur protein (22), but the oxygen generating catalyst is still under investigation.

Instead of using water as a donor and protons as the acceptor we used a simpler reaction, which is depicted in Figure 3. Oxidation of ethylenediamine tetraacetic acid (EDTA) with methyl viologen (MV), suitably sensitized, resulted in oxidation of one of the glycine groups on EDTA to formaldehyde and carbon dioxide, with simultaneous reduction of the MV to a cation-radical (19). This rather simple, slightly uphill chemical reaction produces a deep blue color in one of the reaction products which can be used as an indication of whether or not the reaction is proceeding. We used EDTA as a donor and MV as an acceptor and, instead of using a water-soluble ruthenium bipyridyl complex as the sensitizer, we constructed a surfactant with a C_{16} tail which creates a complex that will locate itself at the water-oil interface when the vesicles are created (Fig. 4). This, in turn, gives rise to vesicles that have EDTA on the outside and EDTA on the inside with the ruthenium complex on both surfaces. With EDTA on both sides of the vesicle, we pass this system through a gel, which has pores too small to accept the vesicle but small enough to accept the EDTA. During gel filtration the EDTA molecules diffuse into the gel pores, and the resulting vesicles have EDTA on the inside and buffer on the outside. We now add the methyl viologen (MV) to the outside of the vesicle, creating an asymmetric vesicle with the donor (EDTA) on the inside and the acceptor (MV) on the outside.

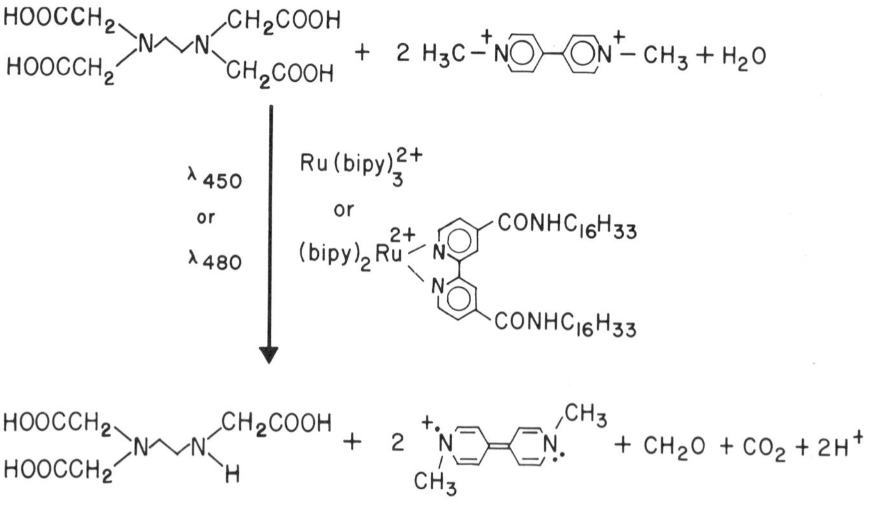

$\Delta G° \sim +20$ Kcal per mole of EDTA

FIGURE 3. Photosensitized reduction of methyl viologen by ethylenediamine tetraacetate.

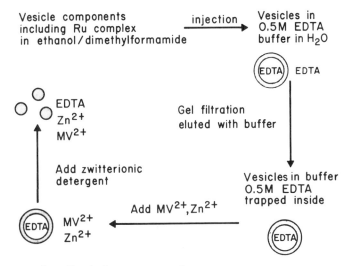

FIGURE 4. Vesicle preparation.

Zinc ion is then added to block any leakage of EDTA from the inside the vesicle.

B. Vesicle Construction

A representation of the method of vesicle construction is shown in Figure 5 (23). At the beginning of the experiment, the vesicles have a thickness of about 50 Å, a simple bilayer with two phospholipid molecules tail to tail and a ratio of phospholipid to ruthenium (as Ru^{2+}) of about 20:1. The ruthenium is on the surface of the interface in a relatively polar environment. On the inside of the vesicle, there is the donor molecule and on the outside, the methyl viologen.

It seemed to us, when the experiments were in the early stages of development, that there should be a carrier to help the electrons across the membrane. Therefore, a quinone was placed in the membrane to carry the electrons through it and also a proton carrier was placed in the membrane to transfer protons. The actual membrane structure (Fig. 6) had plasto-quinone (Q) and a diborane (B) added to it. With the quinone present, the back reaction would eventually take over since the quinone can actually carry the electron back. When the quinone and proton carrier were removed, the reaction was slower, but with very little induction period (13,14).

The reason for the slow rates turned out to be that neither the donor nor the acceptor were anywhere near the sensitizer during the lifetime of its excited state. It appeared

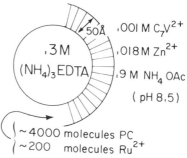

FIGURE 5. Diagram of
vesicle cross-section.

.3 M
(NH$_4$)$_3$EDTA

←50Å

.OOI M C$_7$V^{2+}
.OI8M Zn^{2+}
.9 M NH$_4$ OAc
(pH 8.5)

~4000 molecules PC
~200   molecules Ru^{2+}

desirable to insert something that would trap the electrons
near the surface of the membrane which could then hand the
electrons over to the bulk acceptor.   Therefore, we used
excess heptyl viologen (HV) distributed between the surface of
the membrane and the bulk of the solution.   This provided an
acceptor next to the excited sensitizer which could then
transfer an electron to viologen in the bulk of the solution
very rapidly.   Once the electron leaves the membrane it will
not get back because of the dilution in the bulk viologen.

C. Kinetic Experiments

How did the electron go through the membrane?   Once the
excited electron is trapped by viologen, the external Ru^{2+} has
become Ru^{3+}.   I thought for some time that there had to be a
second quantum absorbed by the internal Ru^{2+} to transfer the
electron across the membrane.   This would make the reaction
dependent upon the square of the light intensity; one quantum
would move the electron from external Ru to viologen and the
other quantum from internal to external Ru.   We then performed
a light-intensity experiment, the results of which are shown
in Figure 7.   The reaction is essentially linear with light
intensity, showing that it is not a two-quantum but a one-
quantum process.   Therefore, the transfer of the electron
through the membrane is not light-dependent.   The exchange
reaction between Ru^{3+} on the outside of the   membrane and Ru^{2+}
inside of the membrane is very fast; the kinetics and the
steps involved in the photoreduction of viologen (C$_7$V^{2+}) are
shown in Figure 8.   The exchange reaction, as far as the
membrane is concerned, is perfectly symmetric, and the two
rate constants are equal but in opposite directions.   All the
other constants can be measured separately,   and thus the rate
of the overall reaction can be measured.   From this the rate
of the exchange reaction can be deduced, and it is very high.
The question was then raised as to whether or not this was
a mass-diffusion reaction, dependent upon the rate of the
molecules on the outside exchanging with the ruthenium ions on

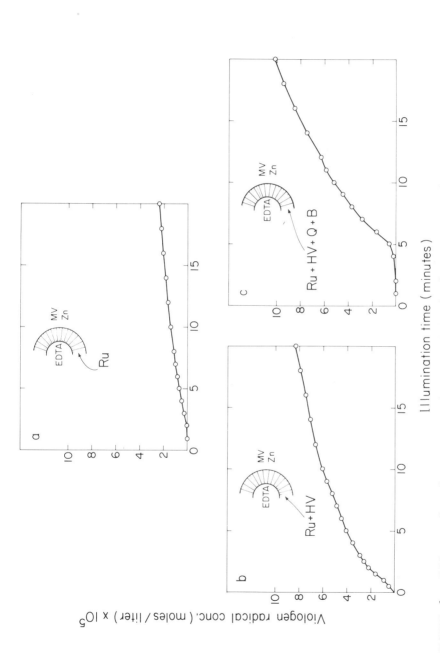

FIGURE 6. Effect of cofactors on photosensitized electron-transfer rates. Progress of the reaction is measured by following the production of viologen radicals.

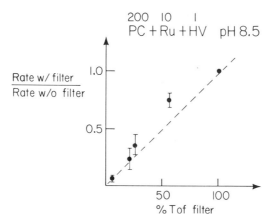

FIGURE 7. Dependence of photoinduced electron-transfer
rate on light intensity.

the inside of the membrane, i.e., an actual migration or
diffusion of the reacting species. If it were, then the diffusion reaction should be independent of ruthenium concentration. If it is a diffusion through the membranes, the
reaction should be unimolecular in ruthenium. However, if the
exchange reaction is light-dependent, it would depend directly
on the ruthenium concentration. The photoreduction kinetics
indicate that when the ruthenium concentration is increased,
the rate is increased. The reaction is thus an isoelectronic
exchange reaction and not a diffusion reaction (20,21). We
have thus demonstrated that it is possible to transfer an
electron across an insulating barrier by an isoelectronic
exchange mechanism by which an electron moves from one side of
the membrane to the other.

Next it was necessary to develop the correct catalytic
system to capture the excited electron on the outside of the
membrane and eventually donate it from the inside. When Ru^{3+}
is on the inside of the vesicle, we have the chemical potential to oxidize water to oxygen, provided a catalyst is
supplied. It takes four electron-transfer events to oxidize
water to molecular oxygen, so the catalyst must be capable of
accumulating four electrons, allowing Ru^{3+} to oxidize water
and be reduced to Ru^{2+}. This type of experiment has been
performed by Grätzel and co-workers (24-26) and by Matsuo et
al. (27), among others, and we discovered that platinum or
cobalt oxide as well as ruthenium oxide could be used for the
catalyst. (See also, Chapters 5 and 6).

This method accumulates four one-electron oxidation steps
on the solid surface, because the Ru^{3+} bipyridyl complex has
the potential for this type of reaction. The other end of the

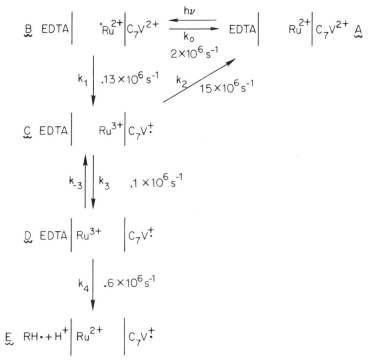

FIGURE 8. Kinetic scheme for photoreduction of heptyl viologen.

reaction is reduced MV (methyl viologen radical cation) which has the potential to generate molecular hydrogen from protons, but only if there is an agent present which will allow the two electrons to accumulate, or to have some other catalyst (such as a rhodium cluster) which will also allow the two electrons to accumulate in the same place. A schematic representation for oxygen evolution from water on a solid catalyst is shown in Figure 9.

We now have developed both catalysts and both systems, i.e., reduction and oxidation. Thus the potential exists to create a device that would perform the same functions as natural photosynthetic quantum conversion. It is, however, difficult to place a solid catalyst on the inside of a vesicle because the solid particles would have to be extremely small to be able to reside there without disturbing the surface. It would be better, if possible, to learn what the natural catalyst might be for oxygen generation.

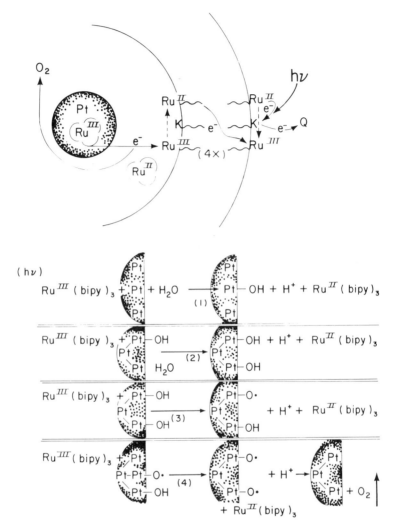

FIGURE 9. Schematic representation of oxygen evolution.

D. Managanese Function

We know a good bit about the natural catalyst for oxygen generation. For example, we know that it is a manganese complex of some sort (10,11). Many years ago, the plant physiologists had determined that when a plant is starved for manganese it fails to have the capacity to generate molecular oxygen, but when manganese ion is added to the plant's environment it recovers this capability very quickly. Many experiments have been done to confirm this idea, i.e., that manganese is somehow involved in an essential role in the first reaction of taking an electron from water to whatever it is in the chlorophyll or other substance to which it is ultimately transferred (28). Also, studies on the determination of the manganese compound have been underway for many years, and it has been unequivocally demonstrated that the compound is not a manganese porphyrin but a weakly bound manganese protein of an unspecified type (29).

Recently, we thought about this manganese problem again and realized that we had been pretty naive to think it was a simple manganese porphyrin. In order to simulate the reactions in green plants, it is necessary to accumulate four electrons. The only way a manganese porphyrin could accumulate these electrons would be two electrons at one time. Originally, I thought there would be two manganese atoms in this reaction, somehow, going in oxidation number from two to three to four, and back again to two, both together in a complex which would create a control system with four electrons. We synthesized a number of binuclear manganese compounds, i.e., containing two manganese atoms in each complex. At this time a spectroscopic method became available to explore the nature of the manganese complex in the living plant material without taking the plant apart.

We used the method of extended X-ray fluorescence spectroscopy (EXAFS) to determine the presence of manganese without interference from the rest of the plant material. Using this spectroscopic method it is possible to examine the manganese atoms alone in a frozen chloroplast sample, which contains about 2 manganese atoms per 1000 chlorophyll molecules. The interference pattern of the scattered X-rays detects only the manganese atoms and its nearest neighbors (30). We also studied the binuclear manganese compound previously synthesized for calibration; the EXAFS of this compound is shown in Figure 10.

From this spectroscopic information it is possible to deduce that in the chloroplasts there are two manganese atoms with a small distance between them. The origin (Fig. 10c) represents the position of one of the manganese atoms, and the second manganese atom shows an intensity peak somewhat over 2 Å distant. A similar picture is observed in the chloroplasts, showing that there are at least two manganese atoms within 2.5 Å of each other. The EXAFS method proved that there is at least a binuclear manganese compound present in the chloroplast material. We can say, therefore, that in the chloroplast there are manganese atoms acting in pairs (or groups larger than pairs) which can perform the same electron transfer reactions that occur in the natural material. We then prepared tetrapyridyl porphyrin (a chlorophyll analogue), with the structure shown in Figure 11, which is a surfactant

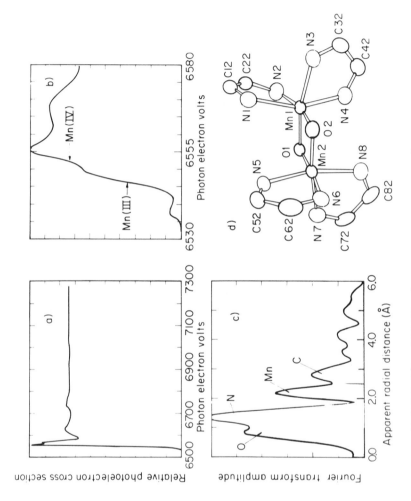

FIGURE 10. EXAFS of manganese compounds.

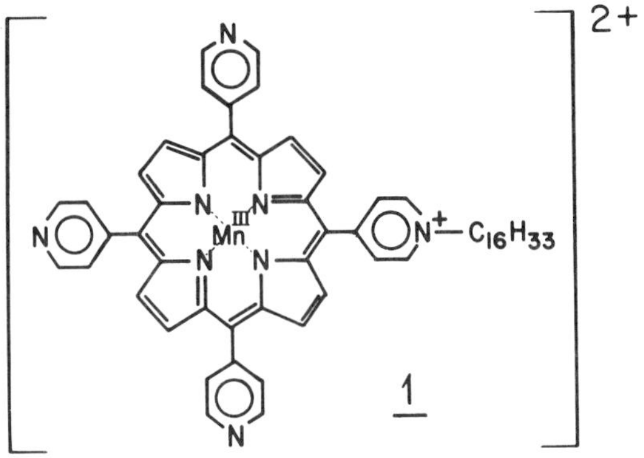

FIGURE 11. Structure of tetrapydridyl Mn^{III} porphyrin.

porphyrin that will lodge at the interface of the vesicle in place of ruthenium.

With that as background, we undertook to use the synthetic vesicles which we had developed previously (19,23) and to replace the ruthenium with this surfactant manganese porphyrin. A general scheme for photochemical generation of oxygen sensitized by manganese compounds is shown in Figure 12. This indicates that there must be two or more manganese atoms (as shown in the last step) acting in concert in the vesicles.

We performed a reaction with manganese porphyrin in a homogeneous solution in methanol (a single phase). The porphyrin, containing manganese as Mn^{III}, and methyl viologen were dissolved in methanol; when the light was turned on, MV was reduced to its cation radical which seemed to be in the right direction energetically. At that point, the reaction did not appear to go backward. However, when we extracted the manganese, instead of finding the expected Mn^{IV}, the substance turned out to be Mn^{II}. Evidently we were raising Mn^{III} porphyrin into an excited state, handing an electron over to the viologen, the radical-cation of which was then visible.

The only way we could understand that result was to assume that two Mn^{IV} atoms were acting together, as shown in Figure 12. When Mn^{III} is excited, and the electron is passed on to methyl viologen, it leaves behind Mn^{IV} to react with water to make an intermediate Mn^{IV} oxide, which can be written as an Mn^{III}-oxygen radical, two of which could come together to make a bis-Mn^{III} peroxide. Finally, one electron from each of the Mn^{III}-oxide bonds would be extracted by Mn^{III} to reduce it to Mn^{II}, liberating molecular oxygen. Therefore, we find Mn^{II} and we should observe molecular oxygen. What appeared to happen was that a very potent intermediate oxidized the methanol, and the resulting reduced product was Mn^{II}. We realized that an incipient oxygen atom bearing only six electrons

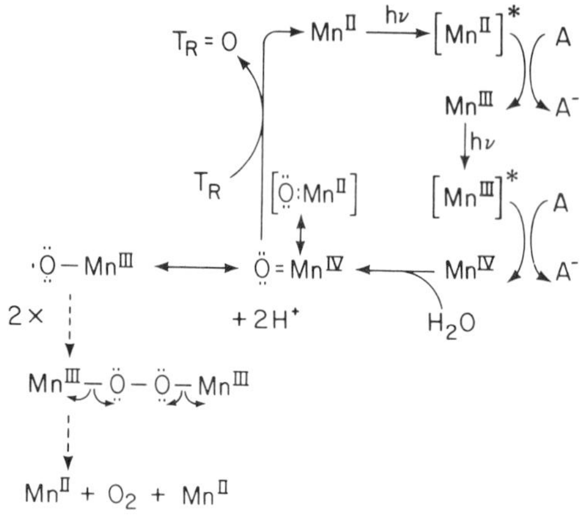

FIGURE 12. Proposed general scheme for photochemical generation of oxygen sensitized by manganese porphyrin.

must be present in the oxidized monomer. If a Lewis base were also present, having an unshared electron pair, it should be possible to hand an oxygen atom on to the base; we should then get Mn^{II}.

Therefore, any compound that has an unshared pair of electrons, or pair of electrons capable of accepting an oxygen atom with six electrons, could be an acceptor. We first used triphenylphosphine (TPP); an oxygen atom (or the Mn^{IV} oxide) with a vacant orbital interacts with the phosphine, leaving behind Mn^{II}. The product formed is triphenylphosphine oxide, in effect, an oxygen-transfer reaction (32). (We are currently conducting an experiment with oxygen-18 water to show that the oxygen atom is from the water and not from some other source). It should be possible to generate more phosphine oxide than total manganese present in this reaction by recycling the Mn compound. If a little platinum is added with the viologen, it is possible to generate any amount of phosphine oxide. If we examine the kinetic curves of Figure 13, we can see that the rate of the reaction is determined by the rate of the initial transfer of an electron from manganese to methyl viologen.

We now have confirmatory chemical evidence that we have converted the Mn^{III} porphyrin to a Mn^{IV} oxide, the crucial intermediate, by simple chemistry. The absorption spectra for three manganese complexes in different oxidation states are shown in Figure 14: Mn^{III} porphyrin, the initial product; Mn^{II} porphyrin, the final product; and Mn^{IV} oxide, the presumed intermediate. All of these compounds can be synthesized separately by routine reactions as shown in Figure 15. We are

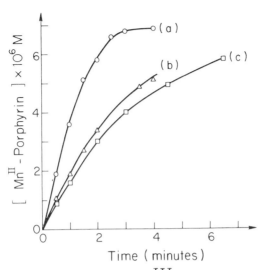

FIGURE 13. Dependence of Mn^{III} porphyrin photoreduction on methyl viologen concentration.

able to oxidize the manganese porphyrins by one- and two-electron steps. Thus, when we oxidize Mn^{II} with iodoso-benzene, we get Mn^{IV}; but if we try to oxidize Mn^{III} with iodosobenzene the reaction does not go. The same product results from Mn^{II} and iodosobenzene, or from Mn^{III} and sodium hypochlorite. If triphenylphosphine is added to this oxidation product, the resulting products are phosphine oxide and reduced manganese; the Mn^{II} formed initially reacts with Mn^{IV} to give back Mn^{III}.

III. DISCUSSION

We have shown photochemically, and confirmed chemically, that a Mn^{IV} oxide intermediate contains a "hot" oxygen atom, probably from water. The specific photochemistry of Mn-sensitized generation of active oxygen atoms is shown in Figure 16. Thus, starting with Mn^{III} and with two quanta we go to Mn^{IV} and then to the phosphine oxide, bringing back the Mn^{II} and on to Mn^{III} again. We have generated two reduced viologen radicals for each oxide, and the net reaction is:

$$PPH_3 + H_2O \xrightleftharpoons[Mn^{III}]{h\nu} PPh_3O + H_2$$

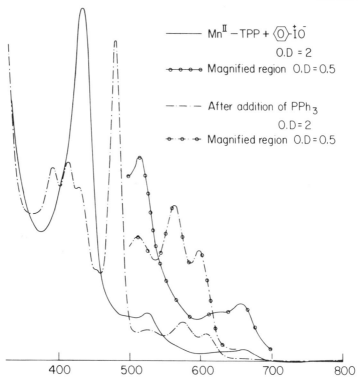

FIGURE 14. Visible absorption spectra of manganese complexes in different oxidation states.

Obviously, it is possible to replace the phosphine with other useful oxygen acceptors, particularly olefins, and the final product will be more useful than phosphine oxide.

It should be emphasized that what has been accomplished here is much more important than getting molecular oxygen. We have generated a "hot" oxygen atom from water -- not from oxygen -- with a cyclic reaction (28) sensitized by absorption of two quanta. Eventually, I predict that the result of these experiments on artificial photosynthetic systems will be a new chemical industry with hydrogen generation (as well as oxygen generation) from water. Although the quantum yield of this reaction is high, so far the energy yield itself is low because the sensitizer has rather weak light absorption.

A diagrammatic representation of the entire reaction is shown in Figure 17. At the bottom of the diagram is what appears to be a cross section of the vesicle containing a manganese dimer to generate oxygen. This drawing should be viewed as a cross section of a tube; the tubes are hung between headers which enclose the space outside the fibers, in this

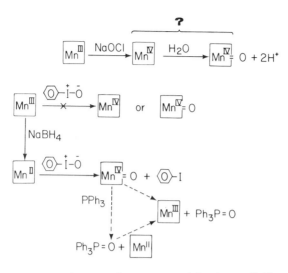

FIGURE 15. One- and two-electron oxidation of Mn porphyrins.

$$2H_2O \longrightarrow O_2 + 4H^+ + 4\bar{e}$$

$$H_2O \longrightarrow \left[\ddot{\underset{\cdot\cdot}{O}}:\right] + 2H^+ + 2\bar{e}$$

$$T_R: + \left[\ddot{\underset{\cdot\cdot}{O}}\right] \longrightarrow T_R = O$$

$$2 \times T_R = O \xrightarrow{Cat.} 2 \times T_R + O_2$$

$$T_R + H_2O \xrightarrow[(h\nu)]{S} T_R = O + H_2$$

FIGURE 16. Specific photochemistry of Mn-sensitized generation of active oxygen atoms.

case containing the quinone-hydroquinone redox couple which will transfer both protons and electrons between the two systems. This couple is similar to the quinone in natural photosynthetic systems and has the same function. Also shown in this diagram are various hydrogen generators (ruthenium, viologen, platinum).

It should be possible to perform these artificial photo-synthetic reactions either in hollow fibers (Fig. 17) or in gels (Fig. 18). These reactions could be made to occur by suspending the vesicles in the gel precursor and then causing the gel to form, either by polymerization or by temperature change. Thus the particles (vesicles) each containing the separate systems could be immobilized and coupled together by the quinone hydroquinone redox system.

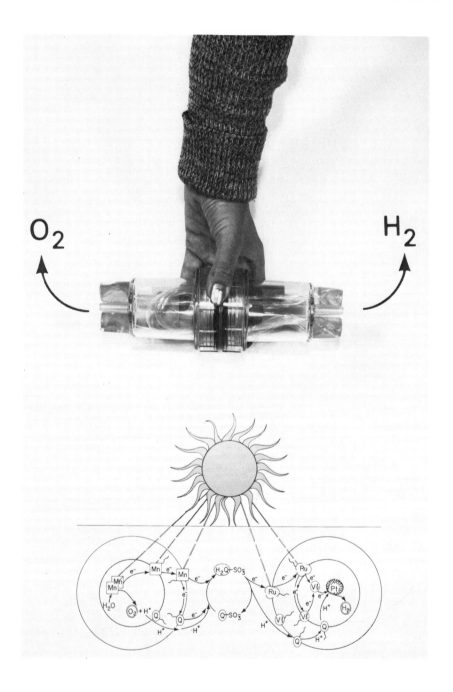

FIGURE 17. Schematic diagram of synthetic chloroplasts: hollow fiber technique.

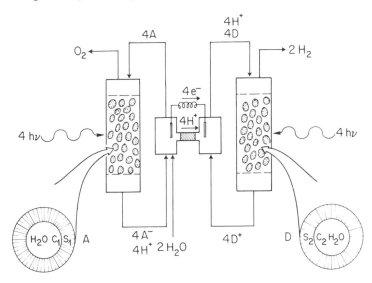

FIGURE 18. Schematic diagram of synthetic chloroplasts: microemulsion technique.

It is my feeling that when we finally construct the artificial system, it will be somewhat different from that depicted in Figures 17 or 18. We will use an oxygen acceptor, instead of the manganese dimer to create a useful oxidized product with hydrogen being produced in the other vesicle. With our artificial system it will be possible to perform exactly the same chemical reactions that occur in natural photosynthetic systems (33).

IV. CONCLUSION

As a result of our efforts to develop artifical photosynthetic systems we have succeeded in designing an oxygen-generating system as well as an oxidation product coupled with a hydrogen generating system. This type of "synthetic chloroplast" will probably be more useful in the chemical industry than would one which more closely reproduces the function of the natural chloroplast.

We have not only designed, but have succeeded in accomplishing, the two separate reactions -- viz., oxidation (or activation) of water and reduction of protons--in two separate systems. What now remains is to couple these two reactions in a continuing (cyclic) and stable manner so that the products generated at both ends of the system may be usefully coupled.

All of the efforts in this direction -- in our laboratory and elsewhere -- are directed to the problem of using water and sunlight as a source of fuels and chemicals in the future. We believe that we are moving in the right scientific direction in our ability to mimic the complex natural photosynthetic process, namely constructing artificial devices for its essential parts. Combining the different components into one comprehensive system will certainly be accomplished in the near future. The fact that this has been achieved long ago in nature gives us encouragement.

REFERENCES

1. J. R. Bolton (ed.), "Solar Power and Fuels", Academic Press, New York, 1977.
2. J. R. Bolton, Annu. Rev. Energy 4, 353 (1979).
3. D. O. Hall, Solar Energy 22, 307 (1979).
4. G. Ciamician, Science 36, 385 (1912).
5. M. Calvin, Bioscience 29, 533 (1979), and references cited therein.
6. Y. Yang and S. C. Trindade, Chem. Econ. Eng. Rev. 11(3), 12 (1979).
7. "GASOHOL: A Technical Memorandum." Office of Technology Assessment, U. S. Congress, Washington, D. C., 1979.
8. W. A. Scheller, "The Nebraska GASOHOL Program: Alcohol Blended Fuel and Protein", 6th National Meeting of Sugar Producers, Campos, Brazil (1978).
9. J. D. Johnson and C. W. Hinman, Science 208, 460 (1980).
10. M. Calvin, Accts. Chem. Res. 11, 369 (1978).
11. M. Calvin, Energy Res. 3, 73 (1979).
12. I. Willner, W. E. Ford, J. W. Otvos and M. Calvin, in "Bioelectrochemistry", (F. Gutmann and H. Kuyzer, eds.), Plenum Press, New York, 1980, pp. 55-81.
13. P. P. Infelta, M. Grätzel and J. H. Fendler, J. Am. Chem. Soc. 102, 1479 (1980).
14. T. Nomura, J. R. Escabi-Perez, J. Sunamoto and J. H. Fendler, J. Am. Chem. Soc. 102, 1484 (1980).
15. For a summary of various experimental approaches, see: "Light-Induced Charge Separation in Biology and Chemistry", (H. Gerischer and J. J. Katz, eds.), Verlag Chemie, Weinheim, 1979.
16. G. Porter, Proc. R. Soc. (London) A362, 281 (1978), and references cited therein.
17. H. Kuhn, J. Photochem. 10, 111 (1979).
18. T. Matsuo, Me. Fac. Eng., Kyushu University 40(1), 26 (1980).

19. W. E. Ford, J. W. Otvos and M. Calvin, Nature 274, 507 (1978).
20. I. Willner, W. E. Ford, J. W. Otvos and M. Calvin, Nature 280, 828 (1979).
21. W. E. Ford, J. W. Otvos and M. Calvin, Proc. Natl. Acad. Sci. USA 76, 3590 (1979).
22. W. Orme-Johnson, Annu. Rev. Biochem. 42, 159, (1973).
23. I. Willner, J. W. Otvos and M. Calvin, in "Solution Behavior of Surfactants", (E. J. Fendler and K. L. Mittal, eds.), Plenum Press, New York, 1981 (in press).
24. J. Kiwi and M. Grätzel, Chimica 33, 289 (1979).
25. K. Kalyanasundaram and M. Grätzel, Helv. Chim. Acta 63, 478 (1980).
26. P.-A. Brugger, P. P. Infelta, A. M. Braun and M. Grätzel, J. Am. Chem. Soc. 103, 320 (1981).
27. T. Matsuo, K. Takuma, Y. Tsutsui and T. Nishijima, J. Coordination Chem. 10, 187 (1980).
28. A. Harriman and G. Porter, J. Chem. Soc., Faraday Trans. II, 1532 (1979); ibid., 1543 (1979).
29. M. Spector and G. D. Winget, Proc. Natl. Acad. Sci. USA 77, 957 (1980).
30. J. A. Kirby, D. Goodin, A. S. Robertson, J. E. Smith, S. Thompson and M. P. Klein, Proc. Natl. Acad. Sci. USA 78, in press (1981).
31. J. Grant, unpublished results from this laboratory.
32. H. G. Mettee, W. E. Ford, T. Sakai, J. W. Otvos and M. Calvin, J. Phys. Chem., in press (1981).
33. W. Haehnel and H. J. Hochheimer, Bioelec. Bioenerg. 6, 563 (1979).

DISCUSSION

Prof. M. Grätzel, Ecole Polytechnique Fédérale, Lausanne:
    What was the chain length of viologen derivatives? Also, would you comment on the specific rates of electron transfer across the microemulsion interface? Are these values available from the quantum yields?

Prof. Calvin:
    We used two or three carbon atoms between the sulfonate and the bipyridine. In answer to your second question, specific rates in terms of electrons per second will depend on surface area, but they have not yet been assessed.

Prof. D. O. Hall, Kings College, London:
    What Fe-S clusters have been used as catalysts for hydrogen evolution?

Prof. Calvin:
  Both the $(Fe-S)_4$, and an $(Fe-S)_2$ cluster similar to Schrauzer's have been used.

Prof. G. Tollin, University of Arizona:
  What do you think are the limiting factors in determining the initial quantum efficiencies in your photosystems?

Prof. Calvin:
  First, the loss from the excited states; and second, the back reaction.

Prof. J. R. Bolton, University of Western Ontario:
  Have you tried to put compounds into the membrane to enhance the efficiency of electron transfer across the membrane?

Prof. Calvin:
  Yes; we have used quinones, but they have had no effect, except to delay the appearance of the electrons on the outside, presumably due to the need to reduce them (i.e., the quinones) before the electrons could appear on the outside.

Prof. P. A. Loach, Northwestern University:
  What ion accompanies electron movement in ruthenium-containing vesicles to maintain electroneutrality?

Prof. Calvin:
  Proton movement accompanies electron transfer.

Dr. P. Ang, Institute of Gas Technology:
  What is your rationale in choosing your redox couples at the inside and outside of your membrane, considering that the difference in redox potentials between the inside and outside of the membrane may influence the electric field on the inside?

Prof. Calvin:
  First of all, we want to have something that "works". Second, the redox couples are chosen so that the chemicals produced have large amounts of energy that can be stored. The fixed charges at the interface also have a very strong influence on the electron-transfer process.

Dr. A. Alberts, TNO, Utrecht:
  Will oxygen diffusion interfere with hydrogen production in a vesicle or fiber system?

Prof. Calvin:
Oxygen can be flushed out faster than diffusion to the membrane. Also, making oxidized products will be useful, at least at the first stage, thus minimizing interference with the hydrogen-production step.

Prof. Sir George Porter, Royal Institution:
Most of the reactions, which you described as occurring with electron transfer across membranes, have also been carried out in homogeneous solutions quite efficiently. Since, in any case, the membranes are permeable to oxygen, what is the advantage of membranes over simple homogeneous solutions?

Prof. Calvin:
It is true that the reactions that I described can be carried out equally well in homogeneous solution. For the complete splitting of water into hydrogen and oxygen, however, membranes will be necessary.

Prof. T. G. Spiro, Princeton University:
In connection with making useful oxidation products you mentioned peroxide as a possibility. Is the Mn(IV)=O in your experiments "hot" enough to generate peroxide?

Prof. Calvin:
When phthalic acid is used as the trap, we believe that peroxy-phthalic acid is generated.

Prof. G. C. Dismukes, Princeton University:
A feature which appears to be important in photosynthetic quantum conversion and which is not represented in the synthetic schemes is the occurrence of a high degree of orientation of the cofactors both in the membranes and, perhaps more importantly, with respect to neighboring electron donors or acceptors. How important will it be to duplicate those features in synthetic schemes to increase the quantum yields, and if important, what are the current prospects?

Prof. Calvin:
The use of a surfactant sensitizer already places it in a special oriented position with respect to the phase boundary. We have shown that placing the primary acceptor has a marked influence on the rate of electron transfer to heptyl viologen vs. methyl viologen. Certainly the geometry of the other components will also be important.

Prof. Porter:

Although models of PSI with sacrificial donors and of PSII with sacrificial acceptors can now be made, the problem of linking the two photosystems through a reversible redox couple without using sacrificial donors and acceptors remains. The green plant uses quinone-hydroquinone couples, and, although quinone has been used as the acceptor and hydroquinone as donor in model systems, the back reaction prevents efficient transfer of charge. How does the chloroplast overcome this difficulty and how can we mimic it in vitro?

Prof. Calvin:

In the scheme of my last slide (Fig. 18) the quinone/hydroquinone couple was part of an electrochemical cell.

Prof. Porter:

I would like to ask Prof. Grätzel to comment on cyclic systems.

Prof. Grätzel:

Our system (see Chapter 5) works on four photons; however, as I understand Prof. Porter's remarks, he alluded to a system in which 4 photons are used to produce O_2 and a reduced chemical while 4 more photons are used to produce H_2 and an oxidized chemical. Could the oxidized and reduced chemicals be combined to reinstate the initial situation?

Dr. A. J. Nozik, Solar Energy Research Institute:

I would like to make a point of clarification with respect to the question raised by Prof. Porter concerning coupled PSI and PSII-like systems for water splitting not requiring sacrificial components. As Prof. Calvin indicated, semiconductor systems can indeed be constructed that couple a p-type semiconductor with an n-type semiconductor through an ohmic contact such that simultaneous illumination of both semiconductors in contact with an aqueous electrolyte produces H^+ reduction to H_2 on the p-type surface and H_2O oxidation to O_2 on the n-type surface. That is, the p-type semiconductor is acting like PSI and the n-type semiconductor is acting like PSII, with the ohmic contact acting like the quinone pool in coupling the excited electron in PSII with the photogenerated hole in PSI. These semiconductor devices, called photochemical diodes, can split H_2O into H_2 and O_2 with visible light without the need for sacrificial reactions. Their present limitations are the low efficiencies ($\sim 1\%$) available from stable devices.

CHAPTER 2

CURRENT STATUS OF BIOMIMETIC SYSTEMS FOR SOLAR ENERGY CONVERSION[1]

Joseph J. Katz
James C. Hindman

Chemistry Division
Argonne National Laboratory
Argonne, Illinois
U.S.A.

I. INTRODUCTION

The energy crises of the 1970's are not the only reason to search for new energy technologies. The enormous expansion in world population predicted by all demographers for the 21st century makes it prudent to devise practical methods for the use of any and all possible energy sources. One of the most attractive of these is solar energy. To be sure, solar energy is used on a vast scale world-wide for growing green plants, but relative to the total amount of energy available in principle from solar insolation, only a miniscule fraction materializes as food stuffs, textiles (cotton, flax), fuel (wood), or polymers (rubber) and other plant products. Consequently, there are strong reasons to devise ways to maximize the use of solar energy, not the least of these being the fact that solar energy is the only truly renewable energy source available on our planet.

A. Biomimetic Artificial Photosynthesis

Several approaches to more effective solar energy utilization are currently under development. Solar energy as a

---

[1]Work performed under the auspices of the Office of Basic Energy Sciences, Division of Chemical Sciences, U.S. Department of Energy.

27

heat source and the direct conversion of light to electricity
by solid-state devices are the farthest along, and these can
be expected to make significant contributions to the energy
budget in the next decades. But other, more visionary approaches are beginning to receive attention, and of these,
biomimetic technologies for solar energy conversion are of
particular interest. Biomimetic approaches to solar energy
conversion, to define the term, are methods that mimic the
essential features of green-plant photosynthesis. The objective is nothing less than artificial photosynthesis.

Green-plant photosynthesis is a process in which electrons
abstracted from water are used to reduce carbon dioxide to
organic compounds containing carbon-hydrogen and carbon-carbon
bonds. Green-plant photosynthesis uses light energy not to
produce heat or electricity but to provide the motive force
for the synthesis of organic compounds. The problems in realizing true artificial photosynthesis are obviously formidable
and intimidating. Unlike the case for direct conversion of
light to electricity, however, in green-plant photosynthesis
nature has provided a complete set of clues which, if interpreted properly, can provide the basis for successful artificial photosynthesis.

Our principal object in this chapter is to describe in as
specific terms as possible the encouraging recent progress in
mimicking the light-energy conversion step in photosynthesis.

B. Biomass and Biomimetic Approaches

A brief discussion of the differences between these two
approaches may be useful. The term "biomass" covers such
varied topics as expanded use of wood as fuel, conversion of
agricultural products or residues to ethanol, gasification of
cellulose, fermentation of animal wastes to methane and the
like, development of much faster growing crops by genetic
selection or genetic engineering, and more efficient utilization of crops, grasses, and forests. Biomass generation,
thus, relies on photosynthesis, but on photosynthesis carried
out by living plants. Biomimetic techniques imitate the sali-
ent features of plant photosynthesis, but the context in which
biomimetic conversion is carried out is purely abiological.
There is much active solar energy conversion research that
makes use of immobilized or encapsulated algae, chloroplasts
or bacterial photoreaction centers isolated from living green
plants or photosynthetic bacteria. For example, bacterial reaction centers have been incorporated into lipid bilayers that
are able to mediate light-energized electron transport (1).
All of these interesting efforts provide useful information
about various aspects of plant or bacterial photosynthesis,

but they do not constitute biomimetic, artificial photosynthesis in the sense we use the term here.

C. Biomimetic Processes

Processes that mimic the essential features of similar processes in living organisms are biomimetic, from the Greek bios, life, and mimesis, imitation. Breslow has used this term to describe organic chemistry that imitates natural reactions and enzymatic processes (2). A biomimetic process or system for solar energy conversion mimics the essential features of natural photosynthesis. Thus biomimetic processes differ in a basic way from biomass production by plants.

Photosynthesis in green plants is often considered to be an inefficient process because of the low ratio of biomass produced to incident solar energy. Natural photosynthesis, if this were correct, would be a poor model. Judged by the efficiency with which a photon is converted to an electron and a positive hole (reducing and oxidizing capacity, respectively), however, the light-energy conversion step in plant photosynthesis is highly efficient. The best available experimental evidence indicates that conversion of one photon produces one electron in the photoreaction centers of both green plants (3) and photosynthetic bacteria (4). It is, of course, true that only that fraction of the energy of a light quantum corresponding to 700 nm radiation in the case of green plants and 865 nm for photosynthetic bacteria, is actually converted, the remainder being discarded as heat. This may not represent inefficiency but rather may serve some purpose of the photosynthetic organism; for example, it may be a way to avoid the formation of too highly excited intermediates, which might be capable of causing tissue damage, or it may be useful for a plant to carry out photosynthesis under conditions where blue (energetic) photons may not be present to a significant extent. As to the overall efficiency of biomass production, clearly the difference between the heat of combustion of a fine piece of furniture and that of an equivalent weight of firewood is not significant. How much energy goes into the ordering of plant cells or in producing an ear of corn can only be surmised, but the energy requirement must be very large, and it must be reflected in a decreased biomass production.

The photosynthetic apparatus in green plants must function in a way that is compatible with the plant's survival. The constituent chemical compounds of the plant are, furthermore, limited in structure by the biosynthetic pathways available to the organism. The components of the photosynthetic apparatus may thus be the most efficient compounds that the plant can

biosynthesize, and need not be the best and most efficient if
optimization of solar energy conversion were the only consideration. Because biomimetic solar energy conversion has only
a single, limited objective, it could be more efficient than
the same process carried out in a living organism which must
pay a high price in energy consumption for growth, reproduc-
tion, and the interruption of insolation during the night. A
biomimetic installation would be immune to viruses, fungal
rusts, and insect predators. Solar energy conversion by a
biomimetic technology could use arid land or land unsuited for
agriculture because of climate or soil. A biomimetic technology for solar energy should ultimately achieve a higher
intrinsic efficiency than its prototype in vivo (5,6) and
could in other ways have significant advantages over green-
plant cultivation.

II. LIGHT ENERGY CONVERSION IN PHOTOSYNTHETIC ORGANISMS

It is self-evident that successful abstraction of the
important features of a biological process will depend on the
extent of our understanding of the process. Critically important is the judgment as to whether a sufficient amount of
information is available to ensure success in the modeling.
In the case of green-plant photosynthesis, it turns out that
in some respects more was known and understood than was generally supposed, but that in other respects less is known than
has been generally assumed.

A. The Photosynthetic Unit

Chlorophyll has long been recognized as the primary photo-
acceptor in photosynthesis and as the principal agent in the
light conversion step itself. Emerson and Arnold (7,8), on
the basis of photosynthesis experiments with flashing light,
concluded that the chlorophyll function in photosynthesis is a
cooperative phenomenon. A large number of chlorophyll molecules, say several hundred, appear to act in concert to convert the energy of a single photon to an electron and a positive hole. These studies have given rise to the concept of
the photosynthetic unit (PSU). In an early version, the PSU
was conceived by E. Katz (9) as a solid-state device in which
the chlorophyll molecules in the unit were considered to be
identical in structure and function, and were highly ordered
in a two-dimensional array. From solid-state considerations,
the chlorophyll "crystal" would be expected to have an electronic conduction band (10), with the result that electrons

and positive holes produced by absorption of light would migrate in such a way as to produce charge separation. There is
no experimental evidence, however, to support the view that
electronic conduction bands and hole migration occur in a
PSU. The arguments against a solid-state description of a PSU
given by Rabinowitch (11) long ago still appear valid and, for
the most part, a solid-state description of the PSU has been
abandoned. The specialized functions of the chlorophyll in
the PSU are now generally considered to be associated with
particular forms or species of chlorophyll.

The chlorophyll in a PSU has three well-defined functions. The large majority of the chlorophyll is used for
collecting light energy and transferring it (in the form of
electronic excitation) to a very few specialized chlorophyll
molecules where charge separation is effected. The chlorophyll species employed in collecting light and transferring
energy is usually referred to as light-harvesting or antenna
chlorophyll, and the chlorophyll species that carries out
charge separation is called photoreaction center chlorophyll.

A photon absorbed by a particular chlorophyll molecule in
the antenna is excited to a higher energy electronic state.
The excitation energy then migrates through the antenna until
it is trapped in the photoreaction center where the light-
energy conversion event actually occurs. To ensure energy
trapping, all photons that are absorbed by the antenna (energy
level at 680 nm) are converted to that of a photon at 700 nm
(the energy level of the reaction center). The energy difference between 680 nm and 700 nm is sufficient to prevent
detrapping of the energy captured in the reaction center. The
conversion of any photon collected by the antenna to a red
photon is a very important part of the antenna function.

All of the chlorophyll in a PSU has optical properties
that are anomalous relative to a solution of pure chlorophyll
in a polar solvent such as acetone or diethyl ether. In such
solutions chlorophyll a (Chl a) occurs in monomeric form and
absorbs light maximally in the red region of the spectrum between 660 nm and 670 nm, depending on the polarizability of
the solvent. All of the chlorophyll in the PSU is red-shifted
and absorbs light maximally at wavelengths longer than 670 nm.
Antenna chlorophyll in green-plants has its absorption maximum
in the red near 678 nm. Photoreaction center chlorophyll absorbs even farther to the red, in green-plants at 700 nm and
in photosynthetic bacteria at 865 nm, hence the conventional
designations for the respective photoreaction center chlorophylls as P700 and P865. The red-shifts in the optical spectra show that the energy level of the lowest excited singlet
state of antenna chlorophyll is lower than that of monomeric
chlorophyll, and that the lowest excited singlet-state energy
level of photoreaction center chlorophyll lies even lower than

that of antenna chlorophyll, which leads to energy trapping in the reaction center.

Although the chlorophyll in the PSU is divided into populations with specialized functions and different optical properties, extraction and examination of the chlorophyll from an organism such as the yellow-green alga Tribonema aequale, or from purple photosynthetic bacteria, reveals that only a single chlorophyll chemical species is present in each instance; Chl a in Tribonema and bacteriochlorophyll a (Bchl a) or b (Bchl b) in the photosynthetic bacteria. The different optical absorptions of chlorophyll in the PSU must, therefore, be due to different species of chlorophyll, which are composed of the same chlorophyll building blocks but which have different architectures and, thus, different physical properties.

A highly schematic representation of a PSU is shown in Figure 1. In this drawing the structure of the antenna is not specified because there is no generally accepted representation. The photoreaction center chlorophyll is usually referred to as P700 (P stands for pigment, the 700 is short-hand for the absorption maximum, 700 nm, of this chlorophyll species). Excitation of the reaction center results in the ejection of an electron:

$$P700 + h\nu \rightarrow P700^{\cdot +} + e^-$$

"Ejection" of an electron is loose usage, for there are no reasons to suppose that free electrons, in the sense of electrons in a metallic conductor, are present at any time in the PSU. The ejected electron produced in a primary conversion act is captured by a primary acceptor (A in Fig. 1) and is then passed along by a series of electron-transfer agents until the ultimate reductant NADPH is formed in the aqueous portion of the chloroplast, where chemical synthesis proceeds by a very complex set of organic reactions powered by NADPH and ATP. Another electron-transport chain conducts electrons removed from water (in green-plants) or some organic compound (in photosynthetic bacteria) to a primary electron donor D generally considered to be a cytochrome that regenerates the oxidized P700$^{\cdot +}$ to its resting state, ready for another light conversion event.

The electron conduits in and out of the reaction center chlorophyll are of the utmost importance to PSU function, for they are so arranged as to achieve a very high degree of directionality in electron flow. The methods used are only now beginning to be understood at the molecular level.

The PSU can, therefore, be regarded as an electron pump in which the energy of light is used to raise low-energy electrons abstracted from water to a high enough level that makes possible the reduction of carbon dioxide.

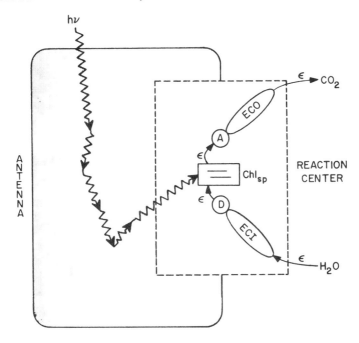

PHOTOSYNTHETIC UNIT

FIGURE 1.   Schematic representation of the photosynthetic unit (PSU) as a light-powered electron pump. Light captured by the antenna is transferred to a photoreaction center where charge separation occurs in a special pair of chlorophyll molecules (Chl_{sp}). The excited Chl_{sp}^* transfers an electron to a primary acceptor A, and the electron is conducted via an electron conduit out (ECO) to the aqueous region of the chloroplast where the chemical reactions of photosynthesis are carried out. An electron abstracted from water is transferred via an electron conduit in (ECI) to D which acts as primary donor to Chl_{sp}^+ to restore the resting state. In this highly simplified PSU only one photoreaction center is shown.

III.   SOME PROPERTIES OF CHLOROPHYLL RELEVANT TO PHOTOSYNTHESIS

The problem of mimicking antenna and photoreaction center chlorophyll is basically a question of forming in the laboratory chlorophyll species that have the optical, redox, and photochemical properties of their natural prototypes.   To accomplish this, much more needs to be known about the properties of chlorophyll on the molecular level than follows only from visible and fluorescence spectroscopy, traditionally the major contributors to chlorophyll chemistry. Such information is now available, derived largely from spectroscopic investigations over the past 20 years by ir (12), nmr (13) and epr (14) spectroscopy. These investigations have revealed new aspects of chlorophyll behavior that are implicit in its molecular structure and that turn out to be highly relevant to chlorophyll functions in vivo.

A. Electrophilic Magnesium in Chlorophyll

Modern spectroscopic investigations make it clear that chlorophyll has an unusual combination of donor-acceptor coordination properties (15). The central Mg atom of chlorophyll (structures of the chlorophylls discussed here are shown in Fig. 2) has coordination properties that figure importantly in the genesis of the various physiologically important chlorophyll species. As shown in the structural formulas, the central Mg atom has a coordination number of 4, i.e., the Mg experiences bonding interactions with the four nitrogen atoms of the pyrrole and pyrroline rings that constitute the chlorophyll macrocycle. From spectroscopic evidence, however, Mg with coordination number 4 in chlorophyll is coordinatively unsaturated. It is electron-deficient and, thus, is an electrophile; there exists a strong driving force to acquire electron-donor groups or functions (nucleophiles) in one or both axial positions of the Mg atom.

When chlorophyll is dissolved in a nucleophilic (polar, Lewis base) solvent such as acetone, diethyl ether or ethyl acetate, the chlorophyll occurs as a monomer with one molecule of solvent in one of the axial Mg positions; the Mg atom then has a coordination number of 5. In a stronger, more basic nucleophilic solvent such as pyridine, the Mg atom has a coordination number of 6 and there are two solvent molecules (L) ligated at the axial positions (16). Monomeric Chl $a \cdot L_1$ or Chl $a \cdot L_2$ has its lowest absorption maximum $(S_0 \rightarrow S_1)$ in the region 660-670 nm, depending on the polarizability of the bulk solvent. Such solutions are intensely fluorescent in dilute solution (fluorescence quantum yield >0.2), in contradistinction to in vivo green-plant chlorophyll, which is only feebly fluorescent with a quantum yield for fluorescence < 0.03 (17).

B. Nucleophilic Functions in Chlorophyll

But how will the electrophilic properties of Mg be expressed in the absence of nucleophiles, if for example chlorophyll is dissolved in nonpolar solvents that lack nucleophilic

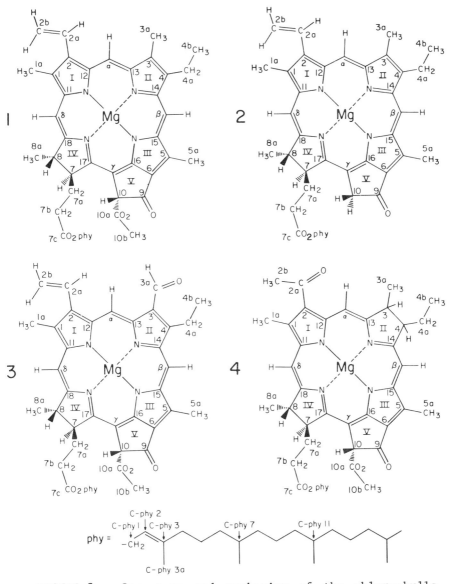

FIGURE 2. Structure and numbering of the chlorophylls. 1, chlorophyll a (Chl a); 2, pyrochlorophyll a (Pyrochl a); 3, chlorophyll b (Chl b); 4, bacteriochlorophyll a (Bchl a). Removal of the central magnesium atom and its replacement by 2H forms the corresponding pheophytin. Chl a (1) is present in all photosynthetic organisms that evolve O_2; Chl b (3) is present in all higher green plants; Bchl a (4) is the characteristic chlorophyll of purple photosynthetic bacteria; Pyrochla a (2) is a synthetic derivative of Chl a.

properties? Chlorophyll itself possesses nucleophilic properties comparable to those of organic solvents containing keto-
carbonyl functions (acetone, cyclopentanone). Chl a (Fig. 2)
contains two ester carbonyl functions at position 7c and 10a
and a keto C=O function at position 9 in Ring V. The weight
of the experimental evidence strongly suggests that the keto
C=O function is by a considerable margin the strongest nucleophile in Chl a. Judging from ir and nmr data (18), the ester
C=O groups are distinctly weaker donors to Mg and this view is
confirmed by the coordination behavior of pyrochlorophyll a,
an important synthetic derviative of Chl a, and the natural
Bchl's c, d, and e all of which lack a carbomethoxy group at
position C-10. Other important chlorophylls contain, in addition to the keto C=O function in Ring V common to all members
of the chlorophyll family, additional nucleophilic functions:
a formyl group, -CHO, in Chl b and Bchl c; an acetyl C=O
function in Bchl a and Bchl b; and an α-hydroxy ethyl
function, -CHOHCH$_3$, in Bchl c, d, and e. In chlorophylls
containing two strong electron donor functions (in the coordination sense) the relative nucleophilic strengths are still
to be determined.

C. Chlorophyll–Chlorophyll Adducts

In the absence of extraneous nucleophiles (and these can
be readily minimized by thorough drying of the solvents and
the chlorophyll), chlorophyll can act as a donor via its keto
C=O group to the Mg of another chlorophyll acting as acceptor
to generate a dimer, (Chl a)$_2$, or an oligomer, (Chl a)$_n$, in
concentrated solutions in aliphatic hydrocarbon solvents (19).
Chl a oligomers absorb in the red maximally at ~678 nm, very
similar to the P680 maximum of green-plant antenna. This similarity has led to proposals that in vivo antenna chlorophyll
may have a structure similar to that of (Chl a)$_n$ in aliphatic
hydrocarbon solvents (20).

D. Chlorophyll Interactions with Bifunctional Ligands

When chlorophyll interacts with a monofunctional nucleophile (ligand) it is evident that a monomeric chlorophyll
species must result. With a bifunctional nucleophilic ligand
containing two donor centers it is likewise evident that a
linear polymer should be formed. Bifunctional ligands that
function as cross-linking agents for chlorophyll include
dioxane (two oxygen atoms), pyrazine (two nitrogen atoms) as
well as many others (21). The oxygen, nitrogen, or sulfur
atoms in a bifunctional ligand can coordinate to the Mg atoms

of two chlorophyll molecules to form species of the structure
-Chl-L-Chl-L-Chl-. If the Mg in such an adduct is allowed a
coordination number of 6, a pyrazine adduct forms with the
probable structure -Chl-N N-Chl-N N-Chl-. In such an adduct
the Mg is 6-coordinated. There is some evidence that suggests
that (Chl a)$_2$ can also be cross-linked to form an adduct of
the composition (Chl-L-Chl)$_n$ in which both 5- and 6-coordin-
ated Mg is present. Adducts with bifunctional ligands can
easily attain colloidal dimensions. Addition of an equimolar
amount of pyrazine (21) or dioxane (22) precipitates chlorophyll even from a good solvent such as carbon tetrachloride.

Chlorophyll-bifunctional ligand adducts have larger red-
shifts than those observed in (Chl a)$_2$ or (Chl a)$_n$. The red-
shift in these aggregates is roughly related to the distance
between the nucleophilic centers in the ligands; the greater
the distance at which the macrocycles are held, the smaller
the red-shift. For a considerable number of bifunctional ligands the absorption maxima are red-shifted to the 680-700 nm
range, making Chl a-bifunctional ligand adducts of interest as
possible models of the photosynthetic reaction center, which
absorbs light maximally at 680 nm and which is involved on the
oxygen side of photosynthesis.

E. Chlorophyll-Water Adducts

There is one class of bifunctional ligands that has particular importance for models of photoreaction center chlorophyll. These ligands have the generic formula RXH, where
R = H or alkyl, and X = O, S or NH. The O, S, and N atoms
have lone-pair electrons not otherwise involved in chemical
bonding that are available for coordination to the central Mg
atom of chlorophyll. The hydrogen atom attached to the
nucleophilic O, S, or NH is available for hydrogen-bonding to
the keto C=O or acetyl C=O functions in Chl a or Bchl a.
Typical of this class of bifunctional ligands are ethanol,
CH_3CH_2OH, ethanethiol, CH_3CH_2SH, or n-butylamine, $n-C_4H_9NH_2$,
and, most important, water, HOH. Coordination of water to a
cation in general increases the positive charge on the coordinated atom, making the attached H atom a much stronger donor
to hydrogen bonds. Water is of exceptional interest because
chlorophyll-water adducts formed by the cross-linking of
chlorophylls by water are strongly red-shifted and are
photoactive.

Addition of water to a solution of (Chl a)$_n$ in n-octane
converts the blue solution (λ_{max} ~ 680 nm) to a yellow-green
colloidal dispersion with an absorption maximum in the red at
740 nm (23). The conversion of P680 to P740 is accompanied by
an equally striking change in the ir spectrum; the absorption

peak at 1652 cm^{-1} assigned to a keto C=O•••Mg vibration in (Chl $\underline{a}$)$_n$ is replaced by a very strong peak at 1638 cm^{-1} that has been assigned to a keto C=O•••O•••Mg hydrogen bond formed
$\qquad\qquad\qquad\qquad\qquad\qquad\qquad\quad$H
between two chlorophyll molecules (24).

An early speculation about the structure of this adduct is shown in Figure 3 (25).   Two possible orientations of the chlorophyll are shown.   In one configuration (Fig. 3a) the chlorophyll molecules are staggered with the phytyl chains on both sides of the stack; any two of the Chl $\underline{a}$ molecules are arranged with C$_2$ symmetry.   In the step configuration (Fig. 3b), all of the phytyl chains are on one side of the micelle, and the Chl $\underline{a}$ molecules in the stack have only translational symmetry.   The more recent X-ray crystal structure of ethyl chlorophyllide $\underline{a}$•2H$_2$O determined by Strouse $\underline{et\ al.}$ (26) and Kratky and Dunitz (27) support a P740 structure in which the phytyl chains are all on the same side.   (Ethyl chlorophyllide $\underline{a}$ is a derivative of Chl $\underline{a}$ in which the large phytyl chain is replaced by the small ethyl group;   this allows the

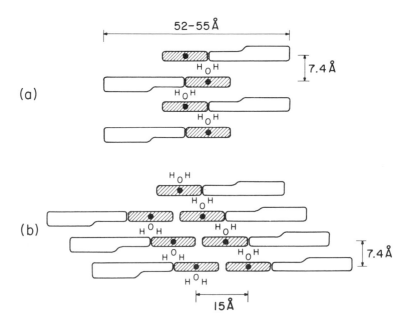

FIGURE 3.   Early representation of the chlorophyll-water adduct P740 (25).   The chlorophyll molecules are cross-linked by water.   The chlorophyll molecules in (a) are arranged with C$_2$ symmetry, in (b) there is only translational symmetry.   The two stacks in (b) are now known to be linked by additional water molecules (26,27).

formation of linear stacks that can form two-dimensional
sheets and a three-dimensional crystal). In Chl $\underline{a}$ itself, the
bulky phytyl chains prevent formation of highly ordered three-
dimensional structures. It is possible that P740 consists of
(portions) of two ordered sheets arranged as in the three-
dimensional structure but with phytyl chains projecting above
and below the two-layer structure. Some longer range periodi-
city might then be introduced by intermingling of the phytyl
chains, so that the P740 structure might be visualized as a
quasi-perodic structure in which ordered bilayers of macrocycles are separated by aliphatic regions consisting of phytyl
chains. [See the P740 structure suggested by Kratky and
Dunitz (27)].

Bchl $\underline{a}$ also forms adducts with water that likewise have
strongly red-shifted electronic transition spectra, and some
Bchl $\underline{a}$-H_2O adducts have been prepared whose optical properties
have a remarkable resemblance to that of $\underline{in\ vivo}$ Bchl $\underline{a}$ in
purple photosynthetic bacteria (28).

Chlorophyll-water adducts have played a crucial role in
the development of photoreaction center models. The exceptionally large red-shifts in the visible absorption spectra,
and the unusual photoactivity of P740 in red-light have provided the point of departure for all of the photoreaction
center chlorophyll models that are under investigation at this
writing.

IV. EPR PROPERTIES OF IN VIVO PHOTOREACTION CENTERS

The most revealing of the probes that can be used to monitor the primary photochemistry of the light conversion act in
photosynthesis is epr spectroscopy. In 1956, Commoner $\underline{et}$
$\underline{al}$. (29) discovered that paramagnetic species are formed in
the light-conversion act in green plants and that these can
readily be detected by an epr experiment. The paramagnetic
species have been shown with high probability to be chlorophyll free-radicals, that is, oxidized chlorophyll with one
unpaired electron. The $\underline{in\ vivo}$ photo-signal is rapidly reversible, disappearing in a short time when the red light is
turned off.

It has been shown by Kok (30,31) that photobleaching
occurs in green plants at ~700 nm, and by Duysens (32,33) that
in purple photosynthetic bacteria photobleaching occurs at
~865 nm during active photosynthesis. From these optical
experiments it was deduced that the photoreaction center
chlorophylls absorb maximally at 700 nm and 865 nm, respectively. The photobleaching was interpreted as an oxidation
process:

$$P700 + h\nu \rightarrow P700^{+}_{\bullet} + e^{-}$$

$$P865 + h\nu \rightarrow P865^{+}_{\bullet} + e^{-}$$

Both the optical transients and the epr signals could be produced by chemical oxidants, reinforcing the view that the reaction center chlorophyll underwent oxidation when irradiated with light, and that the oxidized chlorophyll, containing an unpaired electron was the source of the epr signal. It should be noted that $P700^{+}_{\bullet}$ and $P865^{+}_{\bullet}$ are always written as cations because oxidized Chl a in vivo is known to be a cation (34). However, there is no evidence that establishes oxidized P700 or P865 as charged species; as far as the epr signal is concerned, $P700^{+}_{\bullet}$ and $P865^{+}_{\bullet}$ could be neutral species. Experiments in which the photobleaching was compared to the number of free-radicals detected by epr established that the ratio of light-induced spins in the epr signal to bleached P700 was, within experimental error, 1:1 both in green plants (35,36) and in photosynthetic bacteria (37,38). The evidence is strong that one photon absorbed by P700 or P865 produces one electron. The assignment of the epr signal to $P700^{+}_{\bullet}$ and $P865^{+}_{\bullet}$ is also supported by kinetic studies that show the epr signal and the photobleaching of P700 have similar formation and decay kinetics (35,36). In photosynthetic bacteria P865 is oxidized at room temperature or 4K at the same rate as the photo-epr signal is generated (39). The correlation of this signal in both green-plants and photosynthetic bacteria with a one-electron oxidation of the photo-reaction center is generally regarded as firmly grounded in experiment.

Optical data per se, whatever their other merits, give no structural information. The epr data, however, can be used to provide structural information. The photo-epr signal has a g-value of 2.0025, indicative of a "free" electron (delocalized over a framework composed of carbon and hydrogen atoms), a gaussian lineshape, and a peak-to-peak linewidth of ~7 gauss. The corresponding photoinduced $P865^{+}_{\bullet}$ signal in photosynthetic bacteria is also reversible, gaussian, has a g-value of 2.0025, and a peak-to-peak linewidth of ~9.5 gauss.

Close examination of the otherwise featureless epr signals of $P700^{+}_{\bullet}$ or $P865^{+}_{\bullet}$ reveals significant information of a structural nature. Despite the absence of fine structure in the epr signal that can be used to extract information about the magnitude of the interactions between the unpaired electron and the atomic nuclei that have spin (mainly ^{1}H), it turns out to be possible to extract valuable information even from the line shape of the featureless epr signal. The epr signals of Chl $\underline{a}{\cdot}L_1^{+}_{\bullet}$ and Bchl $\underline{a}{\cdot}L_1^{+}_{\bullet}$ produced by chemical oxidation of monomeric chlorophyll are identical in g-value with $P700^{+}_{\bullet}$ and

P865$^{+}_{\cdot}$, the monomeric free radicals and the oxidized P700$^{+}_{\cdot}$ and P865$^{+}_{\cdot}$ both have gaussian lineshapes. However, there is a very important difference between Chl $\underline{a} \cdot L_1 ^{+}_{\cdot}$ and P700$^{+}_{\cdot}$ or between Bchl $\underline{a} \cdot L_1 ^{+}_{\cdot}$ and P865$^{+}_{\cdot}$, and this is in the linewidth. Chl $\underline{a} \cdot L_1 ^{+}_{\cdot}$ has a linewidth of ~9.3 gauss and the epr signal of Bchl $\underline{a} \cdot L_1 ^{+}_{\cdot}$ has a linewidth of ~13 gauss. The linewidths of P700$^{+}_{\cdot}$ (~7 gauss) and P865$^{+}_{\cdot}$ (~9.5 gauss) are significantly narrower. A comparable narrowing is also observed when the free-radical signals in fully deuterated algae and photosynthetic bacteria are compared to the signals from fully deuterated monomeric Chl $\underline{a} \cdot L_1 \cdot$ and Bchl $\underline{a} \cdot L_1 \cdot$, respectively.

The narrowing of the epr signal in photosynthesizing organisms or chloroplasts relative to the linewidth of monomeric chlorophyll cations prepared by chemical or electrochemical oxidation has been known for a long time. The discrepancy was largely ignored or was attributed to the biological environment in which the _in vivo_ photoreaction center chlorophyll finds itself. The discrepancy in the linewidths makes is impossible to equate P700$^{+}_{\cdot}$ or P865$^{+}_{\cdot}$ with the corresponding monomeric cation free-radicals, and an explanation more satisfying than an otherwise unspecified "biological environment" must be sought.

V. THE CHLOROPHYLL SPECIAL PAIR (Chl$_{sp}$)

The epr properties of the P740 chlorophyll-water adduct radical cation provide the essential clue for an explanation of the epr signal narrowing in P700$^{+}_{\cdot}$ and P865$^{+}_{\cdot}$. The P740 species made by the hydration of (Chl $\underline{a}$)$_n$ in aliphatic hydrocarbon solvents, as mentioned above, is photoactive by the epr criterion when irradiated with red light. Again the g-value of this signal is ~2.0025, indicative of the presence of an unpaired electron delocalized over a carbon-hydrogen framework, but the signal is exceptionally narrow, being on the order of ~1 gauss. Not only is the chlorophyll molecule large, containing 35 protons on the macrocycle, but the P740 adduct is a very large colloidal micelle. It would thus appear at first sight that the epr line should be broadened by the enormous number of protons with which the unpaired electron in P740$^{+}_{\cdot}$ can interact.

The unusually narrow linewidth observed from P740$^{+}_{\cdot}$ can be rationalized by delocalization of the unpaired spin over the entire ensemble of chlorophyll in the P740 micelle. This delocalization can be viewed as a very rapid process of spin migration in which the unpaired electron occupies equivalent sites in the chlorophyll molecules constituting the aggregate. Given a sufficiently high spin-migration rate, the epr

signal collapses to a narrow line. Norris et al. (40) have
carried out a rigorous analysis of the relationship of epr
linewidth to the number of equivalent sites over which spin
migration or delocalization occurs. This analysis requires
that all of the chlorophyll sites be equivalent, and that the
epr lineshape be gaussian. Under these conditions it can be
shown that an unpaired electron delocalized over an aggregate
of N equivalent molecules has a linewidth, ΔH_N, that is related to the linewidth ΔH_M of an unpaired electron delocalized
in a monomer by the relationship:

$$\Delta H_N = \Delta H_M / \sqrt{N}$$

where ΔH_M is the linewidth of monomeric Chl $\underline{a} \cdot L_1^+$ or
Bchl $\underline{a} \cdot L_1^+$. Delocalization over ~100 Chl $\underline{a}$ molecules accounts for the very narrow $P740^+$ signal. Delocalization
over two Chl $\underline{a}$ or Bchl $\underline{a}$ molecules accounts with reasonable
precision for the observed narrowing of the $P700^+$ or $P865^+$
signal (Table I). The narrowing of the in vivo $P700^+$ and
$P865^+$ signals is entirely analogous to that observed in the
epr signals from organic aromatic monocation free-radicals
that undergo spontaneous dimerization to form a monocation
dimer free radical in which the epr linewidth is reduced by a
factor of $1/\sqrt{2}$ [for a review of dimeric organic monocation
free-radicals, see Bard et al. (41)]. Indeed, spontaneous
dimerization of highly aromatic organic monocation free-
radicals is so common that if the in vivo resting photoreaction centers in P700 and P865 did not have optical properties significantly different from those of monomeric Chl $\underline{a}$
or Bchl $\underline{a}$, it would be necessary to consider seriously the
possibility that special-pair formation in the oxidized $P700^+$
or $P865^+$ may occur after the initial oxidation. Recent observations by Norris and Bowman (42) by subpicosecond epr
spectroscopy, however, indicate that the primary electron
donor in the photoreaction center is a pair of chlorophyll
molecules even on the subpicosecond time scale.

The epr linewidths in vivo can thus be rationalized by the
postulate that the primary electron donor is a pair of chlorophyll molecules. The linewidth narrowing appears to hold
equally well for photoreaction centers containing Chl $\underline{a}$ or
Bchl $\underline{a}$, which includes most photosynthetic organisms. We have
called the pair of chlorophylls acting as the primary electron
donor a "special pair" to preclude confusion with the "true"
dimer (Chl $\underline{a})_2$. The true dimer occurs in carbon tetrachloride
or benzene solution, absorbs maximally at 675 nm, is not photoactive, and gives a broad epr signal scarcely distinguish-
able from Chl $\underline{a} \cdot L_1^+$, indicating that spin delocalization does
not occur in this dimer. The (Chl $\underline{a})_2$ dimer is formed by a
keto C=O•••Mg interaction and the two macrocycles are probably

Table I. Comparison of EPR Linewidths of __in vivo__ P700$^{+\cdot}$ and P865$^{+\cdot}$ with __in vitro__ Chl $\underline{a}^{+\cdot}$ and Bchl $\underline{a}^{+\cdot}$

System[a]	ΔH_{obs}[b]	R[c]	N[d]
__in vitro__[e]			
Chl $\underline{a}^{+\cdot}$	9.3 ± 0.3		
^{2}H-Chl $\underline{a}^{+\cdot}$	3.8 ± 0.2		
Bchl $\underline{a}^{+\cdot}$	12.8 ± 0.5		
^{2}H-Bchl $\underline{a}^{+\cdot}$	5.4 ± 0.2		
Photosynthetic Bacteria			
Rhodospirillum rubrum	9.1 ± 0.5	1.41	1.9
Rhodopseudomonas spheroides	9.6 ± 0.2	1.33	1.8
^{2}H-R. rubrum	4.2 ± 0.3	1.29	1.7
^{2}H-Rps. spheroides	4.0 ± 0.2	1.35	1.8
Algae			
Synechococcus lividus	7.1 ± 0.2	1.31	1.7
Chlorella vulgaris	7.0 ± 0.2	1.33	1.8
Scenedesmus obliquus	7.1 ± 0.2	1.31	1.7
^{2}H-S. lividus	2.95 ± 0.1	1.29	1.7
^{2}H-R. rubrum	2.7 ± 0.1	1.41	2.0
^{2}H-S. obliquus	2.7 ± 0.1	1.41	2.0
			Avg. 1.8

[a]Fully deuterated species are designated by the prefix ^{2}H-.

[b]Peak-to-peak linewidth. Gaussian line shapes, g = 2.0025 ± 0.0002.

[c]Ratio of __in vitro__/__in vivo__ epr linewidth.

[d]Number of chlorophyll molecules over which the unpaired spin is delocalized, calculated from $N = (\Delta H_M/\Delta H_{obs})^2$.

[e]Prepared by chemical oxidation with I_2 or $FeCl_3$.

at right angles to each other, so that very little overlap in the two π-systems occurs (43). The chlorophyll special pair consists, like the dimer (Chl $\underline{a}$)$_2$, of two chlorophyll molecules but these are very likely in a parallel rather than perpendicular orientation and in intimate π-π contact. The geometry of the special pair results from the intervention of bifunctional ligands. Because there are so many optical, redox, stoichiometric, and structural differences between the special pair (designated as Chl_{sp}) and the dimer (Chl $\underline{a}$)$_2$, we feel that the Chl_{sp} terminology for the reaction center chlorophyll is necessary to avoid ambiguity. This is especially important because of the possibility that species such as (Chl $\underline{a}$)$_2$ and (Chl $\underline{a}$)$_n$ may be involved in the antenna function (19,20,44).

Because the proposal that the primary donor in light-energy conversion may be a Chl_{sp} has many implications, additional experimental evidence has been sought. Although the epr signals from $P700^{+\cdot}$ and $P865^{+\cdot}$ have no structure and, thus, no information about hyperfine coupling constants can be deduced from them, it still is possible to extract such information by electron-nuclear double resonance (endor) spectroscopy. Endor is a high-resolution variant of epr (45) that makes it possible to extract electron-proton hyperfine coupling constants (hfcs) even when these cannot be deduced directly from a conventional epr experiment. A comparison of hfcs in monomeric Chl $\underline{a} \cdot L_1^{+\cdot}$ and Bchl $\underline{a} \cdot L_1^{+\cdot}$ with $P700^{+\cdot}$ and $P865^{+\cdot}$ is a more stringent test of spin sharing than is line narrowing. A particular hfc in an aggregate in which spin delocalization occurs is related to the hfc in the monomer by the equation:

$$a_{Ni} = a_{Mi}/N$$

where a_{Mi} is the electron-nuclear hyperfine coupling constant for the ith nucleus in the monomer, and a_{Ni} is the coupling constant for that molecular site in an aggregate of size N. For a Chl_{sp}, where $N = 2$, a particular hyperfine coupling constant will be one-half that of the monomer. Coupling constants accounting for >80% of the linewidth of Chl $\underline{a} \cdot L_1^{+\cdot}$ have been assigned by Scheer et al. (46) from endor spectra of a series of isotopically substituted Chl $\underline{a}$ derivatives. Table II compares the hfcs of monomeric Chl $\underline{a} \cdot L_1^{+\cdot}$ and Bchl $\underline{a} \cdot L_1^{+\cdot}$ with those of $P700^{+\cdot}$ and $P865^{+\cdot}$, respectively. The assignment of the endor spectra are more straightforward for $P865^{+\cdot}$ than for $P700^{+\cdot}$, but for both cases the simplest interpretation of the endor data is that the in vivo coupling constants are (approximately) halved relative to the monomer free-radical, thus providing convincing support for the Chl_{sp} model.

Table II. Spin Delocalization as Deduced from a Comparison of in vitro and in vivo ENDOR Hyperfine Coupling Constants[a]

System	Protons[b]	Hyperfine Coupling Constants (MHz)				
		Chl $\underline{a}^{\dotplus}$	Bchl $\underline{a}^{\dotplus}$	in vivo		N[c]
Algae						
Synechococcus lividus	$(\alpha, \beta, \delta, 10)$	0.67		--[d]		
	(1a, 3a, 4a)	3.19		1.68		1.9
Chlorella vulgaris	5a	7.45		3.73		2.0
	7,8	11.8		5.36		2.2
					Avg.	2.0
Photosynthetic Bacteria						
Rhodospirillum rubrum	$(\alpha, \beta, \delta, 10)$		1.4	0.82		1.7
	1a		5.32	2.22		2.4
	5a		9.8	5.67		2.1
	(3,4,7,8)		14.0	7.0		2.0
					Avg.	2.0

[a]Data from ref. (47).
[b]See Figure 2 for proton numbering.
[c]Calculated from N = a_i(in vitro)/a_i(in vivo).
[d]Not assigned.

Bchl $\underline{a}$ triplet states can be detected by epr or by
optically detected magnetic resonance either in intact photosynthetic bacteria or in bacterial reaction centers when the
normal forward path of photosynthesis is blocked (47,48). A
comparison of the properties of in vivo ^{3}Bchl $\underline{a} \cdot L_1$ with ^{3}P865
rules out monomeric ^{3}Bchl $\underline{a}$ as the origin of the in vivo
triplet signal. Both the zero-field splitting parameters
(which characterize the symmetry and the area over which the
triplet state electrons are deployed) and the polarization of
the in vivo triplet epr spectrum strongly suggest delocalization over more than one Bchl $\underline{a}$ molecule in ^{3}P865 [for a
thorough review, see Levanon and Norris (49)].

Perhaps the most convincing evidence for an in vivo Chl_{sp}
comes from recent investigations by electron spin-echo spec-
troscopy on Chl $\underline{a} \cdot L_1^+$ and Bchl $\underline{a} \cdot L_1^+$ as compared to P700$^+$
and P865$^+$, respectively. Electron spin-echo spectroscopy is a
pulsed form of epr spectroscopy in which the unpaired spin is
detected by spin-echo (refocusing of the spins) from the free-
radical excited by an appropriate sequence of high-intensity
radio-frequency pulses. When the hyperfine interaction between the unpaired spin and the nitrogen atoms in the pyrrole
and pyrroline rings that make up the chlorophyll macrocycle
are studied by electron spin-echo spectroscopy, it can be deduced immediately that simple monomeric cation free-radicals
cannot account for the in vivo signals. The spin-echo pulse
envelopes, which have superimposed upon them a modulation pattern produced by interaction of the nitrogen atoms with the
unpaired spin, show differences between the in vitro and in
vivo patterns in these experiments which preclude the
possibility that the respective in vivo signals arise
from Chl $\underline{a} \cdot L_1^+$ and Bchl $\underline{a} \cdot L_1^+$ (50).

VI. MODELS FOR Chl_{sp}

We summarize very briefly the general requirements for a
satisfactory Chl_{sp} model. The lowest energy $S_0 \rightarrow S_1$ electronic transition must be red-shifted relative to the corresponding transition manifold of the antenna chlorophyll to
assure effective trapping of singlet excitation energy that
migrates through the antenna. The model must be able to account for the optical, redox, and spin-sharing properties of
P700 and P865. Detailed discussions of these matters as well
as comparisons of how well various models meet the requirements are available (51-55).

Early models of the Chl_{sp} showed the two chlorophyll
molecules cross-linked by one molecule of water, but in such a
way that the carbonyl functions in both the keto group at

position 9 and the carbomethoxy function at position 10 were involved in the cross-linking (14,56,57). This model placed the two Chl a macrocyles about 6 Å apart. Simultaneous hydrogen-bonding to the carbomethoxy and keto C=O groups prevents a parallel orientation of the transition dipoles in the two macrocycles, and the angles in the hydrogen bonds are not optimum. Consequently, a number of variants has been suggested (51,52,58-60).

The original Chl_{sp} model was asymmetric in the sense that the two macrocycles are not identical, as one of the Chl a macrocycles is acting as donor and the other as acceptor in the coordination interactions that cross-link them. Based on the Strouse (26) and Kratky and Dunitz (27) structures for ethyl chlorophyllide a·$2H_2O$ it is possible to derive models that are asymmetric or have only translational symmetry by using the first two or first three macryocycles of the stacks that are the fundamental units of the crystal structures. Using the first two members of the stack gives an asymmetric model. In the first three, the second and third have translational symmetry (51,52). In these models, as in the stack in the crystal, the macrocycles are cross-linked by one molecule of water whose oxygen atom is coordinated to Mg, and one of whose hydrogen atoms is hydrogen-bonded to the keto C=O function of the next macrocycle in the stack. There is some question whether asymmetric models or models with only translational symmetry provide equivalent sites for spin sharing. As these structures show large red-shifts there is a presumption that the electronic interactions that provide the red shift may also provide sufficient equivalence for spin sharing. Judging from spin sharing in $P740^{+}$ at least, translational symmetry is sufficient. There is no experimental basis at this writing for deciding whether an asymmetric pair will be able to share spins, and further discussion of this point is deferred until such information is forthcoming.

Models with C_2 symmetry have received the most attention. Fong (58) has presented a Chl_{sp} model with C_2 symmetry that resembles an orientation suggested earlier as a possible element in the P740 structure (Fig. 3a). Fong chose, for reasons that are not obvious, to cross-link the two macrocycles by hydrogen-bonding only to the carbonyl functions of the Ring V carbomethoxy groups, the keto C=O functions not being involved in any way in the cross-linking of his special-pair proposal. Such a choice is intrinsically implausible, however, because the keto C=O function is, from all considerations, a stronger donor function (19). Aside from this, the Fong model places the macrocycles ~5.7 Å apart and in a configuration such that the transition dipoles are ~60° to each other. Exciton calculations indicate no significant red-shift for this model. Fong and Koester (61) prepared a low-temperature Chl a-water

species to which they assign a structure involving hydrogen-
bonding to the carbomethoxy groups. Examination of the low-
temperature species by ir spectroscopy shows quite conclu-
sively that it is the keto C=O and not the carbomethoxy
carbonyl groups that are involved in the formation of the low
temperature species (60). The Fong model is also inconsistent
with experimental observations on the covalently linked
special-pair models described below. Linked pyrochlorophyll a
macrocycles, which lack a Ring V carbomethoxy group, assume a
folded configuration even more readily than do linked Chl a
macrocycles. The Fong model is thus inconsistent with all
known experimental evidence. The idea of C_2 symmetry, however, has its attractions.

Shipman et al. (60) and Boxer and Closs (59) have proposed
a Chl_{sp} structure (Fig. 4) that has C_2 symmetry but avoids the
problems posed by the Fong structure. C_2 symmetry in a Chl_{sp}
would assure full equivalence of the highest occupied molecular orbital (HOMO) on each of the macrocycles that make up the
special pair. The interaction of the two HOMO's would generate two supermolecular orbitals with equal contributions from
each of the monomeric units. Removal of an electron from the
supermolecular HOMO of the special pair would leave the

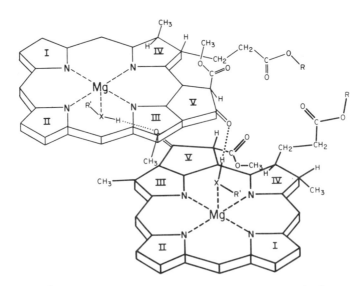

FIGURE 4. Chl_{sp} model of Shipman et al. (60).   The two
chlorophyll molecules are cross-linked by two molecules of a
nucleophile from the class R'XH where R' = H or alkyl, and
X = O, S, or NH.   The nucleophile may be water, ethanol, or
nucleophilic groups present in protein side-chains.   A similar
structure has been proposed by Boxer and Closs (59).

remaining unpaired electron delocalized equally over both macrocycles as required by epr and endor for the oxidized photoreaction center in vivo.

The Chl_{sp} model of Shipman et al. (60) was originally based on low-temperature ir data and on the Kratky and Dunitz (27) and Strouse (26) structures for ethyl chlorophyllide a·$2H_2O$. The cross-linking coordination interactions are provided by two molecules of a bifunctional ligand with the generic formula R'XH, where X is O, S, or NH, and R' is H or alkyl. Thus, the cross-linking agent can be a primary alcohol, amine, or thiol, or simply water. In the Shipman et al. (60) and the Boxer and Closs (59) proposals, the chlorophyll oxygen function used in cross-linking is the keto C=O function of Ring V.

The arrangement in Figure 4 brings the two macrocycles to within π-π stacking distance of 3.6 Å (as deduced from CPK space-filling molecular models), a distance that just brings the π-systems into contact at their van der Waals radii. Optimum π-π overlap is guaranteed in this orientation by the use of the keto C=O group as the cross-linking site. In this orientation the transition dipoles in the two macrocycles are parallel rather than at 60° to each other, as is the case when cross-linking is effected through the carbomethoxy C=O groups.

The optical requirements for a Chl_{sp} model are satisfied on both theoretical and experimental grounds by the Shipman et al. (60) and Boxer and Closs (59) arrangements. The Q_y (lowest excited state) transition dipoles are parallel and, from exciton theory, the red-shifted Q_y exciton transition should have all of the oscillator strength. Exciton calculations based on the recorded optical properties of the ethyl chlorophyllide a·$2H_2O$ crystal and of ethyl chlorophyllide a monolayers on water (62,63) indicate that the strong hydrogen-bonding to the keto C=O function of Ring V in combination with π-π stacking provides an environmental shift for each macrocycle to 686 nm, precisely that required to account for a ~700 nm absorption in the proposed structure.

Shipman et al. (60) have generalized the model by pointing out that nucleophiles other than water can be used for cross-linking. The arrangement of the macrocycles in Figure 4 is such that the cross-linking nucleophile is not restricted only to a small ligand such as water. The proposed structure is sufficiently open at either end so that even very large nucleophilic groups can be inserted without difficulty. Nucleophilic ligands of the class R'XH include nucleophiles characteristically present in protein side chains, and thus the possibility of protein participation in Chl_{sp} presents itself. Based on in vivo experiments the nucleophilic hydroxyl groups of serine and threonine, the $-NH_2$ group of lysine or the -SH group of cysteine could be used for cross-

linking. The amino acids, arginine, tyrosine or proline also
have nucleophilic groups that could be used in the same way.
Nor does there appear to be a rigid requirement that the two
cross-linking nucleophiles be identical. One protein side-
chain nucleophile plus one molecule of water, or two different
side-chain nucleophiles from the protein could be used for
cross-linking purposes. Full equivalence of the two macrocycles would be lost by the use of two dissimilar nucleophiles, but the near equivalence might still be adequate for
spin-sharing purposes. It is even conceivable that a certain
degree of non-equivalence of the two macrocycles might be
desirable in Chl_{sp} function; two such macrocycles would not be
expected to have identical redox properties, and one of them
would be better oxidant, the other a better reductant.
Internal charge transfer might thereby be facilitated as the
first step in charge separation. The possibility that a
variety of photoreaction centers could be present in living
organisms that have essentially the same optical properties
but which differ somewhat in redox or spin-delocalization
properties may need to be taken into account in considering in
vivo photosynthesis.

The possible participation of protein in the formation of
in vivo Chl_{sp} has some features of special interest. Bacterial reaction centers isolated from photosynthetic bacteria
contain Bchl a, bacteriopheophytin a, and three polypeptides;
and numerous chlorophyll-protein complexes have been isolated
from green-plants (64). As there is every reason to suppose
that in chlorophyll-protein complexes the chlorophyll is not
attached to the protein by covalent bonds, the kind of coordination interaction under discussion here provides a rationale
for optical red-shifts in chlorophyll-protein complexes. The
red-shifts could originate not as a result of a direct chlorophyll-protein coordination interaction but because of interactions between chlorophyll macrocycles positioned by protein
side-chain groups.

VII. BIOMIMETIC CHLOROPHYLL SPECIAL PAIRS

Chlorophyll entities with optical and epr properties of
P700 and $P700^{+}$ have been prepared in the laboratory (a) by
self-assembly and (b) by chemically linking two chlorophyll
macrocycles with a covalent bond.

A. Self-Assembled Systems with Optical Absorption Near 700 nm

By self-assembly we mean the formation of molecular
species (loosely termed aggregates) from monomeric units by

the expression of forces implicit in the molecular structure of the monomer. The formation of a molecular crystal is an obvious example of self-assembly. Between a highly ordered, three-dimensional crystal and a gas at low pressure there is a continuum of states ordered to different extents. The extent of the ordering can range from highly ordered monolayer assemblies to small systems where ordering occurs over only two or a few molecules. An example of a highly ordered system is the organized, rigid molecular aggregates (organizates) of specifically ordered architecture prepared by monolayer assembly techniques and described by Kuhn (65). In vivo chlorophyll appears to be in the intermediate range as far as ordering is concerned. From the fact that chlorophyll itself (i.e., without any water or other nucleophilic ligand) has never been crystallized it can be inferred that it has only a small intrinsic tendency, if any, to form highly ordered structures of a crystalline nature. However, it has long been known that aggregates of chlorophyll in colloidal dispersions (66), in films cast from concentrated solutions (67), or films formed by adsorption on solid substrates (68), highly concentrated solutions in nonpolar solvents (44,67), or formed by interaction of chlorophyll with bifunctional ligands (21,69) must be ordered in rather specific ways because of the optical changes resulting from aggregate formation. These chlorophyll aggregates vary in the extent of ordering from local interactions involving two chlorophyll molecules as in the dimer, (Chl a)$_2$, to order extending over hundreds of chlorophyll molecules in colloidal micelles of the chlorophyll a-water adduct absorbing at 740 nm, or in the chlorophyll a-pyrazine adduct absorbing at ~690 nm. The ordering in these systems is an expression or manifestation of both the coordination properties of the Mg atom and the oxygen functions of the chlorophyll molecule, which, under suitable conditions can lead to large-scale order.

The notion that specific coordination interactions between chlorophyll molecules may provide the rationale for the kind of ordering experienced by in vivo chlorophyll has had a surprisingly long gestation period. A pioneer in the study of what can now be described as self-assembled chlorophyll systems was Krasnovsky (67) who noted the resemblance between the optical properties of chlorophyll in concentrated solutions or films and those of native chlorophyll in vivo. Brody and Brody (70) and Stensby and Rosenberg (71) found evidence from absorption and emission spectroscopy of chlorophyll for the formation of ~700 nm absorbing species at low temperatures in polar solvents. Fong and Koester (61) also noted the development of long wavelength absorptions in strongly cooled solutions of Chl a in nonpolar solvents containing water. The stoichiometry of the long-wavelength absorbing species that

they deduced from a somewhat tortured analysis of the optical
data is suspect because they ignored the non-conservation of
oscillator strength when the long wavelength species formed,
and because their experimental system contained much more
water than they assumed in their equilibrium calculations.
The structure of the low-temperature form advanced by Fong and
Koester (61) has been shown to be incompatible with low-
temperature infrared spectra (60). It was generally considered that the red-shifts observed by all of these workers was
a result of aggregation of chlorophyll to dimers, trimers and
perhaps higher aggregates. As will be seen below, a really
convincing structural assignment of the low-temperature chlorophyll species in such systems is still forthcoming. It is
evident, however, that in these systems self-assembly does
occur as the systems are cooled, and, because the species
produced by strong cooling have optical and epr properties
strongly reminiscent of P700 (60,72), the possibility that
similar forces may be operative in the formation in in vivo
P700 has stimulated renewed interest in self-assembled chlorophyll systems (73).

Self-assembly in films may yield mixtures of species as
judged from visible absorption spectra. Because of the immobilization inherent in a solid film, equilibration is all
but impossible. Self-assembly in solution is preferable because the mobility of the systems permits, at least in principle, conversion of all of the chlorophyll to a configuration
of minimum energy. For conversion to a Chl a species absorbing in the vicinity of 700 nm, experience shows that a
nucleophile is required which is capable of coordination to Mg
and simultaneously able to form hydrogen bonds with the keto
C=O function in the Ring V of another molecule of Chl a.
Water has the required properties but its use poses difficulties. Water has limited solubility in nonpolar solvents so
that it is difficult to incorporate stochiometric amounts when
concentrated Chl a solutions (> 0.01 M) are used. Water is a
ubiquitous contaminant of organic solvents and of chlorophyll
preparations, which makes for difficulties in determining the
water content of the system. Finally, when water is used for
cross-linking, one of the two hydrogen atoms is not involved
in hydrogen-bonding, and this is an energetically unfavorable
situation. The second hydrogen atom could perhaps be H-bonded
to the nitrogen atom of one of the pyrrole rings of the macrocycle, but there is no experimental evidence on this point and
the fate of the second water proton is still unclear. The use
of ethanol in place of water considerably simplifies the experimental protocol. Ethanol is miscible with hydrocarbon
solvents and stoichiometric amounts of ethanol equivalent even
to 0.1 M Chl a solutions can be readily incorporated into the
system. The problem of what to do with a free hydrogen atom

on the coordinated oxygen also does not occur with a primary alcohol. The most successful self-assembled systems mimicking P700 thus have been prepared with ethanol as the cross-linking nucleophile. Self-assembly can also be achieved with n-butyl amine and ethanethiol, as well as with other nucleophiles capable of hydrogen-bond formation.

The important variables in the preparation of a self-assembled P700 species are the solvent system, the chlorophyll concentration, the cross-linking nucleophile, and the temperature. When a 0.094 M solution of Chl $\underline{a}$ in toluene containing 0.14 M ethanol is cooled to 170K the absorption maximum at 668 nm at room temperature shifts largely to 702 nm (Fig. 5) (73,74). Infrared spectroscopic investigation shows that at room temperature, where monomeric 670 nm chlorophyll is the predominant species, the keto C=O absorption maximum is at 1697 cm^{-1}. This frequency is characteristic of the free (uncoordinated) keto C=O stretch vibration. At low temperatures, where the ~700 nm absorbing species is predominant, the peak at 1691 cm^{-1} shows a large decrease in area, and a new absorption peak appears at 1658 cm^{-1}, assigned to a bound keto C=O function. The ester C=O absorption maximum at 1737 cm^{-1} shifts slightly to 1733 cm^{-1} with little or no change in area. The infrared data would appear to be incompatible with any

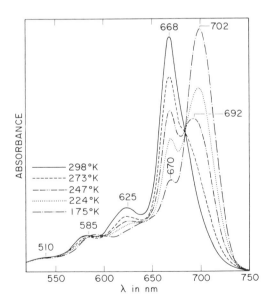

FIGURE 5. Self-assembled Chl$_{sp}$. Visible spectra of 0.094 M Chl $\underline{a}$ and 0.14 M ethanol in toluene solution as a function of temperature. Essentially complete conversion to a ~700 nm absorbing species occurs at 175K (73,74).

low-temperature structure involving participation by the ester C=O.

By choice of solvent, Chl $\underline{a}$ concentration, and molar ratio of Chl $\underline{a}$ to ethanol, it is possible to form systems that are still fluid when a large measure of self-assembly has already occurred. A 0.005 M solution of Chl $\underline{a}$ containing 0.05 M ethanol in methylcyclohexane is substantially converted to a species absorbing at ~700 nm at -100°C, where the system is still fluid and, thus, can be used for nmr structural studies. Systems have also been prepared where there is a substantial concentration of a self-assembled ~700 nm absorbing species at temperatures between -60 and -40°C.

In these self-assembled systems, the absorption peaks are generally broad, which can be taken to indicate structural heterogeneity in the absorbing species. This heterogeneity may stem from a distribution of assembled species, which may have essentially the same composition but in which the orientation of the macrocycles varies. The stoichiometry of these species has not been established. The principal difficulty in using the optical absorbance data is the non-conservation of oscillator strength when the ~662 nm species is converted to a ~700 nm absorbing form(s). Deconvolution of the absorption spectra into gaussian components shows that the oscillator strengths of the assembled species may be as much as 40% greater than that of the monomeric units from which they are composed. While it is reasonable to suppose that the 700 nm species are small entities, there is no evidence that permits an unequivocal choice between a structure with C_2 symmetry, as shown in Figure 4, or a stack of two or three Chl $\underline{a}$ molecules arranged as in the ethyl chlorophyllide $\underline{a} \cdot 2H_2O$ structure (26,27). Exciton calculations indicate that a stack of two will have its absorption maximum in the red at 695 nm and a stack of three at 702 nm (63). All of these structures (as well as a selection of longer stacks) may be present in the assembled system, and until additional structural information becomes available such a possibility cannot be ignored.

The oxidized self-assembled Chl_{sp} show the desired spin sharing properties. A solution of 0.1 mM Chl $\underline{a}$, 1.5 mM ethanol and 100 mM tetranitromethane in toluene gives a suitably narrowed 7.5 gauss epr signal when irradiated with 700 nm light at 96K. When the temperature is raised to 145K, the signal remains gaussian and its width decreases to 7.1 gauss. The epr signal from this system is very similar in line shape, linewidth, and temperature dependence to the epr signal in the green alga Chlorella vulgaris resulting from the photo-oxidation of P700.

The optical properties of the self-assembled systems lend support to the view that the ~700 nm absorbing species in these systems must have much in common with covalently linked pairs in their folded configuration discussed below. The long wavelength emission from these systems has a fluorescence maximum and lasing characteristics essentially the same as the linked pairs in their folded configuration. Self-assembled systems containing "pairs", as well as monomeric Chl $\underline{a}$ and perhaps true dimers or oligomers, have been found to be of considerable interest in energy-transfer studies. When emission is excited with 337 nm light in self-assembled systems, only fluorescence from the longest wavelength absorbing species is emitted at ~730 nm; such systems can be induced to emit coherent laser light at this same frequency (75). Excitation at 661 nm, where only monomeric Chl $\underline{a}$ absorbs, produces fluorescence at ~680 nm, the characteristic monomer fluorescence, and no fluorescence at longer wavelengths. Excitation with 700 nm light produces only ~730 nm fluorescence. Whether energy transfer appears to occur in these systems thus depends on the excitation frequency. This may have important implications for the analysis of fluorescence and energy transfer in vivo.

Depite the problems with stoichiometry and structure, self-assembled Chl_{sp} systems have many uses. Self-assembled systems are easy to prepare, as no chemical syntheses are involved, and, because they are composed of monomeric Chl $\underline{a}$ units that still have phytyl chains, the self-assembled Chl_{sp} have hydrophobic properties more likely to resemble those of the in vivo Chl_{sp}. The ease of preparation suggests that these self-assembled biomimetic systems may occupy an important place in future research in photosynthesis and biomimetic solar energy conversion.

B. Chemically Linked Biomimetic Chl_{sp}

In self-assembled systems, the desired orientation of two chlorophyll molecules required for a biomimetic representation of P700 is achieved spontaneously by making use of the intrinsic coordination properties of the chlorophyll molecule. These coordination forces, however, are not strong enough at room temperature to overcome the unfavorable entropy loss associated with the ordering of the macrocycles. Low temperatures, hostile solvents, a high chlorophyll concentration, and carefully selected chlorophyll-nucleophile ratios are desirable to induce self-assembly. At room temperature in systems that assemble at low temperatures, only minute amounts of assembled Chl_{sp} are present (these can be detected by fluorescence) and the chlorophyll is almost entirely present as monomeric Chl $\underline{a}$·L_1. The entropy loss associated with special-pair ordering at room temperature can be successfully overcome by forging a chemical link between the two macrocycles which

mechanically prevents diffusion from separating the pair
components. This reduction in the entropy of dissociation is
sufficient to permit spontaneous folding to the desired configuration even at room temperature when an appropriate
nucleophile is presented to the linked macrocycles.

Boxer and Closs (59) linked two Chl a macrocycles by an
ethylene glycol link connected at both ends by ester bonds to
the propionic acid side-chains. The chemical manipulations
required for the trans-esterification, at least up to now, are
incompatible with retention of the Mg. Consequently, reinsertion of the Mg is required after the covalent linkage is
effected. The simplest reinsertion procedure uses a solution
of magnesium perchlorate in pyridine, but elevated temperatures are required, and this results in the loss of the carbomethoxy group at position 10 (see Fig. 2). Most of the covalently linked special pair models are, therefore, constructed
of pyrochlorophyll a macrocycles in which the carbomethoxy
group has been replaced by H, considerably simplifying the
subsequent reinsertions of magnesium. Wasielewski et al. have
synthesized covalently linked pairs with both Chl a (76) and
Bchl a (77) macrocycles (Fig. 6) using a new procedure developed by Eschenmoser et al. (78) for reinsertion of Mg.

Chl a or Pyrochl a linked pairs dissolved in a polar
solvent have optical properties essentially identical with
those of the monomers from which they are constituted. Dissolved in a nonpolar solvent such as carbon tetrachloride or
benzene, the linked pairs have optical properties very similar
to those of (Chl a)$_2$ or (Pyrochl a)$_2$ species which form by

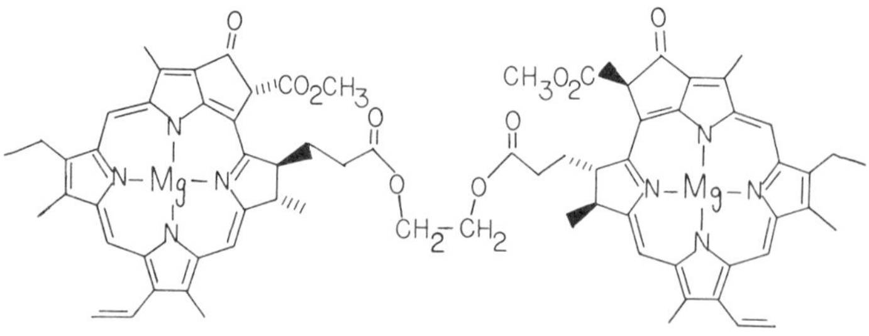

BIS(CHLOROPHYLLIDE a) ETHYLENE
GLYCOL DIESTER

FIGURE 6. Covalently linked Chl$_{sp}$ in its open configuration. The two Chl a (60) or Bchl a (77) macrocycles are
linked by an ethylene glycol bridge. Boxer and Closs (59)
were the first to report a covalently linked special pair
model, which they prepared by linking two Pyrochl a macrocycles in the same way.

either inter- or intra-molecular keto C=O•••Mg coordination bonds. Addition of water or ethanol causes a red-shift to ~700 nm, which is plausibly interpreted as a result of the assumption of the folded configuration shown in Figure 7. In the case of the linked Pyrochl a pairs, the red-shift is complete, whereas for the linked Chl a pairs the shift is partial. Steric problems are the reason for the differences in the folding behavior of the two linked pairs. Chl a in the laboratory, but not in nature, contains ~15% of a diastereo-mer, Chl a', which is formed whenever Chl a is dissolved in a polar solvent of even weak basicity (79,80). The stereoisom-erism at C-10 results in a distribution of pairs in which about 73% is the a-a pair, 25% is the a-a' pair and 2% is the a'-a' pair.

When folding occurs only the a-a and the a'-a' pairs are able to assume the closely spaced C₂ structure shown in Figure 7. Space-filling models show that the a-a' pair cannot assume the folded configuration because of steric compression of one carbomethoxy group between the two macrocycles. Fold-ing of the linked Chl a pair is, therefore, incomplete, but from the optical spectra it appears that all of the pairs that can fold do so. The capacity for complete folding in the linked Pyrochl a pair and the ease with which chemical

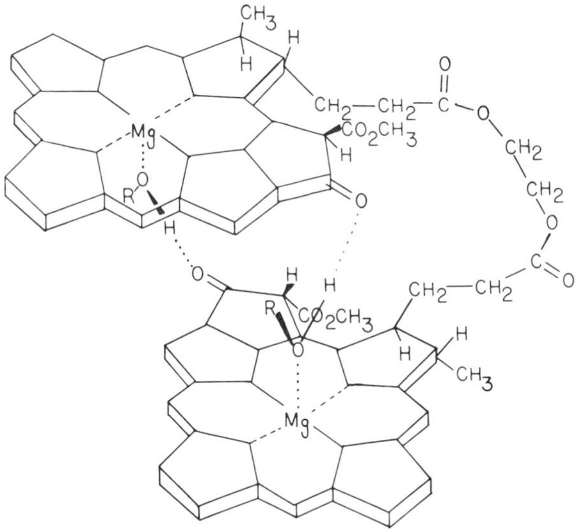

FIGURE 7. A Covalently linked Chl$_{sp}$ model in its folded configuration. The dotted lines indicate the cross-linking hydrogen bonds to the two keto C=O functions. The nucleophile is ethanol. The linkage prevents the two macrocycles from diffusing away from each other.

manipulation and Mg insertion can be carried out are responsible for the choice of Pyrochl a systems for biomimetic linked
pair models of Chl_{sp}.

The conclusion that the red-shift results from folding to
the configuration shown in Figure 7 is buttressed by nmr data.
The proton resonances of the linked Pyrochl a pair dissolved
in dry benzene are extremely broad, but in water-saturated
benzene the lines are narrow and the spectrum becomes well-
defined. The ^1H nmr spectrum of the folded form has only one
resonance for each proton, implying that the two rings are
equivalent on the nmr time scale. The resonance peak associated with the methyl group at position 5 is strongly shifted
upfield, which is exactly what is expected for this group of
protons when positioned over another macrocycle as in
Figure 7. Similar nmr observations on the linked Chl a and
Bchl a pairs all point to the assumption of the configuration
of Figure 4 in these systems when folding is induced.

Whereas the linked Chl a and Pyrochl a pairs in their
folded configurations have the optical properties of P700,
this is not the case for the linked Bchl a pair. Folding
results in an optical shift only to 803 nm rather than to
~865 nm, which is the absorption maximum of special pair
Bchl a in intact photosynthetic purple bacteria or reaction
centers. It would therefore appear that an optical shift to
~865 nm involves interactions with the additional Bchl a and
bacteriopheophytin a known to be present in the bacterial
reaction center, but the nature of the interactions required
for the optical shift is completely obscure.

An optical shift to ~700 nm on folding is not the only way
in which covalently linked pairs mimic P700. When a 0.1 mM
solution of the linked Chl a pair in water-saturated carbon
tetrachloride containing an equimolar amount of I_2 is irradiated with red light in the microwave cavity of an epr spec-
trometer, an intense gaussian epr signal with a linewidth of
7.5 gauss is observed. The narrowed linewidth of the oxidized
synthetic Chl_{sp}^{+} is consistent with expectations for an unpaired spin shared by two macrocycles. For the linked Bchl a
pair, the photooxidized species yields a gaussian epr signal
with a linewidth of 10.6 ± 0.3 gauss and a g-value of 2.0027.
Here again, the folded pair shows the desired ability to de-
localize an unpaired spin.

The redox properties of the linked pairs also mimic in
vivo P700. The oxidation potential of P700 in spinach chloroplasts is generally estimated to be +0.43 V (nhe). Oxidation
of the folded linked Chl a pair indicates an oxidation potential close to 0.5 V and distinctly lower than that of monomeric Chl a.

The optical, epr, and redox properties of the covalently
linked Chl a and Pyrochl a pairs are remarkably faithful

replications of P700 (Table III). The linked Bchl a pair mimics the spin-sharing properties of the in vivo bacterial reaction center but not yet the optical properties. The covalently linked pairs so far prepared lack the hydrophobic phytyl chain which likely plays an important part in making chlorophyll compatible with the lipid components of the photosynthetic membrane. In terms of electronic properties, however, these systems come very close to successful biomimetic replicas for the primary electron donor in the photoreaction center. The success so far attained in biomimetic representations of the primary donor in plant photosynthesis encourages the belief that the development of biomimetic solar energy conversion techniques is a realistic goal.

VIII.   MODELS OF Chl_{sp} ELECTRON TRANSFER

In the initial event of photosynthesis, the excited Chl_{sp} acts as the primary electron donor to a nearby electron acceptor. It is now generally accepted that the initial electron transfer proceeds by a radical-pair mechanism (81). (For detailed discussions of the radical-pair mechanism, see refs. 51,52,54,82). The first electron acceptor (A) is diamagnetic before the initial electron-transfer step. Electron transfer creates a radical-pair state,

$$Chl_{sp} + A \xrightarrow{h\nu} [Chl_{sp}^{\ddagger} + A^{\dot{-}}]$$

Photochemical charge transfer occurs in a few picoseconds apparently from the lowest singlet excited state of the Chl_{sp}. In a laboratory situation when charge transfer occurs there is a high probability that $A^{\dot{-}}$ will return the electron to the original donor. In the photosynthetic apparatus the probability of return after charge separation is, on the contrary, very small. The reasons for the slow back reaction are presumed to be primarily geometric. For example Warshel (83) attributes the fast forward and slow reverse electron transfer to the differential effects of small changes in the distance between donor and acceptor in the ground state and the charge-transfer state. It is generally considered that the exact positioning of the donor and acceptor, and their redox potentials, are critical factors in assuring a rapid forward and slow reverse electron-transfer reaction in the photosynthetic unit. Because of the importance of high directionality in the initial electron transfer step for the efficiency of light energy conversion there is considerable interest in providing reaction center models that mimic the high efficiency of electron transfer in vivo.

Table III. Comparison of Synthetic Photoreaction Centers with P700 and P865

System	Absorption	Fluorescence	Lasing	ΔH_{pp} (gauss)
in vivo				
P700	~700 nm			7.0
P865	~865			9.6
Covalently linked				
Bis(chlorophyllide a)ethylene glycol diester				
open	662	683,734 nm	N.L.[a]	
folded	697	730	733 nm	7.5
Bis(pyrochlorophyllide a)ethylene glycol diester				
open	666	680	N.L.	
folded	696	730	731	
Tris(pyrochlorophyllide) 1,1,1 tris (hydroxymethyl)ethane triester				
open	670	684	N.L.	
folded	670,690	726	735	
Bis(bacteriochlorophyllide a)ethylene glycol diester				
open	760		N.L.	
folded	803		735	10.6
Self-assembled				
Chl a, ethanol at 181K	702	730	732	7.1

[a] No lasing observed.

Pellin et al. (84) have described an ingenious biomimetic model for electron transfer from the Chl_{sp}. A covalently linked Pyrochl a pair is folded with a derivative of pyropheophytin a (Pyropheo a) in which the phytyl chain has been replaced by ethylene glycol. Pyropheo a is used as the acceptor because of the likelihood that pheophytin acts as an intermediary in electron transfer from the in vivo special pair. In the monoglycol ester of Pyropheo a, the unesterified hydroxyl group can act as a cross-linking nucleophile to produce a folded pair in which two Pyropheo a macrocycles are fixed at a known distance from the folded Pyrochl a dimer.

In such an assembly, the characteristic ~730nm fluorescence of the folded pair is almost completely quenched. In the pair folded by ethanol, addition of ordinary Pyropheo a enhances the fluorescence quantum yield at 730 nm. Pellin et al. measured the excited-state difference spectrum and found that the spectral changes occur in less than 6 psec after excitation, and that either a radical-pair or a charge-transfer state has been produced on a very short time scale, very much the same as happens in vivo. The spectral changes, however, persist for 10-20 ns, with subsequent return to the ground state. It thus appears that the reverse reaction in this system is at least a thousand-fold slower than the forward electron or charge transfer. It would also appear that the geometry and redox properties in the assembly provide the required electron-transfer characteristics to mimic the in vivo situation, and in a much simpler fashion than might have been anticipated.

Boxer and Bucks (85) have linked a Pyropheo a molecule to a linked pair of Pyrochl a molecules through the 3-position on one of the linked pair of macrocycles. No optical or charge-transfer properties of this system have appeared in the literature, and, therefore, it is not possible to compare the two reaction center models at this time.

It should be remarked that a variety of laboratory systems involving linked porphyrins (86,87) and porphyrins linked to electron acceptors (88,89) have been described. Study of these systems makes a valuable contribution to the understanding of electron transfer in the primary light conversion step in photosynthesis as well as in electron-transport phenomena in general. Rapid advances in biomimetic photoreaction centers expanded to include electron donor and acceptor chains can be expected. (See Contributed Papers, Sessions I and II.)

IX. ANTENNA CHLOROPHYLL MODELS

Antenna chlorophyll in green plants has its absorption maximum in the red near 680 nm. Green-plant antenna chlorophyll is, thus, red shifted by 10 to 20 nm relative to Chl $\underline{a}$ dissolved in polar solvents such as diethyl ether or pyridine, in which Chl $\underline{a}$ is known to be monomeric. Monomeric Chl $\underline{a} \cdot L_1$ is highly fluorescent, but antenna chlorophyll is only weakly fluorescent. Energy trapping, dissipation, and transfer in the antenna are exceedingly fast events that do not have a convenient monitor as is epr for Chl_{sp} behavior. Thus, little of a specific nature can be said about either the structure or functional aspects of the antenna.

A. Light Harvesting Chlorophyll-Proteins as Models for Antennae

There are two general approaches to the interpretation of the anomalous spectral properties of antenna chlorophyll. The first view holds that chlorophyll occurs in vivo as specific chlorophyll-protein complexes and that it is in chlorophyll-protein interactions that the basis for antenna structure must be sought. Thus, Boardman et al. (90) state "that the specific interactions of the chlorophyll molecules in vivo are influenced to a large extent by some coordination of chlorophyll molecules to amino acid side chains of protein. Even if the spectral properties of chlorophyll in vivo were due to interacting pigment molecules in a hydrophobic environment, it seems very likely that the protein plays an important role in providing the hydrophobic environment and determining the relative orientations of the chlorophylls to one another." In this view, the chlorophyll-protein complexes that can be isolated by detergent extraction of thylakoid membranes are formed by very specific association of chlorophyll and thylakoid membrane proteins, and it is in the properties of these that the key to the structure and function of the antenna must be sought.

The principal evidence in favor of a chlorophyll-protein complex interpretation of in vivo chlorophyll is based on the X-ray crystal structure of a bacteriochlorophyll-protein determined by Matthews and Fenna (91), and the large measure of success achieved in the isolation of bacterial reaction centers (92). The Fenna and Matthews structure (93,94) is of great interest since it is the only chlorophyll-protein for which any structural information exists. The crystal contains red-shifted Bchl $\underline{a}$, but from the crystal structure data there are no Bchl $\underline{a}$-Bchl $\underline{a}$ interactions that could account for the red-shift. This has been taken as evidence that chlorophyll-protein interactions per se are capable of red-shifting the chlorophyll visible absorption spectrum. However, despite many efforts to provide a satisfactory explanation for the red-shift, no convincing interpretation to our knowledge has been advanced, and the optical properties of this bacterio-chlorophyll-protein are still to be explained. This complex is not photoactive and it has no known function, although it has been suggested that it could act as an intermediary in the transfer of excitation energy from the antenna chlorophyll to the photoreaction center (91).

There are also some puzzling structural features of this chlorophyll-protein. For instance, the organism from which the complex is extracted has as its major chlorophyll, chloro-bium chlorophyll, Bchl c, which is present to the extent of 95% of the total chlorophyll content. Bchl c is esterified with farnesol, a C_{15} alcohol (not with phytol as shown in Fig. 1 of ref. 91). The esterifying alcohol of the Bchl a in the protein complex, however, is phytol (J. J. Katz, unpublished work). This makes for a serious biosynthesis problem, for it seems at first glance to require two quite independent and isolated biosynthetic systems to assure the right esteri-fying alcohol for each of the two chlorophylls. The determin-ation of the structure of the Bchl a-protein complex is, to be sure, an important step forward in the understanding of chlor-ophyll-proteins in the photosynthetic membrane, but more will have to be learned about it before it can be accepted as a model for all chlorophyll-protein interactions.

Boxer and Wright (95) have addressed the chlorophyll-protein interaction problem in an ingenious experiment by preparing a complex of methyl pyrochlorophyllide a (Pyrochl a in which the phytyl group is replaced by methyl) and apomyo-globin. The pyrochlorophyllide a is inserted into the cleft left unoccupied by the removal of the heme molecule normally present in myoglobin. In the reconstituted pyrochlorophyl-lide a-apomyoglobin 1:1 complex, the pyrochlorophyllide has optical absorption and emission properties "remarkably similar to those in organic solvents" in which the pyrochlorophyl-lide a is known to be monomeric. The only exception to the similarity in optical properties is in the circular dichroism spectrum, where the CD peaks of the apomyoglobin complex are opposite in sign to those observed on the chromophore in organic solvents. This is interpreted by Boxer and Wright to be a consequence of the chirality of the protein. The protein moiety in all other respects has only a slight effect on the properties of the ground or excited singlet electronic states of the chromophore. While this result again cannot be gen-eralized to all protein-chlorophyll interactions, it does not

provide any indication that close proximity to a protein per-
turbs the electronic states of chlorophyll in any unusual way.

B. Chlorophyll Oligomer as a Model for Antenna

Another way of approaching the optical red shift in
antenna chlorophyll is by way of chlorophyll-chlorophyll
interactions as the basis of the red shift. it was recognized
long ago by Krasnovsky (96), Livingston (97), Brody et
al. (70) and many others that solutions of chlorophyll in
nonpolar solvents, chlorophyll films (67), colloidal dispersions of chlorophyll (66), monolayers of chlorophyll spread on
water (98,99), and chlorophyll adsorbed on solids such as
silica (99) have spectral properties that mimic those of in
vivo antenna chlorophyll to a considerable extent. It is the
interactions between the π-systems of juxtaposed chlorophylls
or the electronic perturbations produced by participation of
the keto C=O (or other nucleophilic groups in the chlorophyll
molecule) in coordination interactions with Mg or in hydrogen-
bonding that are responsible for the red-shifts. The numerous
red-shifted chlorophyll species, some of which closely resemble those found in vivo, which can be prepared in the
laboratory in very simple well-defined systems that contain no
protein, suggest that the basis for in vivo chlorophyll
structures should be sought for in direct chlorophyll-chlorophyll or bifunctional ligand-mediated chlorophyll-nucleophile-
chlorophyll interactions.

There is obviously much overlap in the two approaches as
chlorophyll-chlorophyll interactions are not at all excluded
by concurrent protein participation, and in the end all the
interactions among all of the components of the photosynthetic
membrane will have to be explained. But the two views suggest
different tactics. The red-shifted laboratory systems, complicated as they may be, are nevertheless much simpler to
study, and the reasons for the red-shifts are more readily
isolated for study. We can now consider two such laboratory
systems that point in the direction of a biomimetic antenna.

The chlorophyll "aggregates" found in aliphatic hydrocarbon solvents are red shifted to 680 nm and are only weakly
fluorescent as is antenna chlorophyll in vivo. Detailed
comparisons (44) of the optical properties of chlorophyll solutions in aliphatic hydrocarbon solvents and in vivo antenna
chlorophyll by computer-assisted deconvolution of the absorption peaks into gaussian components suggests that the in vitro
(Chl a)$_n$ oligomers, which are self-assembled by keto C=O$\cdots$Mg
interactions, and in vivo antenna chlorophyll have markedly
similar optical characteristics. The optical spectra of both
species can, for example, be fitted by identical deconvolution

parameters.   A structure for (Chl a)$_n$ deduced from optical,
infrared, and molecular weight measurements is shown in
Figure 8.   This chlorophyll species is, on the basis of its
optical properties, a plausible model for antenna chlorophyll.   Detailed arguments in support of this model have been
given elsewhere (44,55) and will not be repeated here.
    Antenna chlorophyll models in which the antenna is con-
sidered to be composed of monomeric chlorophyll units that do
not experience coordination interactions with other chlorophyll molecules have been proposed.   To our knowledge, only
one structural proposal [other than that of Matthews and
Fenna (91)] for a chlorophyll-protein has been made.   This is
the structure proposed by van Metter and Knox (100) for a
chlorophyll a/b-protein complex considered to have a light-
harvesting function.   The Chl a molecules in the model are
arranged in such a way that no physical contact can occur between them or with the Chl b molecules.   We do not believe
that the optical properties expected of this model are consistent with those of in vivo antenna.
    Beddard et al. (101,102) have proposed an antenna model in
which monomer chlorophylls are prevented from direct contact
with each other by the galactolipids present in the photosynthetic membrane, but again the absorption and fluorescence
properties of such a model do not seem to us to be able to

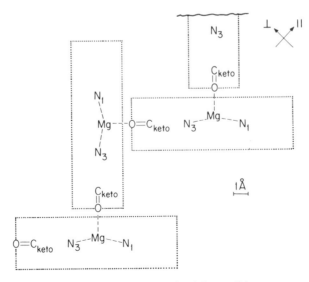

FIGURE 8.   Proposed structure of Chl a oligomer, projected
on a plane containing the oligomer axis.   The arrows in the
upper right hand corner indicate directions parallel (||) and
perpendicular (⊥) to the oligomer axis (44).

account for the optical properties of in vivo antenna chlorophyll.   Shipman (103) has concluded from computer-modeling
studies that for an antenna composed of monomer units to have
its (lowest energy) absorption maximum shifted to 680 nm by
transition-density coupling requires that the local chlorophyll concentration must be much greater (1 M) than the average chlorophyll concentration in the photosynthetic membrane
(~0.1 M). Even in solid films of Chl a the absorption maxima
are not shifted beyond 678 nm, which is the same as the absorption maximum of a Chl a oligomer in n-octane solution. A
pressure of 20 kbars on a Chl a film absorbing at 680 nm
produced no significant red-shift (104).   Red-shifts from
transition-density coupling do not appear to be a promising
route.

Objections have been raised to the oligomer antenna model
on stability grounds and on the ability of $(Chl\ a)_n$ to
function in energy transfer.   The keto C=O•••Mg coordination
interactions in the $(Chl\ a)_n$ oligomer are disrupted by water
to form a Chl a-water adduct absorbing at ~740 nm.   The
question then is: is there a region in the photosynthetic
membrane in green plants sufficiently anhydrous and sufficiently protected from water access to maintain the integrity
of the oligomer? It should be noted that water vapor below a
partial pressure of 10 torr at room temperature cannot compete
successfully for coordination to Mg in competition with the
keto C=O function in Ring V (105).   The annular region of a
lipid bilayer is reasonably free of water, and the hydrophobic
region of the membrane is a possible location of the antenna (28). Even if penetration by water occurred, the activity of the water would have to be substantial for disruption
of the oligomer to occur.   In fact a transition from a 680 nm
Chl a species to a 740 nm species very similar to the transformation observed when $(Chl\ a)_n$ is hydrated in n-octane
solution has been reported to occur in vivo.   Lippincott et
al. (106,107) have shown that tobacco chloroplasts, which absorb maximally at ~680 nm and give a ~7 gauss photo-epr signal
in red light, are transformed by treatment with 65% methanol
or acetone in water to a species absorbing at 740 nm and an
epr signal with a ~1 gauss linewidth, all of which are remarkably similar to the effects of hydration of $(Chl\ a)_n$ in
n-octane.   This transformation possibly can be explained by
extraction of the chlorophyll by the organic solvent and re-
precipitation at the membrane interface. However, we prefer
to think that it provides evidence for the possible existence
of $(Chl\ a)_n$ oligomers in vivo.

Chl a oligomers are essentially non-fluorescent, and it
has been argued that the ability to fluoresce is essential for
energy transfer to occur (101,102). However, chlorophyll fluorescence is a nanosecond phenomenon, whereas energy transfer

in the antenna proceeds on a much faster time scale, perhaps
on the subpicosecond time scale. The loss of energy by nonradiative processes likely is slower than transfer and, therefore, even if a nonradiative channel were available, energy
transfer could occur before the nonradiative channel became
operative. If this were the case, the ability to fluoresce
would not be essential for energy transfer, and the oligomer
may still be able to function in light-harvesting and energy
transfer. In addition, the fluorescence properties of
$(Chl \underline{a})_n$ are not unequivocally established. While the true
Chl $\underline{a}$ dimer has a very low fluorescence yield, it is not
necessarily true that the oligomer is equally nonfluorescent.
We have suggested that quenching traps for the concentration
quenching of chlorophyll fluorescence are likely to be asymmetric, i.e., the chlorophylls constituting the trap may be
required to be nonequivalent (108). The synthetic Chl_{sp} which
is formed of two fully equivalent chlorophyll molecules, is
intensely fluorescent and can in fact be used in a dye
laser (109). In a linear oligomer, the first chlorophyll
molecule in the chain differs from all the rest, and it is
only the first two chlorophyll molecules than can act as a
trap. If the first Chl $\underline{a}$ molecule were to have its keto C=O
perturbed by hydrogen-bonding or by a coordination interaction, then all of the Chl $\underline{a}$ molecules would approach equiva-
lency and the oligomers would have higher fluorescence than
the dimer, $(Chl \underline{a})_2$. In our opinion, Chl $\underline{a}$ oligomer as a
model for antenna has sufficient attractions to justify fur-
ther investigation.

C. A Covalently Linked Chl_{sp}-Antenna Model

Synthetic Chl_{sp} systems prepared by self-assembly or by
chemical synthesis mimic reaction-center chlorophyll in optical and epr properties, but they represent only the core of
the photoreaction center. We can now describe recently developed systems that are rudimentary Chl_{sp}-antenna models.

These systems can be prepared from recently synthesized
triesters of Mg or Zn pyropheophorbide bound covalently
through their propionic acid groups to a trihydric alcohol
(Fig. 9) (110). In polar solvents these compounds are in open
configuration and each of the three macrocycles has nearly
monomer-like properties. The addition of small amounts of
hydrogen-bonding nucleophiles to solutions in nonpolar solvents causes two of the macrocycles to assume a special-pair
configuration while the third macrocycle remains free
(Fig. 10). In this configuration the optical properties of
the triester are of special interest as the system displays

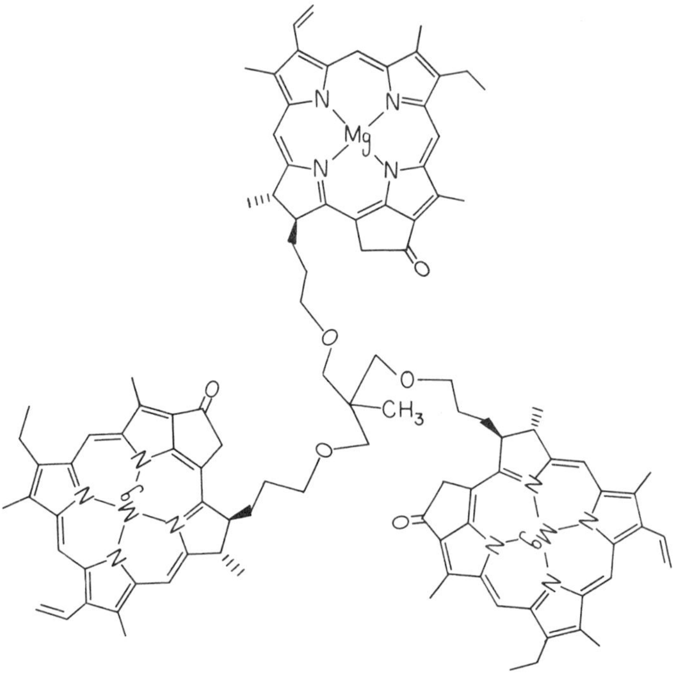

FIGURE 9. Structure of tris(pyrochlorophyllide a) 1,1,1 tris(hydroxymethyl)ethane triester. The covalent link is by way of an ester at the propionic acid side chains of the macrocycles. In this configuration, the optical properties of the triester are almost indistinguishable from a monomeric Pyrochl a molecule.

energy-transfer properties that qualify it as a rudimentary Chl_{sp}-antenna complex.

In the open configuration the Mg- or Zn-triesters are strongly fluorescent, but no lasing action can be detected even when very high pumping power is used. This behavior resembles that of the covalently linked pairs in the same media (109). The triesters in their folded configuration, however, lase readily when pumped by a N_2 laser (337 nm). Lasing in the Mg-triester occurs at 735 nm and in the Zn-triester at 724 nm. In these systems, almost independent of the extent of folding, lasing occurs only at long wavelengths. These systems thus differ in an important respect from self-assembled systems. In the latter, both short (~665 nm) and long (~700 nm) wavelength species are present, and emission from either the monomer or the folded pair can be observed. Fluorescence from the short wavelength chlorophyll

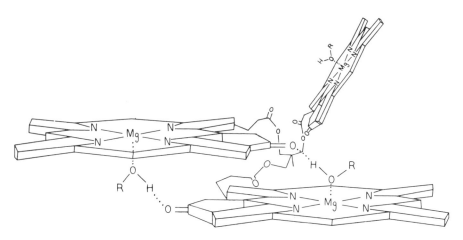

FIGURE 10. Covalently linked triester in its folded configuration. In the presence of a nucleophile ROH, two of the three macrocycles shown in Figure 9 are cross-linked as in Figure 7. The third macrocycle cannot fold for steric reasons, so that in its folded configuration, one macrocycle, which absorbs light maximally at shorter wavelengths, acts as an antenna to the folded pair.

is quenched only when self-assembly to the 700 nm species is essentially complete. Energy transfer is clearly more efficient in the triester system than in the self-assembled systems. The high efficiency of energy transfer can reasonably be expected to be a consequence of the close and fixed attachment of the monomer macrocycle to the paired macrocycles. While such covalently linked species are not known to exist in nature, the energy-transfer properties of these synthetic systems are expected to provided useful information relevant to antenna function in vivo.

X. THE OXYGEN SIDE OF PHOTOSYNTHESIS

Very little progress toward a biomimetic model of the oxygen-evolving side of photosynthesis can be reported. The nature of the photoreaction center P680 presumed is still a considerable mystery. The suggestion by Davis et al. (111) that PSII photoreaction center (also known as P680) might consist of ligated, monomeric Chl a accounts for the redox properties but does not seem to account for the optical properties of P680. Hall has suggested (112) that the enzyme, superoxide dismutase, which catalyzes disproportionation of the free-radical O_2^{-} to molecular oxygen may be involved in oxygen evolution, is interesting but as yet is unexplored. To us, the possibility that oxygen evolution in photosynthesis is essentially the reverse of oxidative metabolism in mitochondria is an attractive one. As O_2^{-} is the first product in oxidative reactions with O_2, this free-radical may be the immediate precursor of O_2 in photosynthesis. Plant superoxide dismutase is a manganese enzyme, which should provide a different role for at least part of the manganese in the chloroplast than the one usually assigned. The recent reconstitution of the oxygen-evolving mechanism by Spector and Winget (113) has aroused much interest and is a promising new approach to the mechanism of oxygen evolution. In general, however, an understanding of oxygen evolution in photosynthesis on a molecular level remains for the future, and until this necessary prerequisite is satisfied, little progress in biomimetic oxygen evolution can be expected.

XI. CONCLUSION

Synthetic systems of well-defined composition and structure have been prepared that mimic many of the properties of the special-pair chlorophyll involved in light-energy conversion in vivo, in energy transfer to the Chl_{sp}, and in electron transfer from the photo-oxidized Chl_{sp} to the primary electron acceptor. While much remains to be done, especially on the oxygen-evolving side of photosynthesis, we can look forward with a reasonable measure of confidence to biomimetic systems that will increasingly approximate the entire PS I photoreaction center.

REFERENCES

1. Physics Today, September 1980, pp. 19-21.
2. R. Breslow, Chem. Soc. Reviews 1, 553 (1972).
3. J. T. Warden and J. R. Bolton, J. Am. Chem. Soc. 95, 435 (1973).
4. C. A. Wraight and R. K. Clayton, Biochim. Biophys. Acta 333, 246 (1973).
5. J. J. Katz and M. R. Wasielewski, Biotechnology and Bioengineering Symp. No. 8, 423 (1978).
6. J. J. Katz, T. R. Janson, and M. R. Wasielewski, in "Energy and the Chemical Sciences" (The 1977 Karcher Symposium, S. D. Christian and J. J. Zuckerman, eds.) Pergamon Press Ltd., Oxford, 1978, pp. 31-57.

7. R. Emerson and W. Arnold, J. Gen. Physiol. 15, 391 (1931–1932).
8. R. Emerson and W. Arnold, J. Gen. Physiol. 16, 191 (1932–1933).
9. E. Katz in "Photosynthesis in Plants" (J. Franck and E. Loomis, eds.), The Iowa State University Press, Ames, Iowa, 1949, pp. 287–292.
10. A. Szent-Györgi, "Bioenergetics", Academic Press, New York, 1957.
11. E. I. Rabinowitch, "Photosynthesis", Interscience Publishers, New York, 1956, Vol. 2, Pt. 2, p. 1299.
12. J. J. Katz, R. C. Dougherty and L. J. Boucher, in "The Chlorophylls", (L. P. Vernon and G. R. Seely, eds.) Academic Press, New York, 1966, pp. 185–251.
13. J. J. Katz and C. E. Brown, Bull. Mag. Resonance, in press (1981)
14. J. J. Katz and J. R. Norris, in "Current Topics in Bioenergetics" (D. R. Sanadi, ed.), Vol. 5, Academic Press, New York, 1973, pp. 41–75.
15. J. J. Katz, Dev. Appl. Spectroscopy 6, 201 (1968).
16. T. A. Evans and J. J. Katz, Biochim. Biophys. Acta 396, 14 (1975).
17. E. Rabinowitch and Govindjee, "Photosynthesis", John Wiley & Sons, Inc., New York, 1969, p. 196.
18. L. L. Shipman, T. R. Janson, G. J. Ray and J. J. Katz, Proc. Natl. Acad. Sci. USA 72, 2873 (1975).
19. J. J. Katz, L. L. Shipman, T. M. Cotton and T. R. Janson, in "The Porphyrins" (D. Dolphin, ed.) Vol. V, Academic Press, New York, 1978, pp. 401–458.
20. T. M. Cotton, A. D. Trifunac, K. Ballschmiter and J. J. Katz, Biochim. Biophys. Acta 386, 181 (1974).
21. J. J. Katz, in "Inorganic Biochemistry" (G. L. Eichorn, ed.) Vol. II, Elsevier, Amsterdam, 1973, pp. 1022–1066.
22. K. Iriyama, N. Ogura and A. Takamiya, J. Biochem. 76, 901 (1974).
23. J. J. Katz and K. Ballschmiter, Angew. Chem. Int. Ed. Engl. 7, 286 (1968).
24. K. Ballschmiter and J. J. Katz, J. Am. Chem. Soc. 91, 2661 (1969).
25. J. J. Katz, K. Ballschmiter, M. Garcia-Morin, H. H. Strain and R. A. Uphaus, Proc. Natl. Acad. Sci. USA 60, 100 (1968).
26. H.-C. Chow, R. Serlin and C. E. Strouse, J. Am. Chem. Soc. 97, 7230 (1975).
27. C. Kratky and J. D. Dunitz, J. Mol. Biol. 113, 431 (1977).
28. J. J. Katz, W. Oettmeier and J. R. Norris, Phil. Trans. R. Soc. London B273, 227 (1976).
29. B. Commoner, J. J. Heise, B. B. Lippincott, R. E. Norberg, J. V. Passoneau and J. Townsend, Science 126, 57 (1957).

30. B. Kok, Nature 179, 583 (1957).
31. B. Kok, Biochim. Biophys. Acta 22, 339 (1956).
32. L. N. M. Duysens, Nature 173, 692 (1954).
33. L. N. M. Duysens, Brookhaven Symp. Biol. No. 11, 10 (1958).
34. D. C. Borg, J. Fajer, R. H. Felton and D. Dolphin, Proc. Natl. Acad. Sci. USA 67, 813 (1970).
35. J. T. Warden and J. R. Bolton, J. Am. Chem. Soc. 94, 4351 (1972).
36. J. T. Warden and J. R. Bolton, J. Am. Chem. Soc. 95, 6435 (1973).
37. J. R. Bolton, R. K. Clayton and D. W. Reed, Photochem. Photobiol. 9, 209 (1969).
38. P. A. Loach and D. L. Sekura, Photochem. Photobiol. 6, 381 (1967).
39. J. D. McElroy, D. C. Mauzerall and G. Feher, Biochim. Biophys. Acta 333, 261 (1974).
40. J. R. Norris, R. A. Uphaus, H. L. Crespi and J. J. Katz, Proc. Natl. Acad. Sci. USA 68, 625 (1971).
41. A. J. Bard, A. Ledwith and H. J. Shine, Adv. Phys. Org. Chem. 13, 156 (1976).
42. J. R. Norris and M. K. Bowman, Abstracts, Fifth International Congress on Photosynthesis, Halkidiki, Greece, p. 419 (1980).
43. K. Ballschmiter, K. Truesdell and J. J. Katz, Biochim. Biophys. Acta 184, 604 (1969).
44. L. L. Shipman, T. M. Cotton, J. R. Norris and J. J. Katz, J. Am. Chem. Soc. 98, 8222 (1976).
45. G. Feher, Phys. Rev. 103, 834 (1956).
46. H. Scheer, J. J. Katz and J. R. Norris, J. Am. Chem. Soc. 99, 1372 (1977).
47. J. R. Norris, H. Scheer and J. J. Katz, in ref. (19), Vol. IV, pp. 159-195.
48. P. L. Dutton, J. S. Leigh and D. W. Reed, Biochim. Biophys. Acta 292, 654 (1973).
49. H. Levanon and J. R. Norris, Chem. Rev. 78, 185 (1978).
50. M. K. Bowman, J. R. Norris, M. C. Thurnauer, J. Warden, S. A. Dikanov and Yu. D. Tsvetkov, Chem. Phys. Lett. 55, 570 (1978).
51. J. J. Katz, J. R. Norris, L. L. Shipman, M. C. Thurnauer and M. R. Wasielewski, Annu. Rev. Biophys. Bioeng. 7, 393 (1978).
52. J. J. Katz, L. L. Shipman and J. R. Norris, in "Chlorophyll Organization and Energy Transfer in Photosynthesis", Ciba Foundation Symposium 61 (new series), Excerpta Medica, Amsterdam, 1979, pp. 1-34.

53. J. J. Katz, in "Advances in Biochemistry and Physiology of Plant Lipids", (L. Appelquist and C. Liljenberg, eds.) Elsevier/North Holland Biomedical Press, Amsterdam, 1979, pp. 37-56.

54. J. J. Katz, in "Light-Induced Charge Separation in Biology and Chemistry", (H. Gerischer and J. J. Katz, eds.) Verlag Chemie, Weinheim, 1979, pp. 331-359.

55. J. J. Katz, J. R. Norris and L. L. Shipman, Brookhaven Symp. Biol. No. 28 (J. M. Olson and G. Hind, eds.),1977, pp. 16-54.

56. K. Ballschmiter and J. J. Katz, Nature 220, 1231 (1968).

57. M. Garcia-Morin, R. A. Uphaus, J. R. Norris and J. J. Katz, J. Phys. Chem. 73, 1066 (1969).

58. F. K. Fong, Proc. Natl. Acad. Sci. USA 71, 3692 (1974).

59. S. G. Boxer and G. L. Closs, J. Am. Chem. Soc. 98, 5406 (1976).

60. L. L. Shipman, T. M. Cotton, J. R. Norris, and J. J. Katz, Proc. Natl. Acad. Sci. USA 73, 1791 (1976).

61. F. K. Fong and V. J. Koester, Biochim. Biophys. Acta 423, 52 (1976).

62. L. L. Shipman, J. R. Norris and J. J. Katz, J. Phys. Chem. 80, 877 (1976).

63. L. L. Shipman and J. J. Katz, J. Phys. Chem. 81, 577 (1977).

64. J. P. Thornber and R. S. Alberte, in "Photosynthesis I", (A. Trebst and M. Avron, eds.) Springer-Verlag, New York, 1977, pp. 574-582.

65. H. Kuhn in ref. (54), pp. 151-169; J. Photochem. 10, 111 (1979).

66. B. B. Love and T. T. Bannister, Biophys. J. 3, 99 (1963).

67. A. A. Krasnovsky and M. I. Bystrova, Biosystems 12, 181 (1980), and references cited.

68. L. I. Nekrasov and R. Kapler, Biofizika 11, 48 (1966).

69. G. M. Sherman and E. Fujimori, Arch. Biochem. Biophys. 130, 624 (1969).

70. S. S. Brody and M. Brody, in "Photosynthetic Mechanisms of Green Plants", Publication 1145, National Academy of Sciences - National Research Council, Washington, D.C., 1973, pp. 455-478.

71. P. S. Stensby and J. Rosenberg, J. Phys. Chem. 65, 906 (1961).

72. F. K. Fong, A. J. Hoff and F. A. Brinkman, J. Am. Chem. Soc. 100, 619 (1978).

73. T. M. Cotton, P. A. Loach, J. J. Katz and K. Ballschmiter, Photochem. Photobiol. 27, 735 (1978).

74. T. M. Cotton, Ph.D. Thesis, Northwestern University, Evanston, Illinois, 1976.

75. M. J. Yuen, L. L. Shipman, J. J. Katz and J. C. Hindman, Photochem. Photobiol., in press (1981).

76. M. R. Wasielewski, M. Studier and J. J. Katz, Proc. Natl. Acad. Sci. USA 73, 4282 (1976).
77. M. R. Wasielewski, U. H. Smith, B. T. Cope and J. J. Katz, J. Am. Chem. Soc. 99, 4172 (1977).
78. H. P. Isenring, E. Zass, K. Smith, H. Falk, J. L. Luisier and A. E. Eschenmoser, Helv. Chim. Acta 58, 2357 (1975).
79. J. J. Katz, G. D. Norman, W. A. Svec and H. H. Strain, J. Am. Chem. Soc. 90, 6841 (1968).
80. P. H. Hynninen, M. R. Wasielewski and J. J. Katz, Acta Chem. Scand. B33, 637 (1979).
81. M. C. Thurnauer, J. J. Katz and J. R. Norris, Proc. Natl. Acad. Sci. USA 72, 3270 (1975).
82. J. R. Norris and J. J. Katz, in "The Photosynthetic Bacteria", (R. K. Clayton and W. R. Sistrom, eds.) Plenum Press, New York, 1978, pp. 397-418.
83. A. Warshel, Proc. Natl. Acad. Sci. USA 77, 3105 (1980).
84. M. J. Pellin, K. J. Kaufmann and M. R. Wasielewski, Nature 278, 54 (1979).
85. S. G. Boxer and R. R. Bucks, J. Am. Chem. Soc. 101, 1883 (1979).
86. J. L. Y. Kong and P. A. Loach, in "Frontiers of Biological Energetics - Electrons to Tissues", Vol. 1 (P. L. Dutton, J. S. Leigh, and A. Scarpa, eds.) Academic Press, New York, 1978, pp. 73-82.
87. T. L. Netzel, P. Kroger, C. K. Chang, I. Fujita and J. Fajer, Chem. Phys. Lett. 67, 223 (1979).
88. T.-F. Ho, A. R. McIntosh and J. R. Bolton, Nature 286, 254 (1980).
89. J. Dalton and L. R. Milgrom, J. Chem. Soc., Chem. Comm., 609 (1979)
90. N. K. Boardman, J. M. Anderson and D. J. Goodchild, in "Current Topics in Bioenergetics" (D. R. Sanadi and L. P. Vernon, eds.) Vol. 8, Pt. B, Academic Press, New York, 1978, pp. 35-109.
91. B. W. Matthews and R. E. Fenna, Accts. Chem. Res. 13, 309 (1980).
92. G. Gingras, in ref. (82), pp. 119-131.
93. R. E. Fenna and B. W. Matthews, Nature 258, 573 (1975).
94. B. W. Matthews, R. E. Fenna, M. C. Bolognesi, M. L. F. Schmid and J. M. Olson, J. Mol. Biol. 131, 259 (1979).
95. S. G. Boxer and K. Wright, J. Am. Chem. Soc. 101, 6791 (1979).
96. A. A. Krasnovsky, in "Progress in Photosynthesis Research" (Proc. First Int. Congress, Freudenstadt, 1968, H. Metzner, ed.), Lichtenstern, Munich, 1969, pp. 763-770.
97. R. Livingston, Quart. Rev. Chem. Soc. (London) 14, 174 (1960).
98. B. Ke, in ref. (12), pp. 253-279.

99. N. A. Mamleeva and L. I. Nekrasov, Zh. Fiz. Khim. 50, 1794 (1976). (Chem. Abstr. 85, 119796 (1976)).
100. R. S. Knox and R. L. van Metter, in ref. (52), pp. 177–186.
101. G. S. Beddard and G. Porter, Nature 260, 366 (1976).
102. G. S. Beddard, S. E. Carlin and G. Porter, Chem. Phys. Lett. 43, 27 (1976).
103. L. L. Shipman, J. Phys. Chem. 81, 2180 (1977).
104. J. R. Ferraro and J. J. Katz, unpublished work.
105. K. Ballschmiter and J. J. Katz, unpublished work.
106. J. A. Lippincott, J. Aghion, E. Porcile and W. F. Bertsch, Arch. Biochem. Biophys. 98, 17 (1962).
107. J. A. Lippincott, and B. B. Lippincott, Arch. Biochem. Biophys. 105, 359 (1964).
108. M. J. Yuen, L. L. Shipman, J. J. Katz and J. C. Hindman, Photochem. Photobiol. 32, 281 (1980).
109. J. C. Hindman, R. Kugel, M. R. Wasielewski and J. J. Katz, Proc. Natl. Acad. Sci. USA 75, 2076 (1978).
110. M. J. Yuen, G. L. Closs, J. J. Katz, J. A. Roper, M. R. Wasielewski and J. C. Hindman, Proc. Natl. Acad. Sci. USA 77, 5598 (1980).
111. M. S. Davis, A. Forman, and J. Fajer, Proc. Natl. Acad. Sci. USA 76, 4170 (1979).
112. D. O. Hall, in "Bioinorganic Chemistry II", (K. N. Raymond, ed.), Advances in Chemistry Series, No. 162, American Chemical Society, Washington, D.C., 1977, pp. 227–250.
113. M. Spector and G. D. Winget, Proc. Natl. Acad. Sci. USA 77, 957 (1980).

DISCUSSION

Prof. J. R. Bolton, University of Western Ontario:
Do you interpret the variation of emission spectra with excitation wavelength as indicating a variety of absorbing species?

Dr. Katz:
We believe that the variation in emission as a function of excitation wavelength is indicative of a multiplicity of chlorophyll species in the self-assembled systems.

Dr. J. S. Connolly, Solar Energy Research Institute:
We have some recent data (Contributed Paper III-3) which show that the fluorescence lifetime of Bchl a is longer in solvents in which the magnesium is hexacoordinated (e.g., neat pyridine, THF) than in pentacoordinating solvents

(e.g., ether, acetone), and that it is shorter still in
strongly H-bonding solvents (e.g., ethanol, methanol).
This trend does not appear to occur as strongly in
Chl a. First, would you correct this impression if wrong
[Ed. note: it is]; and second, would you care to offer an
explanation for these solvent effects on fluorescence
lifetimes of chlorophylls?

Dr. Katz:
I am not aware that anyone other than yourself is studying
this aspect. A change in the coordination number of magnesium in chlorophyll affects principally the Q_x transition and only marginally the Q_y. Because Q_x and Q_y are
more clearly resolved in Bchl a than in Chl a, the effect
may be more evident in the former. That the fluorescence
lifetime may also be a function of coordination number
seems a reasonable possibility to me.

Dr. K.-H. Grellmann, Max-Planck-Institut für biophysikalische
Chemie, Göttingen:
The derivative which contains three Chl molecules (Figs. 9
and 10) mimics the antenna and the reaction center as
well. You did not show us an absorption spectrum of this
compound. Is it just an overlap of the two component
spectra?

Dr. Katz:
In the folded tri-ester, two pyrochlorophyll a molecules
constitute a special pair, the third is free. The free
molecules absorb at shorter wavelengths, and the emission
is only from the folded pair. Consequently this system
may be thought of as a primitive antenna-special pair
model.

Dr. T. L. Netzel, Brookhaven National Laboratory:
Chl a monomers lase and "folded" dimers lase. However
"open" dimers don't lase. Can you rationalize this behavior? Secondly, "folded" Chl a dimers absorb where P700
does, but the "folded" Bchl a dimer absorbs at 803 nm, not
where P865 does. Can you explain the large shift for P865
relative to the Bchl a monomeric absorption at 770 nm?

Dr. Katz:
As far as I know, lasing action has not been induced in
covalently linked pairs in their open configurations.
Lasing has been observed to occur only from the folded
configurations. Whether there are additional relaxation
paths (vibrational modes or charge transfer) available to
the open configuration, or whether present excitation

techniques are inadequate to achieve a population inversion is still unclear.

In answer to your second question: Folded Bchl a covalently linked pairs have spin-sharing properties but do not have the optical properties of P865. Presumably the Bpheo a and additional Bchl a molecules present in the reaction center are responsible for the additional red-shifts.

Prof. D. O. Hall, Kings College, London:
The role of Mn and superoxide dismutase may be important in the O_2 evolution mechanism, as you mentioned; we have published on this. I also think that the role of super-oxide in other water-splitting/oxygen-evolving systems should be examined.

Dr. Katz:
I strongly agree. O_2 evolution in photosynthesis may be the reverse of O_2 uptake in mitochrondria where superoxide is involved.

Dr. H. Tributsch, C.N.R.S., Laboratoires de Bellvue:
You said that photosynthesis is by far the most important and efficient solar energy conversion mechanism which nature has evolved. This is accepted today as a fact which nobody would like to challenge. However I wish to call your attention to a mechanism, which, in my opinion, could be nearly as important. Approximately 40% of the solar energy incident on a forest is used for evaporation of water from leaves, which is, in fact, an evaporation of water from thin capillaries and drives the transport of water in plants. We believe, and this is the topic of our poster (Contributed Paper XI-12), that such a mechanism -- which involves the evaporation of molecules from intimate contact with a suitable substrate -- could possibly be put into practice more easily than a biomimetic system.

Dr. Katz:
I have stated that photosynthesis is extremely efficient, mainly because one photon produces one electron in the photoreaction center. Using solar energy to evaporate water in photosynthesis requires a large input of energy to compensate for the high heat of evaporation of water, and is thus to a degree parasitic.

Dr. A. F. Haught, United Technologies Research Center:
The last question has raised an interesting point on the efficiency of photoconversion. If the sunlight is

absorbed in a quantum process it can be converted to useful work with an efficiency of ~31% (unconcentrated
sunlight). If on the other hand, the sunlight is absorbed
as heat and used thermally, as in the evaporation of
water, thermal conversion can have a conversion efficiency
of ~54% with the same unconcentrated sunlight. This point
is addressed in Contributed Paper XI-2.

Prof. H. Linschitz, Brandeis University:
It may be helpful to clarify the relationship between
Dr. Katz' fluorescence and lasing observation and our
experiments demonstrating unfolding of the dimer upon
excitation, with the resulting formation of covalently
linked triplet and ground state moieties. We concluded
that the reaction proceeds via initial transition of the
excited dimer to its triplet state, followed by unfolding. As we have explicitly pointed out, the initial
singlet-to-triplet conversion requires sufficient time to
yield significant fluorescence, or, under suitable conditions, to lead to lasing action. Thus there is no contradiction nor incompatability between these two sets of
observations.

CHAPTER 3

QUANTUM HARVESTING AND ENERGY TRANSFER

Christian K. Jørgensen

Département de Chimie minérale
analytique et appliquée
Université de Genéve
Geneva, Switzerland

I. INTRODUCTION

Our closest star provides all the energy used by plants and animals by emitting a standard continuous spectrum of an opaque object ("black body") at a temperature T of 5750K. Not only is this twenty times higher than room temperature, but it is also important that the emission maximum of the energy-density spectrum (W m^{-2} nm^{-1}) occurs at ~5.0 kT (k ≃ 0.7 cm^{-1} K^{-1}), although the maximum of such a spectrum plotted as W m^{-2}/cm^{-1} occurs at ~2.8 kT [see ref.(1)]. This Conference is concerned with utilization of photons having energies about a hundred times kT (for T = 300K) prevailing at the Earth's surface.

Other authors in this Volume have discussed sophisticated systems modelled more or less on photosynthesis. It might be worth mentioning, however, that the excited state of the uranyl ion (occuring at 10^{-8} M as a carbonate complex in sea water, and sometimes at higher concentrations in rivers and lakes) (2,3) is almost as oxidizing as fluorine (E^{o} ~ 2.6 V) and performs complicated photochemical reactions with organic molecules, starting with abstraction of a hydrogen atom. Since the turnover rate in bright sunshine at the ocean's surface involves an excitation (on the average) every ten minutes, the excited state of this ion may remove organic molecules (i.e., those not biodegradable) that would otherwise accumulate from the great forests arriving, e.g., via the Amazon and Zaire rivers.

II. ATOMIC AND MOLECULAR SPECTRA

Over the past quarter century, quantum chemistry has developed a broad rationalization of the origin of the absorption bands in several categories of transitions. It should be emphasized that this understanding has been obtained by comparative induction (4,5) from the numerous spectra which became available after 1950, when reliable spectrophotometers for the visible and near-ultraviolet became standard equipment. In close analogy to the development of atomic spectroscopy (until it was abandoned after 1930 by the physicists, who were attracted by the more fashionable nuclear properties), induction from facts has been much more fruitful than numerical calculations. In condensed matter (liquids, amorphous and crystalline solids) the p, d and f orbitals used in the description of monatomic entities persist to a remarkable degree for the higher ℓ values. Thus, lanthanide compounds (1,6) containing from one to thirteen 4f electrons have excited J-levels at energies only $\leqslant$ 6% lower than the corresponding gaseous ion, and the "ligand field" effects representing deviations from spherical symmetry correspond to the (2J+1) states of a given level being separated, at most, a few hundred wavenumbers.

The situation is quite different in non-metallic compounds of the Fe (3d), Pd (4d) and Pt (5d) groups, where one or more of the five d-like orbitals is antibonding with energies, dependent in a regular way on the central atoms and on the ligands in the spectrochemical series (4), some 4,000 to 40,000 cm^{-1} above the remaining non-bonding orbitals. Contrary to statements in text books, this is not due to the tiny non-spherical part of the local Madelung potential, but is an effect of (weaker or stronger) covalent bonding, as is also known outside the transition metals. A closer analysis shows that the electron configuration (classifying the J-levels of the monatomic entities) presents an opportunity to define the oxidation state in compounds showing certain analogies with Russell-Saunders coupling (4,7). Whereas, L of this approximation is not usually a well-defined quantum number in d-group compounds, the total spin S has the same possibility of being well-defined in compounds as in monatomic entities. In d-group complexes containing from two to eight d-like electrons, the choice between different S values for the ground state is determined by the competition between the presence of antibonding electrons and the lower interelectronic repulsion in the partly filled shell concomitant with higher S.

It may be worthwhile to point out a difference between organic and transition-group chemistry. Any species containing an odd number of electrons (S is a positive, half-

odd integer) is called a "free-radical" by organic chemists,
and can be expected to dimerize to a diamagnetic (S = zero)
product. Exceptions to this rule are known outside the transition groups (NO, O_2^-, O_2^+, O_3^-, ClO_2, etc.) but it should be
mentioned that they are often connected with the presence of
anti-bonding electrons. More remarkable, in gaseous O_2 which
contains an even number of electrons, the triplet (S = 1)
ground state contains two of the four possible electrons in
two anti-bonding molecular orbitals (MO). Although these MO's
are filled in O_2^{2-} and in F_2, it is generally true that most
colorless species outside the d groups contain non-bonding and
bonding electrons, but not typically antibonding electrons.
The quantum chemist is strongly motivated by the Franck–Condon
principle that optical transitions, and ionization processes
studied by photoelectron spectra (8–10), conserve the nuclear
positions of the system before excitation or ejection of an
electron by a high-energy photon, whereas the chemist (by
definition) is more interested in processes that modify the
internuclear distances. Unfortunately, systems containing
three or more (n) nuclei have their electronic energy levels
determined by (3n – 6) spatial variables, creating diverging
difficulties for deductive quantum chemistry of polyatomic
entities. Frequently, the addition of one or two antibonding
electrons, e.g., to linear ONO^+, which is isoelectronic with
CO_2, modifies the stereochemistry, say to bent NO_2 and NO_2^-;
or planar SO_3 to pyramidal SO_3^{2-}; or B_2H_6 with two hydride
bridges to ethane-like $B_2H_6^{2-}$.

As long as the absorption spectra of f and d group compounds (Fig. 1) are treated by analogy to atomic spectroscopy,
the inner-shell (Rydberg) transitions can be described as
$4f \rightarrow 5d$, as in Ce(III), Pr(III), Tb(III) and in many 4f group
divalent systems or $5f \rightarrow 6d$ in transthorium M(III) and M(IV)
states (1). Whereas it is certain that the first four excited
J-levels of the gaseous mercury atom belong to the configuration [78]6s6p it is less certain that the strong characteristic transitions in the ultraviolet of Tℓ(I), Pb(II) and Bi(III)
can be described to a good approximation as $6s \rightarrow 6p$ (4,11).
The weak absorption band of $Fe(H_2O)_6^{2+}$ at 40,500 cm^{-1} is
ascribed to an excited configuration [18]$3d^5 4s$ (though it is
not certain whether S = 3 or 2). Like the xenon atom, iodide
ion has $5p \rightarrow 6s$ transitions at even higher energy (4,12),
although this is of less relevance to our subject because of
the atmospheric ozone cut-off at 33,000 cm^{-1}. In the latter
examples, the excited state containing one 4s or 6s electron
can evolve rapidly in time, dissociating to a solvated electron, and perhaps ultimately liberating H_2 from the solvent.

However, for our purposes, the most interesting absorption
bands are due to electron transfer. Such transitions are also
known between extended organic molecules with relatively small

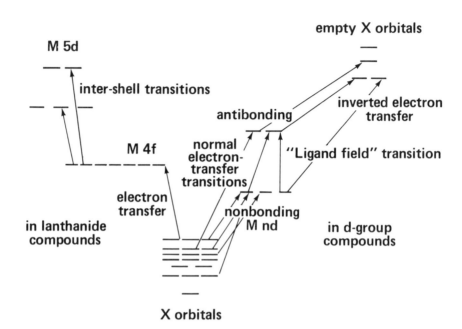

FIGURE 1. Electronic transitions among various orbitals. The ligating neighbor atoms X provide the 18 orbitals in the middle of the figure (adapted to octahedral chromophores MX_6). To the left, M is a lanthanide with negligible energy differences between the seven 4f orbitals, and inter-shell transitions 4f → 5d as well as electron transfer to 4f are shown. To the right, M contains a partly filled 3d (Fe group), 4d (Pd) or 5d (Pt group) shell with two anti-bonding and three approximately non-bonding orbitals. Normal electron transfer from the reducing X to the oxidizing M is shown, as are a typical "ligand field" transition between the d-like orbitals and inverted electron transfer from the reducing M to low-lying empty orbitals of the ligands. It should be emphasized that the energy levels of many-electron systems cannot be exhaustively described by orbital energy differences, that the energetic effects of interelectronic repulsion can be large, and in particular, that under equal circumstances, the total energy tends to be lowest for the highest total spin quantum number S compatible with the distribution of electrons in the orbitals.

ionization energies (13) and oxidizing molecules having low-lying empty MO's such as tetracyanoethylene, iodine, and titanium tetrachloride (14). An interesting case is gaseous, yellow IrF_6 which, when condensed with xenon by cooling, yields a deep purple mixture having excited states due to $Xe^+IrF_6^-$, thereby transferring an electron between the spatially isolated Xe5p and Ir5d shells (15). As shown by Bartlett in 1962 (16) the corresponding mixture of gaseous red PtF_6 and xenon produces yellow crystals (later shown to be $XeF^+PtF_6^-$).

The electron-transfer spectra of transition group complexes most frequently involve immediately adjacent atoms. It was recognized early that $3d^5$ Fe(III) (17), $3d^6$ Co(III) (18), $3d^9$ Cu(II), and $4d^6$ Pt(IV) with reducing ligands exhibit such transitions, which are not observed in the comparable $3d^{10}$ Zn(II) and Ga(III) complexes (19,20), and such electron-transfer bands were studied extensively (e.g., by Fromherz before 1930) long before the internal transitions in the partly filled d shell were rationalized by "ligand field" theory. It was realized after 1956 (21,22) that the major categories of transition group electron-transfer spectra are due to one or more reducing ligands collectively losing an electron from an MO to an empty or partly filled shell, thus lowering the oxidation state of the central atom by one unit; or the other way around, a filled or partly filled shell can lose an electron to a low-lying empty MO of the ligand(s). The latter situation is called "inverted electron transfer" because it occurs only in a rather specific class of conjugated organic ligands (among others, pyridine and its derivatives, cyclic $C_5O_5^{2-}$, bipyridyl and phenanthroline). Acetylacetonate (23) shows both normal and inverted electron-transfer bands.

It has been possible (4,20,24) to establish a proportionality between the energy of the lowest allowed transition and the difference between the optical electronegativity, χ_{opt}, of the most reducing ligand and χ_{uncorr} (i.e., before corrections for spin-pairing energy and antibonding character of the d-like orbitals) of the central atom in a definite oxidation state. If the proportionality constant is chosen to be 30,000 cm^{-1}, χ_{opt} of the halide ligands are remarkably close to the Pauling values 3.9, 3.0, 2.8, and 2.5 for F, Cl, Br and I, respectively. It has been possible to establish χ_{opt} of polyatomic ligands, e.g., 3.5 for H_2O, ~ 3.4 for Xe, 3.2 for SO_4^{2-}, 2.8 for N_3^-, 2.7 for acetylacetonate, and 2.6 for several RS^-. In contrast, it is very difficult to give a definite value for oxides (25-27). One reason is the highly varying Madelung potential and another is that covalent bonding makes the central atom look less oxidizing, e.g., in chromate and permanganate compared with the chemically far less stable CrF_6, and no fluoride of Mn(VII) is known. A third kind of electron-transfer band is due to interactions between atoms of differing metallic elements, as in the case of red Ag_2CrO_4, blue $AgMnO_4$ and strongly colored silver(I) and thallium(I) salts of 5d group hexahalide complexes, in which the electron is transferred from the filled Ag4d or Tℓ6s shell to the d-like orbitals of the anion (28). Certain instances of "mixed oxidation state colors" fall in this third category (29), such as Prussian Blue, $K[Fe(CN)_6Fe]$, which contains distinct Fe(II) C_6 (S=0) and Fe(III) N_6 (S=5/2) and is thus not essentially different from the fox-red copper(II) and uranyl ferrocyanides.

III. FLUORESCENT CONCENTRATORS AND OTHER APPLICATIONS

One of the most important uses of solar energy is in photovoltaic cells of p- and n-type silicon or other adamantoid semiconductors, such as GaAs. Many such devices were developed by NASA to supply electricity for various space vehicles. Obviously, two major criteria were low weight and long lifetimes, but it seems that silicon cells may be adapted to the terrestrial economy if we are willing to pay a price for electricity (in "indexed" units) of the same order of magnitude as prevailed between the two World wars (30,31). Some optimists hope for technological breakthroughs and "auto-catalyzed" price decreases that result from economiés of scale as in the electrolytic production of metallic aluminum or in the computer industry. Nevertheless, it would be of tremendous help if the amount of silicon needed could be cut by a factor of ten. Among the more trivial reasons is that the production of silicon itself is quite energy-consuming, and another being that the dispersed nature of sunlight (peak insolation on the order of 1 kW/m^2) requires large areas of coverage, viz., 10^4 m^2 (2.47 acres) for 1 MW at the exceptionally high overall efficiency of 10 percent.

A possible solution to this problem may be the Reisfeld concentrator (Fig. 2) (32,33) which makes use of the isotropic fluorescence of the uranyl ion in its first excited state. This transition corresponds to electron transfer from an MO (with odd parity delocalized on the two oxo ligands in linear OUO^{2+}) to the empty 5f shell of uranium, thereby decreasing its oxidation state from U(VI) to U(V). This idea originated with Goetzberger and Greubel (34) who suggested using the total reflection inside a glass plate with refractive index n. If the angle θ between a line perpendicular to the two strictly parallel surfaces going through the fluorescent species and the direction of the photon it emits is larger than a critical value $θ_0$ given by the relation sin $θ_0$ = 1/n (approximating the

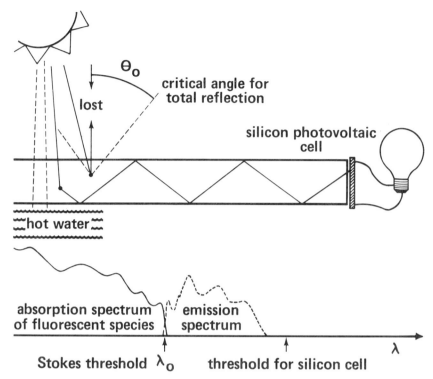

absorption spectrum | emission
of fluorescent species | spectrum

Stokes threshold λ_0 threshold for silicon cell

FIGURE 2. Fluorescent concentrator according to Reisfeld et al. (32,33). At the upper left is shown how photons with lower energy than the Stokes threshold of the fluorescent, transparent plate (seen from the rim) go through the glass and can be used for heating water in a black container. In addition some of the isotropic fluorescent radiation emitted inside the cone (determined by the critical angle θ_0 for total reflection) is lost by escape from the glass surface. Finally, light emitted outside this cone is trapped inside the plate (on the drawing, it undergoes five consecutive total reflections) before hitting the silicon-covered rim. The lower part of the figure is a schematic illustration of typical absorption and emission spectra of a fluorescent glass.

refractive index of air to unity) the emitted light is trapped between the two surfaces by a series of consecutive total reflections, until it hits the rim of the plate. The part of the luminescent light trapped this way is $\cos \theta_0$; as shown in the following numerical examples:

$$
\begin{array}{rcccc}
n & = & 1.41 & 1.54 & 2.00 \\
\theta_o & = & 45° & 40° & 30° \\
\cos \theta_o & = & 0.707 & 0.766 & 0.866
\end{array}
$$

Thus, typical glasses may readily trap ~75% of the luminescent light. However, three major requirements constrain such devices: (i) there should be no absorption bands, even weak ones, in the region of fluorescence, either initially or form-ing reversibly (e.g., triplet-triplet absorption in organic molecules, photochromic rearrangements) or irreversibly during exposure to solar light; (ii) there can be no air bubbles or other scattering defects (much like optical glass fibers) in the glass; and (iii) strict parallellism must be maintained between the surfaces, on which there can be no scratches or refractive index gradients.

Reisfeld and Neuman (32) satisfied some of these criteria in their experiment with uranyl glasses. However, this absorp-tion band has several inconvenient features (2,3,21). The worst is the low molar extinction coefficient ($\varepsilon \sim 10$); if 90 percent of incident blue light is to be absorbed, according to Beer's law the product of the concentration (moles/liter) and plate thickness ℓ (cm) has to be 0.1. Because of self-quenching of fluorescence, it would be unwise to attempt uranyl concentrations as high as 1 M, so a useful combination may be ℓ=1 cm with 23.8 g/liter of uranium in the glass. One may then proceed along one of two alternatives, as shown in Figure 2. About 80 percent of the solar light below about 20,500 cm^{-1} is allowed to pass through the glass, through a narrow air gap (since direct contact between the glass plate and cooling water would be deleterious for total reflection) and then hit a black surface, producing hot water as in con-ventional installations for solar heating; or one may cover the lower surface with a very good mirror. The low-energy photons are lost (since those in the far infrared are dis-sipated in the glass) but only half the optical thickness dictated by Beer's law is required.

The other undoubted disadvantage of the uranyl ion is the much higher energy of its luminescence compared with the band-gap (9,000 cm^{-1}) of the silicon photovoltaic cell. Actually, a given amount of current (corresponding at most to one elec-tron per photon) is not increased by photon energies consider-ably above the energy gap, but rather is decreased by unfor-tunate side-effects of heating of the silicon by the excess energy (30; see also Chapter 11, this Volume). Reisfeld and Kalisky (33) suggested that this problem could be circumvented by energy transfer from the excited uranyl ion to lanthanides (1) such as Nd(III) or Ho(III) homogeneously dispersed in the glass. These $4f^3$ and $4f^{10}$ systems show luminescence with high quantum yields of excited J-levels much closer to, but lying above, the energy gap of silicon in the near infrared. In the contributed program at this Conference, Reisfeld et al. (35) discussed the use of $3d^3$ Cr(III) and $3d^5$ Mn(II) for similar purposes.

Much more research is needed to find a material with both a high luminescence quantum yield and strong absorption throughout most of the solar spectrum above 10,000 cm^{-1}. An obvious argument might be that fluorescent dye-stuffs have much higher extinction coefficients (in the 10^4 range) which allow much smaller values of the product of molecular concentration and optical path; however, they have great difficulties satisfying the unconditional requirements stated above. Most dye-stuffs have to be dispersed in organic polymers which are susceptible to degradation by UV radiation, surface scratches and plastic deformations of the plane parallel surfaces. There is also a general tendency of fluorescent dyes to induce photochemical reactions in the surrounding plastic which cause discoloration with time. Reisfeld et al. (35) have argued that thin fluorescent films may be intercalated between two glass plates, which may also filter out the undesirable UV radiation.

A practical example of the uranyl glass system might involve glass plates in squares (or regular hexagons, which would present a slight advantage) with sides of 200 cm, 1 cm thick and weighing ~40 kg times the specific gravity of the glass. They can be covered by an 800 cm^2 rim of silicon ribbon, i.e., 50 times less than the surface area of the plate. The efficiency will thus be about 50 times greater than that of a plate fully covered with silicon, multiplied by the product of the luminescence quantum yield, the fraction of solar light (say 0.2) above the Stokes threshold (~20,500 cm^{-1} in the uranyl ion), and the loss factor (due to absorption, scattering and refractive index losses) which will be slightly more than cos θ_0. The product of these four factors may lie somewhere between 0.06 and 0.12, thus reducing the amount of silicon needed by a factor of 3 to 6.

From a purely geometrical point of view, a large part of the silicon on the rim might be replaced by mirrors, but this would probably be unwise in view of the very long light paths, corresponding typically to 20 or 100 total reflections. On the other hand, if problems with thermal expansion could be overcome, the amount of silicon might be halved by using both sides of the ribbon fixed between adjacent plates. If we estimate that an output of 1 MW could be provided by 10^4 m^2 of silicon, we would need perhaps 2 × 10^3 m^2 of silicon combined with 10^5 m^2 of uranyl glass weighing about 2 × 10^3 tons and containing 23.8 tons of uranium. At present, the isotope U238 is a relatively inexpensive by-product of isotope separation performed by diffusion of UF_6, and reasonably large (but not

impossibly large) amounts can be obtained for $10,000 a ton.
Hence, the investment in uranium would be about $0.24 per
watt, probably of the same order of magnitude as the cost of a
0.1 m^2 glass plate.

One of the great advantages of the Reisfeld concentrator
is that it does not need expensive tracking equipment of the
kind needed, for example, in solar furnaces (30). If the
solar radiation does not fall absolutely perpendicular on the
glass plate, but under an angle θ, the first-order result is a
multiplication of the photovoltaic yield by cos θ since the
luminescence produced by the entering radiation is isotropic.
If one is willing to accept a decrease of the average yield by
some 10 to 20 percent, one does not need daily adjustment following the motion of the sun. However, if one lives at a considerable latitude (i.e., far from the equator) it may be
worthwhile to make monthly adjustments according to the
seasons. Under slightly cloudy conditions the same arguments
about scattering within an angle around 30° of the solar light
can be applied (36).

A. Some Aspects of Solar Photochemistry

Besides photovoltaic applications, photochemistry induced
by solar light has many fascinating aspects (37,38), as discussed elsewhere in this Volume. This application is appealing not only in desert areas but also, for example, in Switzerland which has considerable mountainous regions unsuited for
agricultural productivity. It is important to note that the
Gibbs energy of a reasonably long-lived excited state is increased by the Stokes threshold (the origin of the absorption
band). This may very well have the consequence that the excited species is simultaneously more reducing and more oxidizing (see Chapter 4). Such behavior is well known for thermal
ground states, such as H_2O_2 (which is not only an oxidant, but
is also more reducing than water in acidic solutions toward
cerium(IV) or permanganate), and the asymptotic limit is disproportionation, e.g., of MnO_4^{2-} in alkaline solution forming
two-thirds MnO_4^- and one-third MnO_2 by acidification. A good
illustration of the idea of simultaneously enhanced reducing
and oxidizing characteristics is the widely investigated
ruthenium(II) bipyridyl complex, $Ru(bipy)_3^{2+}$, which has a
ground-state belonging to [36]$4d^6$. We can write the excited
states due to inverted electron transfer as $4d^5(\pi^*)$. Although
the maximum of the rather structured absorption spectrum of
this red-orange species occurs at ~22,000 cm^{-1}, the Stokes
threshold of the weak shoulder representing the first excited
state is closer to 17,700 cm^{-1} or 2.19 eV. The standard
oxidation potential E^o to form the $4d^5$ $Ru(bipy)_3^{3+}$ is not

known with high precision, but is close to +1.2 V (nhe) (see Chapter 4). Combining these two numbers we see that E^o of the excited $4d^5(\pi^*)$ state is not far from -1.0 V, i.e., much more reducing than Eu(II) or Cr(II) aquo ions, and almost as reducing as Yb(II). This aspect, however, is complicated by the fact that hydrated electrons react with the $(4d^6)$ ground state to form then species, Ru(bipy)$_3^+$ (39), and it has been noted previously (40) that excited Ru(bipy)$_3^{2+}$ is reduced the same way by Eu(II). The same reaction is obtained with Co(I), Zn(I) and Cd(I) produced as short-lived species in aqueous solution from hydrated electrons and M(II) (39). The question whether Ru(bipy)$_3^{1+}$ contains collectively reduced ligands[1] or is a genuine $4d^7$ Ru(I) may perhaps be decided if the strong band observed (41) at ~20,000 cm^{-1} is not accompanied by a strong band in the infrared. If the $4d^7$ configuration is octahedral, it contains one strongly antibonding electron which should give rise to inverted electron transfer bands at very low energies.

Another interesting complication is provided by the intense absorption bands in the near infrared of mixed-valence Ru(II) and Ru(III) connected with a bidentate ligand such as pyrazine (42); this situation is comparable to cyanide-bridged Fe(II,III) cyano complexes in solution (43). Quite fascinating spectra have been obtained recently on Ru(II) complexes containing one or two 2(2'-pyridyl)quinoline ligands and two or one bipyridyl ligands (44). The excited state due to the absorption seems localized on a definite ligand, but the emission is from the pyridine-substituted quinoline. Transient bleaching of the phenanthroline complex Fe(phen)$_3^{2+}$ induced by picosecond pulses at 532 nm (~19,000 cm^{-1}) has been shown to go through two intermediate stages, the initial $3d^5(\pi^*)$ with a lifetime of ~30 ps and another state with a lifetime of ~700 ps, probably a triplet state of $3d^6$ (45). Transient bleaching has also been studied in several Cr(III), Fe(II) and Ru(II) complexes (46).

It is likely that fluorescent collectors using total reflection will soon move to elements more abundant than uranium, and that solar photochemistry will move away from ruthenium. Two recent photogalvanic studies of the rubidium anion in tetrahydrofuran (47,48) and certain oxide mixtures used for photoelectrochemical decomposition of water (49) are

---

[1]In other cases of low oxidation numbers, bipyridyl is known not always to be an "innocent" ligand (4), i.e., allowing the oxidation state of the central atom to be easily defined; examples are the strongly colored complexes of collectively oxidized mercaptoaniline or maleonitrilodithiolate.

good examples. Surface chemistry can sometimes be profitably studied by X-ray induced photoelectron spectra (9,50).

Fortunately for living organisms some, but not all, conceivable chemical reactions establish equilibrium in a given system. Like agriculture, the title "Quantum Harvesting" suggested for this session by the Chairman, refers to fortunate kinetic barriers against the dull equilibrium mixtures predicted by severely consistent thermodynamics, which would also turn the stars into white dwarfs, neutron stars or black holes, according to their masses in the 10^{30} to 10^{31} kg range. For chemists working with condensed matter below 1000K, chemical polarizability (51) is highly significant; this phenomenon is distinct from the (approximately additive) electric polarizabilities of molecules, cations and anions derived from measurements of the refractive index. Chemical polarizability has perfectly observable consequences such as the large contributions of relaxation to ionization energies in photoelectron spectra (52) and chemical behavior, including the hydration energies of gaseous ions (4,6).

A related problem is the origin of band intensities (oscillator strengths), especially for transitions inside the same partly filled shell. Whereas inter-shell transitions (such as $4f \rightarrow 5d$ and $6s \rightarrow 6p$) have mechanisms like those in monatomic entities, and the electron-transfer bands can be rationalized qualitatively by their transition dipole moments (20), it is striking that certain d-group complexes such as Cu(II) and Pd(II) have band intensities one to two orders of magnitude higher than expected (51), and at the same time tend to deviate from the highest symmetry available to the complex, at least on a short time scale (5). Such behavior is rationalized in text books as lone-pairs occurring in compounds outside the transition groups, but there are severe difficulties in such a description (9,53). The theory for intensities in the 4f group for transitions (also luminescent) between J-levels has been expressed by Judd (54) and by Ofelt (55), based on arguments by Broer, Gorter and Hoogschagen (see ref. 56), as three parameters of each material multiplying predetermined matrix elements between the two J-levels; but again, chemical polarizability plays a conspicuous role (1).

It is worthwhile to note that <u>any</u> excited state decays by luminescence if competing photochemical reactions or nonradiative relaxation is not more rapid, and the latter process is well understood as being favored by high phonon energies, i.e., vibrational normal-mode frequencies (1,57). There is little doubt that energy transfer (1,57-59), a process involved in many lasers, is also going to be very important for collectors of solar quanta and for many of the other subjects discussed in this Volume.

IV. CONCLUSIONS

Thirty years ago, there were only fragmentary (and in part illusory) explanations available for the visible colors of inorganic compounds. It is interesting to ask the general question of the origins of color in condensed matter (60) and to survey the many different mechanisms. Indeed it is also a kind of "harvesting" to see how the early rationalization by induction of the multitudinous absorption spectra (4,19-26) contained the grains of utility for capturing solar energy a generation later, even though most of these studies were motivated exclusively by the desire to understand nature. The electrostatic model (i.e., non-spherical part of the Madelung potential) of ligand field theory has now been completely replaced by the angular overlap model in which antibonding effects are proportional to the squares of the overlap between orbitals of the central atom and those of the ligating atoms (5,61,62). However, the fact remains that the energy levels of the constituent atoms grow more and more difficult to recognize in the compounds, progressing in the series: inner shells $<$ 4f $<$ 5f $<$ d and p orbitals, at which point we approach the conventional covalent bonding picture in organic molecules (63).

ACKNOWLEDGMENT

These long-range, subtle interrelations in spectroscopy make it practical to be a generalist rather than to be confined to a narrow and special field. Even so, I must confess that I never would have had the interest to develop this conceptual network of slightly abstract categories for the purpose of capturing solar energy had I not been so fortunate as to meet Professor Renata Reisfeld and initiate a lasting and fruitful collaboration.

REFERENCES

1. R. Reisfeld and C. K. Jørgensen, "Lasers and Excited States of Rare Earths," Springer-Verlag, Berlin, 1977.
2. C. K. Jørgensen, Rev. Chim. Min. (Paris) 14, 127 (1977).
3. C. K. Jørgensen, J. Luminescence 18, 63 (1979).
4. C. K. Jørgensen, "Oxidation Numbers and Oxidation States," Springer-Verlag, Berlin, 1969.

5. C. K. Jørgensen, "Modern Aspects of Ligand Field Theory",
    Elsevier/North-Holland, Amsterdam, 1971.
6. C. K. Jørgensen, "Handbook on the Physics and Chemistry
    of Rare Earths", Vol. 3, Elsevier/North-Holland,
    Amsterdam, 1979, pp. 111-169.
7. C. K. Jørgensen, Adv. Quantum Chem. 11, 51 (1978).
8. D. W. Turner, C. Baker, A. D. Baker and C. R. Brundle,
    "Molecular Photoelectron Spectroscopy", Wiley-Inter-
    science, London, 1970.
9. C. K. Jørgensen, Structure and Bonding 24, 1 (1975);
    ibid., 30, 141 (1976).
10. H. Bock, Angew. Chem., Int. Ed. 16, 613 (1977).
11. G. Boulon, C. K. Jørgensen and R. Reisfeld, Chem. Phys.
    Lett. 75, 24 (1980).
12. C. K. Jørgensen, Structure and Bonding 22, 49 (1975).
13. L. E. Orgel, Quart. Rev. (London) 8, 422 (1954).
14. H. H. Perkampus, "Wechselwirkung von π-Elektronensystemen
    mit Metallhalogeniden", Springer-Verlag, Berlin, 1973.
15. J. D. Webb and E. R. Bernstein, J. Am. Chem. Soc. 100,
    483 (1978).
16. N. Bartlett, Proc. Chem. Soc., 218 (1962).
17. E. Rabinowitch, Rev. Mod. Phys. 14, 112 (1942).
18. M. Linhard and M. Weigel, Z. anorg. Chem. 266, 49 (1951).
19. C. K. Jørgensen, Mol. Phys. 2, 309 (1959).
20. C. K. Jørgensen, Prog. Inorg. Chem. 12, 101 (1970).
21. C. K. Jørgensen, Acta Chem. Scand. 11, 166 (1957).
22. C. K. Jørgensen, Adv. Chem. Phys. 5, 33 (1963).
23. C. K. Jørgensen, Acta Chem. Scand. 16, 2406 (1962).
24. C. K. Jørgensen, "Orbitals in Atoms and Molecules",
    Academic Press, London, 1962.
25. J. L. Ryan and C. K. Jørgensen, Mol. Phys. 7, 17 (1963).
26. A. Müller, E. Diemann and C. K. Jørgensen, Structure and
    Bonding 14, 23 (1973).
27. J. A. Duffy, Structure and Bonding 32, 147 (1977).
28. C. K. Jørgensen, Acta Chem. Scand. 17, 1034 (1963).
29. M. B. Robin and P. Day, Adv. Inorg. Radiochem. 10, 248
    (1967).
30. H. Kelly, Science 199, 634 (1978).
31. R. Reisfeld, Naturwiss. 66, 1 (1979).
33. R. Reisfeld and S. Neuman, Nature 274, 144 (1978).
33. R. Reisfeld and Y. Kalisky, Nature 283, 281 (1980).
34. A. Goetzberger and W. Greubel, Appl. Phys. 14, 123
    (1977).
35. R. Reisfeld, E. Greenberg, A. Kisilev and Y. Kalisky, in
    Book of Abstracts, Third Int. Conf. Photochem. Conv.
    Stor. Solar Energy (J. S. Connolly, ed.), Boulder,
    Colorado, August 1980, SERI/TP-623-797, pp. 95-97.
36. A. Goetzberger, Appl. Phys. 16, 399 (1978).
37. E. Schumacher, Chimia 32, 193 (1978).

38. G. Calzaferri, Chimia 32, 241 (1978).
39. D. Meisel, M. S. Matheson and J. Rabani, J. Am. Chem. Soc. 100, 117 (1978).
40. C. Creutz and N. Sutin, J. Am. Chem. Soc. 98, 6384 (1976).
41. Q. G. Mulazzani, S. Emmi, P. G. Fuochi, M. Z. Hoffman and M. Venturi, J. Am. Chem. Soc. 100, 981 (1978).
42. T. J. Meyer, Accts. Chem. Res. 11, 94 (1978).
43. G. Emschwiller and C. K. Jørgensen, Chem. Phys. Lett. 5, 561 (1970).
44. S. Anderson, K. R. Seddon, R. D. Wright and A. T. Cocks, Chem. Phys. Lett. 71, 220 (1980).
45. A. J. Street, D. M. Goodall and R. C. Greenhow, Chem. Phys. Lett. 56, 326 (1978).
46. A. D. Kirk, P. E. Hoggard, G. B. Porter, M. G. Rockley and M. W. Windsor, Chem. Phys. Lett. 37, 199 (1976).
47. S. Goldstein, S. Jaenicke and H. Levanon, Chem. Phys. Lett. 71, 490 (1980).
48. U. Eliav and H. Levanon, Chem. Phys. Lett. 72, 213 (1980).
49. J. Augustynski, J. Hinden and C. Stalder, J. Electrochem. Soc. 124, 1063 (1977).
50. C. K. Jørgensen, Fresenius Z. analyt. Chem. 288, 161 (1977).
51. C. K. Jørgensen, Topics Current Chem. 56, 1 (1975).
52. C. K. Jørgensen, Adv. Quantum Chem. 8, 137 (1974).
53. C. K. Jørgensen, Chimia 31, 445 (1977).
54. B. R. Judd, Phys. Rev. 127, 750 (1962); J. Chem. Phys. 70, 4830 (1979).
55. G. S. Ofelt, J. Chem. Phys. 37, 511 (1962).
56. R. D. Peacock, Structure and Bonding 22, 83 (1975);
57. R. Reisfeld, Structure and Bonding 13, 53 (1973); ibid., 22, 123 (1975); ibid., 30, 65 (1976).
58. R. Englman, "Non-Radiative Decay of Ions and Molecules in Solids", Elsevier/North-Holland, Amsterdam, 1979.
59. R. C. Powell and G. Blasse, Structure and Bonding 42, 43 (1980)
60. M. V. Orna, J. Chem. Ed. 55, 478 (1978).
61. C. E. Schäffer, Structure and Bonding 5, 68 (1968); ibid., 14, 69 (1973).
62. C. Linares, A. Louat and M. Blanchard, Structure and Bonding 33, 179 (1977).
63. C. K. Jørgensen, Israel J. Chem. 19, 174 (1980).

DISCUSSION

Prof. M. Grätzel, Ecole Polytechnique Fédérale, Lausanne:
What is the quantum yield for fluorescence of UO_2^{2+}?
Also, would you care to comment on the system of Prof.
Wagner (University of Freiburg) based on organic dyes?

Prof. Jørgensen:
An indirect estimate of the quantum yield from the average
lifetime compared with the radiative lifetime evaluated
with Einstein's 1917 formula is not particularly reliable
for the uranyl ion, since the (rather small) relevant part
of the oscillator strength occurs in a shoulder within the
first 1500 cm^{-1} above the Stokes threshold. The aquo ion
has a lifetime as short as 10^{-5} s, but as seen in the many
studies by my colleague, Marcantonatos, the dramatic
increase in D_2O indicates nonradiative de-excitation connected with O-H and O-D vibrations, and indeed, the lifetime is in the millisecond range in many solids not having
short uranyl-hydrogen contacts. The quantum yield in typical glasses is 0.3 to 0.8, but this is a point needing
further research.

In reply to your second question: In general, organic
dye-stuffs allow much higher optical density (product of
molar extinction coefficient and path length) but they
have their problems, too. It is difficult to tell whether
organic or inorganic luminescent materials are going to
"win the prize" as the most efficient for fluorescent concentrator plates.

Dr. A. Heller, Bell Laboratories:
Are fluorescent collectors competitive with Fresnel
lenses? Also, are there any practical fluorescent systems
that meet the Stokes' shift, absorption, and emission
criteria?

Prof. Jørgensen:
Good optical quality glass Fresnel lenses have the same
characteristics as parabolic mirrors and other helio-
stats. However, they do not collect slightly scattered
light arriving at small angles from the direct light; this
can amount to about half the total effect in the case of
thin clouds as pointed out by Goetzberger (36). I am
afraid that plastic Fresnel lenses may deform with time,
and round off the sharp edges. I have no strong arguments
about relative price; if you tell me that 1 ft^2 of Fresnel
lens costs \$3, and if it corresponds to 10 W, I agree that

it seems marginally less than the cost of 10 ft^2 of glass, though you lose the low-energy photons for heating, and warm the silicon to an unnecessarily high temperature in the former case. As you know better than I, gallium arsenide would be more favorable at high luminosities, but since gallium is a byproduct of aluminum isolation, its yearly production rate is not readily extensible, but is of the order of magnitude of a few tons.

In answer to your second question, resonant fluorescence with quantum yields less than unity cannot be tolerated. The typical Stokes overlap only costs some 20% once, whereas quantum yields of total reflection each time are multiplied by the cosine of the characteristic angle. If isotropic resonant fluorescence takes place 10 times, even $(0.8)^{10}$ is only 0.1. This argument is somewhat attenuated, however, in vitreous materials with large emission bandwidths (due to inhomogeneous broadening). There is no doubt that the uranyl glasses, with or without energy transfer to trivalent lanthanides, work. There is also no doubt that there are rational hopes for choosing systems with lower Stokes thresholds.

CHAPTER 4

**PHOTOCHEMICAL ELECTRON TRANSFER REACTIONS IN
HOMOGENEOUS SOLUTIONS**

Vincenzo Balzani

Istituto di Fotochimica e
Radiazioni d'Alta Energia del CNR, and
Istituto Chimico "G. Ciamician" dell'Università
Bologna, Italy

Franco Scandola

Centro di Fotochimica del CNR, and
Istituto Chimico dell'Università
Ferrara, Italy

I. INTRODUCTION

An electronically excited state induced in a molecule by
absorption of light is virtually a new species with its own
chemical and physical properties different from those of the
corresponding ground-state molecule. One of the most important consequences of light absorption is that of increasing
the electron affinity and decreasing the ionization potential. Thus, an electronically excited state is expected to be
both a better oxidant and a better reductant than the ground
state of the molecule. It follows that those electronically
excited states which, in fluid solution, live long enough to
encounter a molecule of another solute can often be involved
in electron-transfer reactions (1,2).
   In recent years, light-induced electron-transfer processes
have been extensively investigated with the aim of understanding the primary mechanisms of the natural photosynthetic process and of designing artificial systems for conversion and
storage of solar energy. Although it is generally agreed that
any practical system for solar energy conversion and storage
will involve heterogeneous reactions at some stage of the

overall process, elucidation of the factors that govern homogeneous electron transfer remains a fundamental need for any
further progress in this field.

In this paper we will discuss the role played by thermodynamic and kinetic factors in determining the rate of
electron-transfer processes involving electronically excited
states. The problems of the decrease of the rate constants
with increasing exergonicity as predicted by the Marcus theory
and of adiabatic vs. nonadiabatic behavior will be analyzed.
Most of the reactions discussed will concern transition metal
complexes, not only because of our own interests, but also
because these molecules have been investigated so extensively.

II. ELECTRONICALLY EXCITED STATES AS REDOX REACTANTS

As previously mentioned, electronically excited states
are, in principle, excellent reactants for redox processes.
In practice, however, there may be severe kinetic limitations
to their efficient use. It is well known that an excited
state formed by light absorption may undergo several kinds of
deactivation processes (Fig. 1). Some of these processes do
not need specific interactions with other molecules and are
therefore called intramolecular processes. Only when such
intramolecular processes are not too fast, i.e., when the
lifetime of the excited state is sufficiently long, will the
excited molecule have a chance to meet a molecule of another
solute and eventually to undergo a quenching process. Simple
kinetic considerations show that even when high quencher concentrations are used only those excited states that have lifetimes longer than about 0.1-1 ns can be involved in encounters
with a quencher molecule (3,4).

For many organic molecules the lowest spin-forbidden
excited state, and sometimes also the lowest spin-allowed

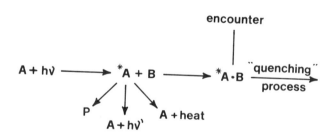

FIGURE 1. Competition between intramolecular and
intermolecular deactivation of an electronically excited
molecule.

excited state, satisfy this condition. For transition metal complexes, however, the lowest spin-allowed excited states in fluid solution generally have lifetimes much shorter than 1 ns because intersystem crossing to the lowest spin-forbidden excited state is very fast due to spin-orbit coupling induced by the metal atom. Thus, for transition metal complexes the only excited state that can be involved in bimolecular processes is usually the lowest spin-forbidden excited state; but, as we will see later, even these states are often too short-lived owing to their high reactivities, lack of rigidity, and strong spin-orbit coupling.

Once an excited state is involved in an encounter with another molecule, different processes may occur depending on the nature and the magnitude of the interaction (Fig. 2). Weak interaction may lead to energy transfer via an exchange mechanism, to electron transfer, or to simple deactivation of the excited state, catalyzed in some way by the quencher. Strong interactions lead to the formation of new chemical species called exciplexes which may undergo radiative or non-radiative deactivation, or may also lead to energy or electron transfer as well as to other kinds of chemical reactions. When the excited state and/or the reaction partner are transition metal complexes and the solvent is polar, the interaction is generally weak. In this paper we will deal

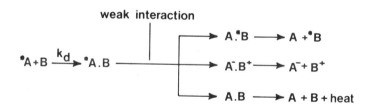

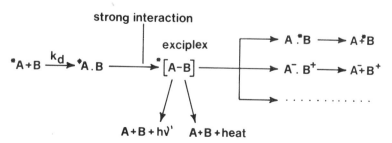

FIGURE 2. Schematic diagram of the possible consequences of weak and strong interactions between an excited state and a quencher.

only with electron- and energy-transfer processes which occur as a consequence of weak interactions in encounters between excited states and other molecules. Weak interaction processes are amenable to theoretical interpretations and we will compare the available experimental results with the proposed theoretical treatments.

The redox reactions of electronically excited states can be used for three fundamental types of applications (Fig. 3): (a) for catalytic purposes when the exceptional propensity of excited states to undergo redox processes is used to obtain products that would have been obtained with much lower efficiency by a thermal reaction; (b) for photochemical transformation of lower energy reactants into higher energy products, i.e., for conversion of light into chemical energy; and (c) for chemical generation of excited states in highly exergonic reactions in order to convert chemical energy into light. Although we will discuss electron-transfer reactions in general, the aim of this paper is, of course, that of offering some guidelines for selecting systems to be used for solar energy conversion.

It should be pointed out at the beginning that the excited states involved in electron- and energy-transfer processes in fluid solution are thermally equilibrated species, so that these processes can be dealt with in the same way as any other chemical reaction, that is, by using thermodynamic and kinetic arguments. In particular, electron-transfer reactions involving ground state (Eq. [1]) or excited state (Eq. [2]) molecules:

$$A + B \rightarrow A^- + B^+ \tag{1}$$

$$^*A + B \rightarrow A^- + B^+ \tag{2}$$

are processes identical in nature, and thus any notion, theory or formalism that can be used for one of them can also be used for the other. At first sight, electronic energy transfer, Eq. [3], might appear to be a process of different type:

$$^*A + B \rightarrow A + {}^*B \tag{3}$$

In reality, in the weak interaction limit there is a close analogy between electron-transfer and exchange energy-transfer processes. In both cases, spatial overlap of the orbitals of the two reactants is required, no bond-making or bond-breaking processes take place, and Franck-Condon restrictions have to be obeyed because electronic rearrangement with (Eq. [2]) or without (Eq. [3]) the net transfer of an electron occurs in a time short compared to that required for nuclear motions.

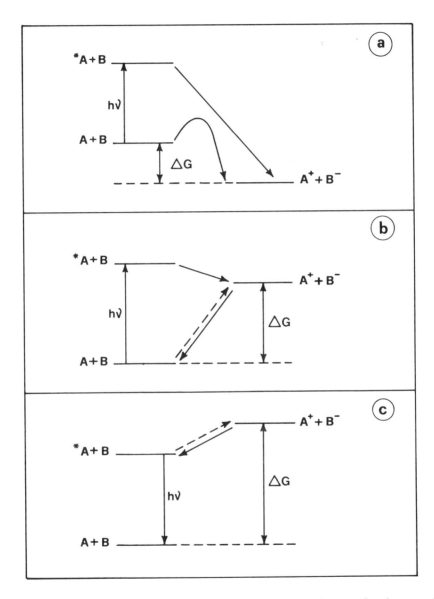

FIGURE 3. Schematic diagram of the three fundamental types of applications of excited state electron-transfer processes. (a) Photocatalysis (see also Chapter 10, this Volume); (b) conversion of light into chemical energy; (c) conversion of chemical energy into light (chemiluminescence).

From an energetic point of view, the only difference between reactions [1] and [2] is that the excited state reactant of [2] possesses some additional energy with respect to the ground state reactant of [1]. If both the excited and ground state molecules lie in their zero vibrational levels, the extra energy content of *A is the 0-0 spectroscopic energy, E^{oo}. Passing to thermodynamic quantities, if the vibrational partition functions of the two states are not very different, the enthalpy difference between excited and ground state molecules is, for practical purposes, equal to E^{oo} and the entropy difference is negligible (1). Thus the excited states have an extra free-energy content which is approximately equal to E^{oo}. This means that the redox potentials of the excited molecule are fairly well approximated by the following equations:

$$E^o(A^+/{}^*A) \simeq E^o(A^+/A) - E^{oo}({}^*A,A) \qquad [4]$$

$$E^o({}^*A/A^-) \simeq E^o(A/A^-) + E^{oo}({}^*A,A) \qquad [5]$$

The standard free-energy change of excited state reactions, e.g., Eq. [2], can be calculated using the appropriate redox potentials. As for energy-transfer processes, Eq. [3], it can be shown similarly that if the partition functions are not very different for the ground and excited states of the same molecule, the standard-free energy change of the process is given approximately by the difference in the 0-0 spectroscopic energies of the two excited states (5).

An excited state should meet several specific requirements in order to be a useful redox reactant for solar energy conversion. It should have a reasonably high energy content, but at the same time it should be induced with photons of reasonably low energy; it should be formed with unit efficiency upon light absorption, regardless of the excitation wavelength; it should have a sufficiently long lifetime; it should not be strongly distorted with respect to the ground state; it should be a good oxidant and/or reductant; and it should be stable towards photodecomposition. Moreover, the resulting oxidation and reduction products should themselves be good oxidants and good reductants, respectively, and they should be stable towards side reactions.

Many of these requirements are more easily met by transition metal complexes than by organic molecules or simple metal ions. For example, simple metal ions do not exhibit intense absorption bands in the visible and, even worse, their lowest excited states often lie at very low energy and are strongly distorted and very short-lived. On the other hand, the lowest spin-allowed excited state in organic molecules often can be reached only with high-energy photons and are usually rather short-lived, while the lowest spin-forbidden excited state, which is long-lived, can be obtained only via singlet absorption sometimes with low efficiency. Transition metal complexes of metals of the second and third transition rows with suitable organic ligands exhibit intense charge-transfer bands in the visible region (which of course are not shown by the simple metal ion or by the free ligand); moreover, their lowest excited states are undistorted because they involve promotion of an electron to a delocalized π^* orbital. They can be populated very efficiently because of the spin-orbit coupling induced by the heavy metal atom, and can still be sufficiently long-lived to be involved in bimolecular processes even though the lifetimes are shorter than the lowest triplet of the free ligand because of enhanced intersystem crossing to the ground state. Another important advantage of transition metal complexes is the presence of redox sites on both the metal and the ligands, which offers additional redox possibilities not available for either simple metal ions or organic molecules.

A. Complexes Containing Polypyridine Ligands

In the last few years it has been found that transition metal complexes containing aromatic ligands such as 2,2'-bipyridine (bpy), 1,10-phenanthroline (phen) or their derivatives are attractive candidates for excited state electron-transfer processes in fluid solution (1,2,6-10). These complexes have octahedral structure and can be obtained with a variety of transition metal ions. The best known among these complexes is $Ru(bpy)_3^{2+}$. Under normal chemical conditions the stable oxidation state of this complex is 2+, but it can be oxidized or reduced without disruption of the molecular structure (Fig. 4a). Its absorption spectrum extends to the visible with a very broad and intense band whose maximum is at 450 nm. This complex is quite stable in aqueous solution and does not undergo appreciable photodissociation. Its lowest excited state has metal-to-ligand charge transfer (MLCT) orbital character and, formally, triplet multiplicity. The excited state has a 0-0 spectroscopic energy of 2.12 eV, is only slightly distorted with respect to the ground state, is obtained with unit efficiency regardless of the excitation wavelength, and exhibits luminescence even in aqueous solution at room temperature where its lifetime is fairly long (0.62 μs). Comparison of the properties of $Ru(bpy)_3^{2+}$ with the requirements discussed at the end of the previous section shows that this complex is an almost ideal candidate as an electron-transfer photosensitizer for artificial photosynthetic systems. It is worthwhile noting that the excited

Table I. Properties of the Lowest Excited States
of Some Tris-1,10-Phenanthroline Complexes[a,b]

*A	Spin and orbital nature[c]	τ[d]	E^{00}, eV[e]	$*E°(*A/A^-)$[f]	$*E°(A^+/*A)$[f]
*Cr(phen)$_3$$^{3+}$	^{2}MC ($t_{2g} \rightarrow t_{2g}$)	270 μs	1.71	1.46	>-0.1[g]
*Fe(phen)$_3$$^{2+}$	^{3}MC ($t_{2g} \rightarrow e_g$)	0.80 ns[h]	≤0.9[i]	-	≥0.1[i]
*Ru(phen)$_3$$^{2+}$	^{3}MLCT ($t_{2g} \rightarrow \pi^*$)	0.81 μs	2.18	0.82	-0.92
*Os(phen)$_3$$^{2+}$	^{3}MLCT ($t_{2g} \rightarrow \pi^*$)	84 ns	1.78	0.57	-0.96
*Rh(phen)$_3$$^{3+}$[j]	^{3}LC ($\pi \rightarrow \pi^*$)	~0.25 μs[k]	2.75	2.00[l]	-
*Ir(phen)$_3$$^{3+}$	^{3}LC ($\pi \rightarrow \pi^*$)	2.9 μs[m]	2.79[m]	1.8[n]	~-0.1[n]

a From ref. (1) and references therein, unless otherwise noted.
b Room temperature, aqueous solution, unless otherwise noted.
c MC = metal-centered; MLCT = metal-ligand charge transfer; LC = ligand centered.
d Excited state lifetime in deaerated solution, unless otherwise noted.
e 0-0 excited state energy.
f v(nhe), unless otherwise noted; calculated from Eqs. [4] and [5].
g By analogy to Cr(bpy)$_3$$^{3+}$.
h Ref. (9).
i By analogy to Fe(bpy)$_3$$^{2+}$ (ref. 9).
j Ref. (10); all data refer to CH$_3$CN solutions.
k Aerated solution.
l v(sce).
m Methanol solution.
n The ground state A/A$^-$ and A$^+$/A potentials are -1.0 and ~+2.8 V (11).

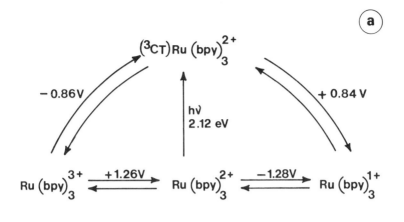

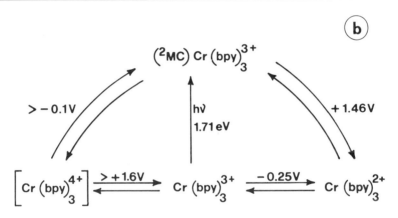

FIGURE 4. Redox potentials for the ground and lowest excited states of $Ru(bpy)_3^{2+}$ and $Cr(bpy)_3^{3+}$.

state of $Ru(bpy)_3^{2+}$ is <u>both</u> a moderately strong oxidant <u>and</u> a very strong reductant (Fig. 4a), so that from a thermodynamic point of view it can both reduce and oxidize water.

The analogous $Cr(bpy)_3^{3+}$ complex (Fig. 4b) in its lowest excited state is a much stronger oxidant than the ruthenium complex, but is a very poor reductant. For both practical and theoretical reasons it is important to have a series of complexes which cover a wide range of values of the various excited state properties. In the family of the polypyridine complexes this "tuning" can be done in three ways: (i) by changing the nature of the central metal (Table I); (ii) by replacing some of the pyridine ligands with other ligands; and (iii) by substituting suitable groups onto the polypyridine rings. The search for other families of transition metal complexes to be used in artificial photosynthetic processes is very active in several laboratories, but it will be difficult to find complexes having more favorable properties than those belonging to the polypyridine family.

III. KINETIC TREATMENT

As previously mentioned, there is always competition among the various intra- and intermolecular deactivation pathways of an excited state. Thus, even if electron transfer or energy transfer is thermodynamically allowed, it can take place only if it is fast enough to win the competition with other intra- and intermolecular deactivation processes. It follows that to make progress in the area of energy migration and energy conversion and storage, where energy- and electron-transfer processes both play fundamental roles, the factors governing the rates of these processes must be elucidated.

The formalisms developed by Marcus (12), Hush (13), Sutin (14) and others for thermal electron-transfer processes (Eq. [1]) can be applied to both excited state electron transfer (Eq. [2]) (1,2) and energy transfer (Eq. [3]) (5). For an excited state electron-transfer reaction, the following kinetic scheme applies:

$$*A + B \underset{k_{-d}}{\overset{k_d}{\rightleftarrows}} (*A \cdot B) \underset{k_{-el}}{\overset{k_{el}}{\rightleftarrows}} (A^- \cdot B^+) \overset{k_s}{\longrightarrow} products \qquad [6]$$

where k_d is the diffusion rate constant, k_{-d} the dissociation rate constant of the precursor complex, k_{el} and k_{-el} are the unimolecular rate constants for the electron-transfer step, and k_s includes all the steps (except k_{-el}) which cause the disappearance of the successor complex. A simple steady-state treatment shows that the experimental rate constant of Eq. [2], k_q, can be expressed as a function of the rate constants of the various steps in Eq. [6] as follows:

$$k_q = \frac{k_d}{1 + \dfrac{k_{-d}}{k_{el}} + \dfrac{k_{-d}}{k_s} \cdot \dfrac{k_{-el}}{k_{el}}} = \frac{k_d}{1 + \dfrac{k_{-d}}{k_{el}} + \dfrac{k_{-d}}{k_s} \exp(\Delta G/RT)} \qquad [7]$$

where ΔG is the standard Gibbs free-energy change of the electron-transfer step. For slow electron-transfer processes, i.e., when $k_{-el} \ll k_s$ and $k_{el} \ll k_{-d}$, Eq. [7] reduces to the pre-equilibrium formulation of outer-sphere electron-transfer reactions given by Sutin (14):

$$k_q = \frac{k_d}{k_{-d}}\, k_{el} = K_o k_{el} \qquad\qquad [8]$$

A completely analogous kinetic treatment can be used for an energy-transfer process (5).

The key step in electron (or energy) transfer is the conversion of the precursor into the successor complex during the encounter. This step can be dealt with using either classical or quantum mechanical treatments.

A. Classical and Quantum Mechanical Approaches

In the classical approach the conversion of the precursor into the successor complex is treated according to the activated complex formalism (14). The rate constant for the electron-transfer step is given by:

$$k_{el} = k_{el}^o\, \exp\!\left(-\Delta G^{\ddagger}/RT\right) = \underset{\sim}{k}\, \frac{k_B T}{h}\, \exp\!\left(-\Delta G^{\ddagger}/RT\right) \qquad [9]$$

where k_{el}^o is the frequency factor, $\Delta G^{\ddagger}$ the standard free-energy of activation, $\underset{\sim}{k}$ the transmission coefficient and $k_B T/h$ is a universal frequency factor (6×10^{12} s^{-1} at 25°C). In order to understand the meaning of $\underset{\sim}{k}$ and $\Delta G^{\ddagger}$ in Eq. [9], consider the potential energy surfaces of the initial and final states of the system as functions of the nuclear configuration (Fig. 5). The electron-transfer step must obey the Franck-

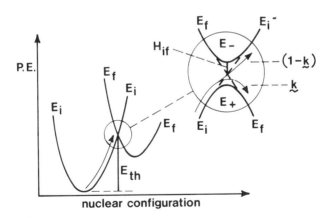

FIGURE 5. Schematic representation of an electron-transfer process according to the classical approach. E_i and E_f are the potential energy surfaces of the initial and final states of the system, respectively. See text for details.

Condon principle which states that the nuclear positions and nuclear velocities remain essentially unchanged during the electronic transition. Thus, the only possibility for electron transfer to occur is near the crossing point of the two curves. This requires the adjustment of the inner vibrational coordinates of the molecules and of the outer solvation spheres to some nonequilibrium configuration prior to electron transfer. The reorganization energy E_{th} is taken to be equal to the reorganization free-energy $\Delta G^{\ddagger}$, which appears in Eq. [9], on the assumption that the entropy contribution related to the reorganization is negligible. Once a suitable nuclear configuration has been reached, whether or not electron transfer occurs is then a matter of electronic factors.

If there is no electronic interaction, the initial system cannot be transformed into the final system and thus the reaction cannot take place. If there is interaction, the degeneracy at the intersection will be removed and two new surfaces will be formed. The separation between the two surfaces is equal to $2H_{if}$, where H_{if} is the interaction energy (electronic coupling matrix element). In the classical theories (like the Marcus theory) the interaction energy is assumed to be small enough so that is can be neglected in calculating E_{th} but large enough so that the reactants are converted into products with unit probability in the intersection region. More generally, in the weak interaction limit the second assumption has to be removed, and the possibility must be considered that a system arriving from the initial state may continue along the original zero-order curve and then return without undergoing reaction. This possibility is taken into account by the transmission coefficient $\underset{\sim}{k}$ which measures the probability that a system will remain on the lower curve, thus giving rise to products. When the transmission coefficient is unity the reaction is said to be adiabatic; when it is less than unity the reaction is said to be nonadiabatic or diabatic.

Theoretical approaches show that when the interaction energy is larger than about 100 cm^{-1} the reaction is expected to be adiabatic. The interaction energy may be much smaller when the separation between the reactants is large or when the reaction is spin- or symmetry-forbidden. When the interaction energy is very small, both the semiclassical Landau–Zener theory (15) and quantum mechanical treatments (16) show that the probability of electron transfer is proportional to the square of the interaction energy H_{if}. Thus, the transmission coefficient contains important electronic information concerning the electron-transfer step and, as we will discuss below, provides a useful link between the classical and quantum mechanical equations.

In the quantum mechanical approach, conversion of the precursor into the successor complex is considered as a nonradiative transition between two electronic states of a "super-molecule" consisting of the donor, the acceptor and the solvent (16,17). According to Fermi's golden rule the probability per unit time that a system in a single vibronic level of the initial state $|iv\rangle$ will pass to a manifold of vibrational levels of the final state $\{|fw\rangle\}$ is given by:

$$W_{iv} = \frac{4\pi^2 H_{if}^2}{h} \sum_w |\langle X_{iv}|X_{fw}\rangle|^2 \delta(E_{iv} - E_{fw}) \qquad [10]$$

where H_{if} is the previously mentioned electronic coupling matrix element; the summation gives the density of final states weighted by the Franck-Condon factor $|\langle X_{iv}|X_{fw}\rangle|^2$; E_{iv} and E_{fw} are the zero-order vibronic levels; and δ is the Dirac delta function which ensures energy conservation. The transition probability from all the vibrational levels of the initial state is given by a summation of single terms given by Eq. [10], normalized according to the Boltzman distribution:

$$W_i = \sum_v \{W_{iv} \exp(-E_{iv}/RT)\}/\sum_v \exp(-E_{iv}/RT) \qquad [11]$$

This approach is valid only when two electronic states (one each for the reactant and product) need to be considered, the density of final levels is large, and the transition probability is small (i.e., a nonadiabatic process) (16,17).

In summary, under the above conditions the rate constant of the electron-transfer step is given by the product of an electronic and a nuclear term:

$$k_{el} = E \cdot N \qquad [12]$$

The electronic term, which may be viewed as the quantum mechanical counterpart of the classical transmission coefficient, is related to the coupling of the initial and final states. The nuclear term, which can be viewed as the quantum mechanical counterpart of the classical activation term, is a thermally averaged summation over all the individual vibronic transitions of the system, each proportional to its Franck-Condon factor. In order to get an analytical expression for the nuclear term, it is usually convenient to separate the vibrational modes of the system into two categories: (i) high frequency ($h\nu > kT$) vibrational modes which have to be treated as quantum modes, and (ii) low frequency ($h\nu < kT$) vibrational modes which can be treated within the framework of the classical approximation (17). Unfortunately, the application of the quantum mechanical approach to practical cases is generally precluded by the lack of knowledge of most of the relevant molecular parameters, and it has been attempted only in model systems (16,17) and in a few real cases (18-20), some of which concern the analogous energy-transfer and spin-conversion processes. Thus, in the following discussion we will use the classical approach, trying to incorporate in it the major conclusions of its quantum mechanical counterpart.[1]

B. Free-Energy Relationships

As we have seen above, the conclusions of the quantum mechanical approach concerning the role played by electronic interactions are embodied in the classical equation by the transmission coefficient. As for the nuclear part, according to the classical Marcus theory (Fig. 6) the free-energy of activation is given by the following equation (see 12,14):

$$\Delta G^{\ddagger} = \Delta G^{\ddagger}(0)\left(1 + \frac{\Delta G}{4\Delta G^{\ddagger}(0)}\right)^2 \qquad [13]$$

where $\Delta G^{\ddagger}(0)$ is the so-called intrinsic reorganizational parameter (i.e., the free-energy of activation for $\Delta G = 0$), which is related to the changes in the nuclear coordinates that must

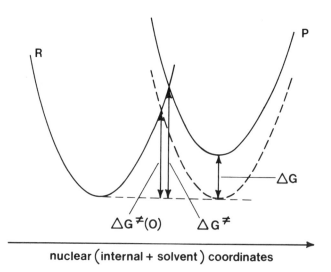

nuclear (internal + solvent) coordinates

FIGURE 6. Schematic diagram of the quantities involved in free-energy relationships such as Eq. [13].

---

[1]A semiclassical treatment which combines various components of the classical and quantum mechanical approaches has been recently put forward by Brunschwig et al. (21).

occur prior to electron transfer, as required by the Franck-Condon principle. A very peculiar aspect of Eq. [13], to be discussed later, is that the free-energy of activation is expected to increase (and thus the reaction rate to decrease) when the reaction becomes strongly favored thermodynamically.

Relations like Eq. [13], where the free-energy of activation is expressed as a function of the net free-energy change together with some intrinsic parameter, are very common in chemical kinetics and are quite useful in understanding homogeneous (i.e., homologous) sets of data (22). Consider, for example, electron transfer between an excited molecule (e.g., $*A$) and a series of structurally related molecules (B_1, B_2 ... B_n) which have variable oxidation potential but the same size, shape, electric charge and electronic structure. It can be assumed that throughout this homogeneous series of reactions:

$$*A + B_1 \rightarrow A^- + B_1^+$$
$$*A + B_2 \rightarrow A^- + B_2^+ \qquad\qquad [14]$$
$$\cdots\cdots\cdots \qquad \cdots\cdots\cdots$$
$$*A + B_n \rightarrow A^- + B_n^+$$

the parameters k_d, k_{-d} and k_s which appear in Eq. [7] are constant as are the transmission coefficient k and the intrinsic reorganizational parameter $\Delta G^{\ddagger}(0)$. From Eqs. [7] and [9] and a free-energy relationship (e.g., Eq. [13]) it follows that, within the homogeneous reactions in Eq. [14], k_q is a function only of the oxidation potentials of the members of the homologous series B_i (see also the next section).[2]

When Eq. [13] is explicitly used in the above treatment, Eq. [7] predicts that $\log k_q$ should increase with decreasing endergonicity, reach a plateau value and then decrease again when $\Delta G < -4\Delta G^{\ddagger}(0)$. The region in which the Marcus model predicts a drastic decrease in the rate constant, as the reaction becomes more and more thermodynamically favored, is called the Marcus "inverted region". Experimental evidence for or against this inverted region cannot be obtained from conventional (i.e., ground-state) kinetic experiments because, in general, ΔG is not sufficiently negative. The use of electronically excited states as reactants and of flash photolysis as a fast relaxation technique (Fig. 7) have made it possible

---

[2]This line of reasoning applies, of course, to any homogeneous series of redox reactions, regardless of whether the constant partner is an oxidant or a reductant in a ground or an excited electronic state.

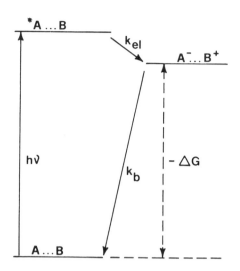

FIGURE 7. Scheme showing that the back reaction (k_b) between two species generated in excited-state electron-transfer experiments can involve large negative ΔG.

in recent years to explore highly exergonic reactions in a systematic way (23-31), showing that the dramatic decrease in reaction rate predicted by the Marcus model does not, in general, take place (see, for example, Fig. 8) (31). However, in some of the systems studied, alternative reaction mechanisms (e.g., H-atom transfer) may actually be operative.

It should be noted also that the expectations of the Marcus model are not consonant with those of quantum mechanical treatments. Although, as we have seen before, the general expression derived from theoretical nonadiabatic approaches can hardly be applied to any practical case, the features emerging from simple calculations on model systems indicate that a noticeable decrease in the rate constant is not expected (at least for moderately exergonic reactions) if all the distorted vibrations of the system are taken into account (16,17). Moreover, as discussed later, in very highly exergonic systems other reaction channels leading to excited state products may come into play, so that a decrease in the experimental rate constant is not likely to be observed in most cases.

An empirical way to account for the above effects is to use other free-energy relationships in place of Eq. [13]. The empirical relationships proposed by Rehm and Weller (23), Eq. [15], and Agmon and Levine (32), Eq. [16]:

$$\Delta G^{\ddagger} = \frac{\Delta G}{2} + \left\{ \left(\frac{\Delta G}{2}\right)^2 + \left(\Delta G^{\ddagger}(0) \right)^2 \right\}^{1/2} \tag{15}$$

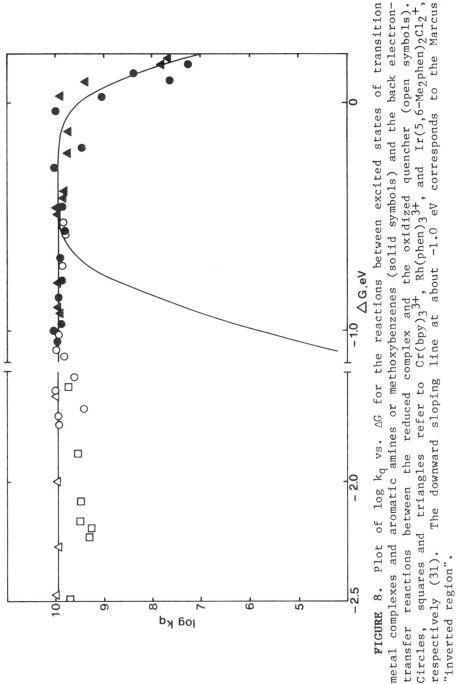

FIGURE 8. Plot of log k_q vs. ΔG for the reactions between excited states of transition metal complexes and aromatic amines or methoxybenzenes (solid symbols) and the back electron-transfer reactions between the reduced complex and the oxidized quencher (open symbols). Circles, squares and triangles refer to $Cr(bpy)_3^{3+}$, $Rh(phen)_3^{3+}$, and $Ir(5,6-Me_2phen)_2Cl_2^+$, respectively (31). The downward sloping line at about -1.0 eV corresponds to the Marcus "inverted region".

$$\Delta G^{\ddagger} = \Delta G + \frac{\Delta G^{\ddagger}(0)}{\ell n2} \; \ell n \left\{ 1 + \exp\left[- \frac{\Delta G \; \ell n2}{\Delta G^{\ddagger}(0)} \right] \right\} \qquad [16]$$

account fairly well for the available results. Although Eqs. [15] and [16] behave quite the same in all ΔG ranges, we will adopt Eq. [16] because it seems of more general use (32). Thus, we propose to use Eqs. [7], [9], and [16] as a formalism which describes the behavior of electron transfer (or energy transfer) in terms of classically defined parameters. Such a formalism can be used to rationalize the available data and to predict the behavior of systems not yet studied.

C. General Kinetic Equation and Log k_q vs. ΔG Plots

Taken together, Eqs. [7], [9], and [16] give the following general expression for the experimental rate constant of reaction [2]:

$$k_q = \frac{k_d}{1 + \dfrac{k_{-d}}{k_s} \exp\left(\dfrac{\Delta G}{RT}\right) + \dfrac{k_{-d}}{k_{el}^o} \exp\left\{ \dfrac{\Delta G + \dfrac{\Delta G^{\ddagger}(0)}{\ell n2} \ell n \left[1 + \exp\left(\dfrac{-\Delta G \; \ell n2}{\Delta G^{\ddagger}(0)}\right) \right]}{RT} \right\}} \qquad [17]$$

As we have seen in the previous section, for a homogeneous series of reactions (e.g., Eq. [14]), k_q is a function only of ΔG, which is given by:

$$\Delta G = E^o(^*A/A^-) + E^o(B^+/B) \qquad [18]$$

where $E^o(^*A/A^-)$ is given by Eq. [5]. Since $E^o(^*A/A^-)$ is constant along the series of reactions, k_q is a function only of $E^o(B^+/B)$. In this case, Eq. [17] predicts that a plot of log k_q vs. ΔG (or vs. $E^o(B^+/B)$) will have the features shown in Figure 9, viz., (i) an Arrhenius type linear region for sufficiently endergonic reactions; (ii) a more or less wide intermediate region (depending on $\Delta G^{\ddagger}(0)$) in which log k_q increases in a complex but monotonic way as ΔG decreases; and (iii) a plateau region for sufficiently exergonic reactions. The intermediate region is centered at $\Delta G = 0$, where the slope of the curve is $0.5/(2.3\ RT)$. As shown in Figure 9, the value of the intrinsic barrier $\Delta G^{\ddagger}(0)$ strongly influences the values of the rate constant in the intermediate nonlinear region. However, this barrier does not affect the plateau value given by Eq. [19], which is equal either to k_d or to $k_d \cdot k_{el}^o/k_{-d}$,

$$k_q^p = k_d k_{el}^o/(k_{el}^o + k_{-d}) \qquad [19]$$

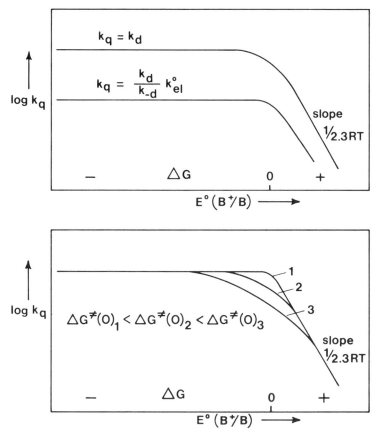

FIGURE 9. Influences of various parameters on $\log k_q$ vs. ΔG plots (see text).

depending on whether k_{el}^o is much larger or much smaller than k_{-d}. Thus, a lower than diffusion-controlled value of k_q^p is related to a low value of the frequency factor, i.e., to a transmission coefficient $\underset{\sim}{k}$ much lower than unity.[3]

---

[3]In particular cases, pre-equilibrium changes on one of the reactants may also cause rate saturation below the diffusion-controlled limit (33). Other factors that probably decrease the pre-exponential term (14) (i.e., orientational factors, use of nuclear frequencies other than the classical $k_B T/h$) are not likely to cause $k_{el}^o < k_d$. Note also that $\underset{\sim}{k}$ values $\geqslant 10^{-3}$ cannot appreciably decrease k_q^p below k_d and thus cannot be revealed by this approach.

When, with increasing driving force, the rate constants of a homogeneous series of outer-sphere reactions tend to be slower than the diffusion-controlled limit because of nonadiabaticity (Fig. 9), it is to be expected that other, electronically more efficient, channels become important at higher exergonicity (34). For example, the formation of an excited state product $^*B^-$ may become thermodynamically allowed. If such a reaction has a more favorable electronic factor (which will usually be the case because formation of an excited state generally involves the transfer of the electron to an outer orbital), the rate constant k_q, as schematized in Figure 10,

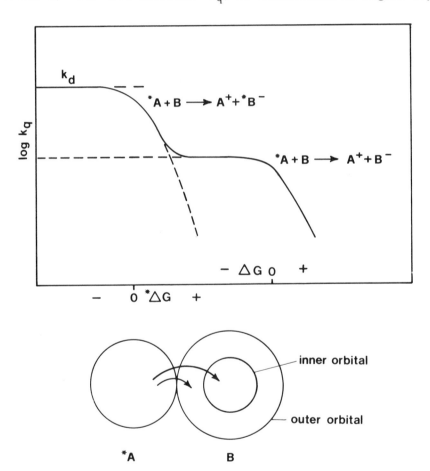

FIGURE 10. Schematic representation of the possible shape of a log k_q vs. ΔG plot when the reaction leading to ground state products is strongly nonadiabatic and a more adiabatic reaction (leading to an excited product) becomes thermodynamically possible at more negative ΔG values.

will increase over the initial nonadiabatic plateau and will reach a higher plateau through another nonlinear region centered at $\Delta G = E^o(^*A/A^-) + E^o(B^+/^*B)$ and depending on the intrinsic barrier $^*\Delta G^{\ddagger}(0)$ of the redox reaction leading to the excited state. In these cases a stepwise behavior of the log k_q vs. ΔG plot is expected, although the observation of distinct plateaus is possible only if the reaction leading to $^*B^-$ becomes important when that leading to B^- is close to saturation, and if the nonadiabatic character (i.e., the electronic factor) of the two processes is sufficiently different. Even when these conditions are not satisfied, different reaction channels may nevertheless appear since an increase of log k_q with decreasing ΔG will not follow Eq. [17].

It should be noted that the above treatment refers only to outer-sphere (weak interaction) electron-transfer processes. It is well known for some metal complexes, that redox processes may also take place via inner-sphere (strong interaction) pathways. For reactions of one of the latter complexes with a homogeneous series of reactants, the contribution of outer- or inner-sphere paths may change with ΔG. For example, it may happen that a reaction is predominantly inner-sphere at small ΔG values and becomes mainly outer-sphere at large and negative ΔG values, or vice versa. These mechanistic changes will also affect the log k_q vs. ΔG plot which, again, will not obey Eq. [17].

Figure 11 shows log k_q vs. ΔG plots of the literature data for electron-transfer reactions of $Ru(NH_3)_6^{n+}$, $Fe(H_2O)_6^{n+}$ (n = 2,3), and $Eu(H_2O)_n^{2+}$ with a "homogenized" family of redox reactants (34).[4] It can be seen that for the $Ru(NH_3)_6^{n+}$ reactions, the rate constants approach the diffusion-controlled limit for slightly negative ΔG values, indicating an adiabatic or nearly adiabatic behavior and a small intrinsic barrier. For the $Fe(H_2O)_6^{n+}$ reactions, the nonlinear part of the plot is considerably broader, as expected for reactions having higher intrinsic barriers. The curves through the experimental points have been obtained using Eq. [17] with $\Delta G^{\ddagger} = 10.6$ and 17.4 kcal/mol for the $Ru(NH_3)_6^{3+/2+}$ and $Fe(H_2O)_6^{3+/2+}$ exchange rates, respectively; [ref. (34) gives the values of the other parameters]. For the oxidation of $Eu(H_2O)_n^{2+}$, the plot in Figure 11 is indicative of stepwise behavior, with a reaction channel that becomes saturated for $k_q^o \sim 10^6$ $M^{-1}s^{-1}$ and another channel that becomes important only at very negative values of ΔG. The former channel is thought to be due to a strongly

---

[4]The results obtained with nonhomologous reactants can be "homogenized" by appropriate procedures. For details see ref. (34).

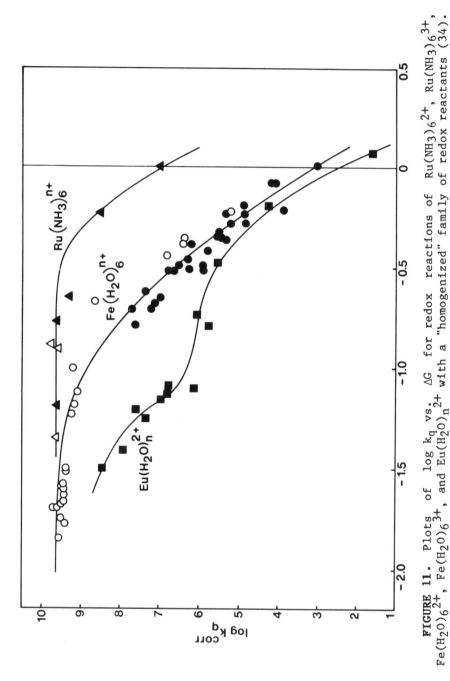

FIGURE 11. Plots of $\log k_q$ vs. ΔG for redox reactions of $Ru(NH_3)_6^{2+}$, $Ru(NH_3)_6^{3+}$, $Fe(H_2O)_6^{2+}$, $Fe(H_2O)_6^{3+}$, and $Eu(H_2O)_n^{2+}$ with a "homogenized" family of redox reactants (34).

nonadiabatic reaction involving the transfer of an electron from an f orbital via an outer-sphere mechanism, while the latter reaction is thought to proceed by an inner-sphere path via a charge-transfer intermediate (34). It cannot be excluded, however, that the peculiar log k_q vs. ΔG plot for oxidation of Eu^{2+} could be due to the participation of different Eu(II) species (e.g., aquo ions, ion-pairs between Eu^{2+} and anions) to the redox process. The log k_q vs. ΔG plot for electron-transfer quenching of aromatic molecules by Eu^{2+} in acetonitrile also shows a step-wise behavior (35).

In conclusion, the log k_q vs. ΔG plots obtained for several homogeneous series of reactions may give useful information on the factors governing the redox processes. When the reaction occurs via a simple outer-sphere mechanism it is possible in favorable cases to disentangle the effects of nonadiabaticity (electronic term) and intrinsic reorganizational barrier (nuclear term) on the rate constant and also to evaluate the corresponding $\underset{\sim}{k}$ and $\Delta G^{\ddagger}(0)$ parameters.

An exactly analogous treatment can also be given for exchange energy-transfer processes such as Eq. [3] (5).

IV. COMPARISONS AMONG TYPICAL EXCITED STATE REDOX REACTANTS

Now that we have discussed both the thermodynamic and kinetic factors which govern excited state electron-transfer reactions, it is instructive to compare the properties of some typical excited state redox reactions. The properties of two reductants $^*Ru(bpy)_3^{2+}$ and $^*Fe(bpy)_3^{2+}$ are listed in Table II. The reasons why $^*Fe(bpy)_3^{2+}$ is so much worse than the analogous $^*Ru(bpy)_3^{2+}$ are related to the different orbital natures of the excited states. For the first transition-row elements the ligand field is rather small and thus the lowest excited state is metal-centered (MC), while for complexes of the second and third transition rows with polypyridine ligands, the MC excited states lie at higher energies and the lowest excited state turns out to be either ligand-centered (LC) or metal-to-ligand charge transfer (MLCT). Metal-centered excited states of d^6 complexes like those of Fe(II) involve promotion of an electron from an approximately nonbonding t_{2g} orbital to a strongly σ antibonding e_g orbital[5] (36). This causes a severe distortion between the ground and excited states, which carries a number of adverse consequences:

---

[5]We have used the symbols corresponding to octahedral symmetry.

Table II. Comparisons between Two Different Excited-State Reductants[a]

$^*A$	Orbital nature[b]	τ, ns[c]	$^*E^{oo}$, eV[d]	$^*E^o(A^+/^*A)$[e]	$\Delta G^{\ddagger}$[f]	k[g]
$^*Fe(bpy)_3^{2+}$	MC ($t_{2g} \to e_g$)	0.81[h]	≤0.9[h]	≥0.1[h]	~0.57[i]	~1
$^*Ru(bpy)_3^{2+}$	MLCT ($t_{2g} \to \pi^*$)	624[j]	2.1[j]	−0.86[j]	~0.2[k]	~1

[a] Room temperature, aqueous solution.
[b] MC = metal-centered; MLCT = metal-to-ligand charge transfer.
[c] Excited state lifetime in deaerated solution.
[d] 0-0 excited-state energy.
[e] V(nhe).
[f] Reorganizational barrier for the self-exchange reaction (eV).
[g] Transmission coefficient for the self-exchange reaction (34).
[h] Ref. (9).
[i] Made up of ~0.35 eV inner-sphere and 0.22 eV outer-sphere contributions (9).
[j] Ref. (1).
[k] Nearly all outer-sphere contribution (9).

(i) radiationless deactivations via strong coupling mechanisms are favored and thus the excited state lifetime is very short; (ii) the free-energy content of the excited state (E^{oo}) is much smaller than the "vertical" excitation energy E_v; and (iii) since $Fe(bpy)_3^{3+}$ does not possess an electron in the $\sigma^*(e_g)$ orbitals, oxidation of $^*Fe(bpy)_3^{2+}$ involves large inner-sphere reorganizational barriers which slow down the reaction unless it is very exergonic (see Section III). In contrast, the analogous $^*Ru(bpy)_3^{2+}$ complex is not distorted with respect to its ground state because the promoted electron lies in a delocalized π^* orbital, and thus none of the above mentioned adverse consequences occurs.

Table III compares the properties of three excited state oxidants. The excellent properties of $^*Cr(bpy)_3^{3+}$ derive from the following circumstances: (i) the excited state is metal-centered but intraconfigurational (i.e., its formation involves only a spin-flip into the t_{2g} orbitals, not promotion to $\sigma^*(e_g)$ orbitals) (36). As a consequence, the excited state is not distorted with respect to the ground state and thus the excited state can be long-lived and E^{oo} is almost equal to E_v; (ii) the ground state can be very easily reduced (Fig. 4b), so that the excited state turns out to have a very positive reduction potential (Eq. [5]); (iii) reduction involves addition of an electron into the nonbonding t_{2g} orbitals, so that there is almost no inner-sphere reorganizational barrier; and (iv) although both excitation and reduction involve "metal" t_{2g} electrons, there is sufficient mixing between these orbitals and the outermost π and π^* orbitals of the ligand to cause adiabatic behavior (28).

The lowest excited state of Eu^{3+} is an intraconfigurational f-f excited state (36). The 4f orbitals are strongly shielded by the outermost 5s and 5p metal orbitals and thus they are really nonbonding orbitals, strongly localized on the metal. This fact has two consequences: (i) the excited state is not distorted with respect to the ground state; and (ii) the interaction between these orbitals with those of the solvent or of other species is very small. Both (i) and (ii) preclude efficient radiationless transitions and thus cause a long excited state lifetime. However, (ii) also results in a very small value of the transmission coefficient $\underset{\sim}{k}$, so that the potentially good properties of $^*Eu^{3+}$ (long lifetime, strong oxidizing power) cannot be profitably used. For reduction of $^*Ru(bpy)_3^{2+}$ the same considerations previously mentioned for oxidation of this species apply. Of course, high intrinsic barriers and, especially, lower transmission coefficients may turn out to be useful properties whenever an electron-transfer reaction needs to be slow, as in the case of dissipative back electron-transfer reactions of cyclic redox processes for solar energy conversion.

Table III. Comparisons among Three Different Excited-State Oxidants[a]

*A	Orbital nature[b]	τ, ns[c]	*E_{00}, eV[d]	*$E°(*A/A^-)$[e]	$\Delta G^{\neq}$[f]	$\underset{\sim}{k}$[g]
*$Cr(bpy)_3^{3+}$	MC (d)	77[h]	1.7[h]	+1.5[h]	~0.2[i]	~1
*$Eu(H_2O)_n^{3+}$	MC (f)	~100[j]	2.1[j]	+1.7[k]	0.43–0.86[ℓ]	~10^{-11}
*$Ru(bpy)_3^{2+}$	MLCT ($t_{2g} \rightarrow \pi^*$)	0.62[h]	2.1[h]	+0.8[h]	~0.2[m]	~1

[a]Room temperature, aqueous solution.

[b]MC = metal-centered; MLCT = metal-to-ligand charge transfer.

[c]Excited state lifetime in deaerated solution.

[d]0–0 excited-state energy.

[e]V(nhe).

[f]Reorganizational barrier for the self-exchange reaction (eV).

[g]Transmission coefficient for the self-exchange reaction (34).

[h]Ref. (1) and references therein.

[i]Ref. (2).

[j]Ref. (37).

[k]Ref. (35).

[ℓ]Ref. (34).

[m]Ref. (9).

V. CONCLUSIONS

An excited state should meet several spectroscopic, thermodynamic and kinetic requirements in order to be a useful reactant in solar energy conversion systems based on electron-transfer reactions. Many of these requirements are more easily met by transition metal complexes than by organic molecules or simple metal ions. The structural and electronic factors which determine the spectroscopic and thermodynamic properties of the various excited states of transition metal complexes are reasonably well known. The parameters which govern the kinetic behavior, namely, the electronic transmission coefficient k and the intrinsic barrier $\Delta G^{\ddagger}(0)$, can be evaluated by using the semi-empirical approach described in Section III; these parameters are also related to structural and electronic factors. By appropriate combination of metals and ligands it is thus possible to design transition metal complexes having very different excited-state properties. The complexes containing polypyridine ligands (particularly, $Ru(bpy)_3^{2+}$) seem to be at present the best candidates as electron-transfer photosensitizers for artificial photosynthetic systems.

ACKNOWLEDGMENTS

We wish to thank Drs. N. Sutin, G. Orlandi, N. Sabbatini and M. T. Indelli for helpful discussions.

REFERENCES

1.   V. Balzani, F. Bolletta, M. T. Gandolfi and M. Maestri, Top. Curr. Chem. 75, 1 (1978).
2.   N. Sutin, J. Photochem. 10, 19 (1979).
3.   V. Balzani, L. Moggi, M. F. Manfrin, F. Bolletta and G. S. Laurence, Coord. Chem. Rev. 15, 321 (1975).
4.   N. J. Turro, "Modern Molecular Photochemistry", W. A. Benjamin, Menlo Park, Calif., 1978, p. 628.
5.   V. Balzani, F. Bolletta and F. Scandola, J. Am. Chem. Soc. 102, 2152 (1980).
6.   N. Sutin and C. Creutz, in "Inorganic and Organometallic Photochemistry" (M. S. Wrighton, ed.), Advances in Chemistry Series, No. 168, American Chemical Society, Washington, D. C., 1978, pp. 1-27.

7. T. J. Meyer, Israel J. Chem. 15, 200 (1977).
8. P. J. DeLaive, C. Giannotti and D. G. Whitten, J. Am. Chem. Soc. 100, 7413 (1978).
9. C. Creutz, M. Chou, T. L. Netzel, M. Okumura and N. Sutin, J. Am. Chem. Soc. 102, 1309 (1980).
10. R. Ballardini, G. Varani and V. Balzani, J. Am. Chem. Soc. 102, 1719 (1980).
11. J. L. Kahl, K. W. Hanck and K. De Armond, J. Phys. Chem. 83, 2613 (1979).
12. R. A. Marcus, Faraday Disc. Chem. Soc. 29, 21 (1960); Annu. Rev. Phys. Chem. 15, 155 (1964); J. Chem. Phys. 43, 679 (1965); Electrochim. Acta 13, 995 (1968); in "Tunneling in Biological Systems" (B. Chance, D. C. DeVault, H. Frauenfelder, R. A. Marcus, J. R. Schrieffer and N. Sutin, eds.), Academic Press, New York, 1979, pp. 109-127.
13. N. S. Hush, Electrochim. Acta 13, 1005 (1968).
14. N. Sutin, in "Inorganic Biochemistry" (G. Eichorn, ed.), Elsevier, Amsterdam, 1973, pp. 611-653; in "Tunneling in Biological Systems" (B. Chance, D. C. DeVault, H. Frauenfelder, R. A. Marcus, J. R. Schrieffer and N. Sutin, eds.), Academic Press, New York, 1979, pp. 201-224; G. M. Brown and N. Sutin, J. Am. Chem. Soc. 101, 883 (1979).
15. L. D. Landau, Phys. Z. Sov. Union 2, 46 (1932).
16. N. R. Kestner, J. Logan and J. Jortner, J. Phys. Chem. 78, 2148 (1974); M. Bixon and S. Efrima, Chem. Phys. Lett. 25, 34 (1974); R. P. Van Duyne and S. F. Fischer, Chem. Phys. 5, 183 (1974).
17. J. Ulstrup and J. Jortner, J. Chem. Phys. 63, 4358 (1975).
18. E. Buhks, M. Bixon, J. Jortner and G. Navon, Inorg. Chem. 18, 2014 (1979).
19. G. Orlandi, S. Monti, F. Barigelletti and V. Balzani, Chem. Phys. 52, 313 (1980).
20. E. Buhks, G. Navon, M. Bixon and J. Jortner, J. Am. Chem. Soc. 102, 2918 (1980).
21. B. S. Brunschwig, J. Logan, M. D. Newton and N. Sutin, J. Am. Chem. Soc., 102, 5798 (1980).
22. F. Scandola and V. Balzani, J. Am. Chem. Soc. 101, 6140 (1979).
23. D. Rehm and A. Weller, Ber. Bunsenges Phys. Chem. 73, 834 (1969); Isr. J. Chem. 8, 259 (1970).
24. R. Ballardini, G. Varani, F. Scandola and V. Balzani, J. Am. Chem. Soc. 98, 7432 (1976).
25. E. Vogelman, S. Schreiner, W. Rauscher and H. E. Kramer, Z. Phys. Chem., Neue Folge 101, 321 (1976).
26. N. Sutin and C. Creutz, J. Am. Chem. Soc. 99, 241 (1977).
27. V. Breyman, H. Dreeskamp, E. Koch and M. Zander, Chem. Phys. Lett. 59, 68 (1978).

28. R. Ballardini, G. Varani, M. T. Indelli, F. Scandola and V. Balzani, J. Am. Chem. Soc. 100, 7219 (1978).
29. M. T. Indelli and F. Scandola, J. Am. Chem. Soc. 100, 7732 (1978).
30. J. K. Nagle, M. J. Dressick and T. J. Meyer, J. Am. Chem. Soc. 101, 3993 (1979).
31. F. Scandola, M. T. Indelli, R. Ballardini and G. Varani, Book of Abstracts, VIII IUPAC Symposium on Photochemistry, Seefeld, Germany, July 1980, p. 314, and unpublished work.
32. N. Agmon and R. D. Levine, Chem. Phys. Lett. 52, 197 (1977).
33. R. A. Marcus and N. Sutin, Inorg. Chem. 14, 213 (1975).
34. V. Balzani, F. Scandola, G. Orlandi, N. Sabbatini and M. T. Indelli, J. Am. Chem. Soc. 103, 3370 (1981).
35. N. Sabbatini, A. Golinelli, M. T. Gandolfi and M. T. Indelli, Inorg. Chim. Acta. 53, L213 (1981).
36. V. Balzani and V. Carassiti, "Photochemistry of Coordination Compounds", Academic Press, London, 1970, p. 432.
37. J. G. Bunzli and J. R. Yersin, Inorg. Chem. 18, 605 (1979).

DISCUSSION

Prof. Sir George Porter, Royal Institution:
   The otherwise favorable ruthenium tris(bipyridyl) complex has a threshold absorption which means poor energy efficiency as a solar energy absorber. Substitution on the ligand has relatively small effects. How can this situation be improved?

Prof. Balzani:
   Large changes in the absorption spectrum can be obtained by changing the metal. This, however, usually causes changes also in other excited-state properties.

Prof. D. G. Whitten, University of North Carolina:
   In response to the question by Prof. Porter and Prof. Balzani's answer concerning the wavelength limitations of $Ru(bipy)_3^{2+}$: By substituting the bipyridyl ligand, primarily with meta- or para-carboxyester groups (on one or more of the ligands), the absorption maximum can be greatly red-shifted, in some cases with threshold excitation beyond 600 nm.

Prof. C. H. Langford, Concordia University, Montreal:
Because of the economic advantages of first transition series metals like Fe, I'd like to suggest some mitigation of your pessimism with respect to their use. The $Fe(bipy)_3^{2+}$ lifetime near 1 ns was first estimated by "supersensitizer" experiments at an optically thin electrode. This kinetic scavanging technique used higher concentrations than those for $Ru(bipy)_3^{2+}$, but still did not require inconvenient values. I think the distortion produced by $\pi \rightarrow \sigma^*$ ligand field transitions can be over emphasized. A rigid ligand can help. Consequently, there are a very large number of well-known and cheap systems that remain unexplored. On another tack, charge-transfer states can be moved to the visible by changes of ligand. Little work has been done on S donors as compared to N and O donors. I think it would be quite premature to assume that the best metal complexes must use 4d or 5d metals, which tend to be rare and expensive. Indeed, use of intraconfiguration transitions is not well explored.

Prof. Balzani:
I partially agree on some of the points you raised, for example that rigid ligands may help to reduce distortion of metal-centered excited states. Concerning the complexes with S ligands, they tend to display very low-energy excited states, which is clearly a disadvantage. Intraconfigurational MC excited states may exhibit very poor kinetic properties, as seems to be the case for Eu ions.

Dr. H. Tributsch, C.N.R.S., Laboratoires de Bellvue:
Referring to your comment on ruthenium-sulfur complexes, I would like to say that we have, in collaboration with several scientists of our laboratory, identified and studied a semiconducting ruthenium compound which has the composition RuS_2. It has an energy gap of approximately 1.5 eV and thus absorbs visible light. We did not observe photocorrosion into sulphur and ruthenium ions, and the system apparently combines the favorable catalytic properties of other ruthenium compounds for water oxidation with the capacity to harvest photons for the generation of reactive holes. It might therefore be reasonable to study ruthenium-sulfur complexes.

Prof. Balzani:
Thank you for this piece of information. However, I would like to point out that the properties of solid ruthenium sulfides may not be related to those of monomeric Ru-S complexes to be used in homogeneous processes.

Prof. R. J. Watts, University of California at Santa Barbara:
   I would like to comment that our studies indicate that the
   lowest excited state of $IrCl_2(phen)_2^+$ is metal-centered
   rather than charge-transfer in nature. Therefore, if that
   complex is to be used as a photocatalytic agent, that
   drawback would have to be taken into account.

Prof. Balzani:
   The slide on $Ir(phen)_2Cl_2^+$ is an old one and does not take
   into account your most recent work which shows that two
   emissions (CT and MC) are obtained from this complex.
   However, the general concept behind that slide, that
   tuning of the lowest excited state can be obtained by
   changing the type of liquids, remains valid.

Dr. A. F. Haught, United Technologies Research Center:
   If I followed your analysis correctly, I believe the ΔG
   you used in your model is the standard free-energy change
   for the molecules (i.e., ΔG^0). In view of its significance for the energetics of back reactions, it is important to distinguish between the standard ΔG^0 and the
   actual free-energy change in the reaction, which includes
   the effects of the relative concentrations of the species
   included.

Prof. Balzani:
   Of course, I agree with you that the actual free-energy
   change in an electron-transfer reaction depends on concentrations. In our approach, the experimental rate constant
   is given in terms of rate constants of single steps, equilibrium constants and free-energy relationships. Thus,
   the standard free-energy change has to be used as the relevant parameter.

Dr. H. Kisch, Institut für Strahlenchemie, Max-Planck-Institut, Mülheim:
   Complexes containing sulfur may well be suited for catalysis of H_2O cleavage. We found that metal dithiolenes are
   very active catalysts for H_2 generation.

Prof. Balzani:
   Which metals are you using?

Dr. Kisch:
   One can use any transition metal, but Zn is the most
   effective.

Prof. G. C. Dismukes, Princeton University:
   It may have come across to the audience during your discussion that the Marcus theory fails to work when the exothermicity of the electron-transfer reaction becomes very
large.   Although this may be the case in the excited
states of the transition metal complexes you have discussed, there are systems where qualitative agreement with
Marcus theory at large exothermicity is observed.   The
work by Miller et al. at Argonne National Laboratory
(J. Chem. Phys., Feb. 1980) is one example. These studies
involve thermally activated electron transfer from trapped
electrons in organic glasses to organic acceptors. A fall
in the rate of electron transfer is observed at negative
ΔG for many acceptors.   However, a plateau in the rate
constant at negative ΔG is found for those acceptors with
low-lying excited states which fall below the ΔG of the
reaction.   Evidently the participation of these excited
states can permit constant electron-transfer rates, although the rate predicted for the ground state is falling
with more negative ΔG.

Prof. Balzani:
   I am aware of Miller's results.   However, I don't think
that his systems can be compared with those discussed in
my lecture.   The electron-transfer reactions studied by
Miller are very peculiar since they involve: (i) solvated
electrons as one of the reactants; (ii) electron transfer
between reactants trapped at long distances; (iii) low
temperature.   From an experimental point of view, such
processes always have very small rate constants.   The
electron-transfer reactions that I have discussed involve:
(i) large molecules as reactants; (ii) electron transfer
within an encounter in fluid solutions; (iii) high temperature.   Experimentally, the rate constants of these
systems are always very high.

Prof. Whitten:
   You mentioned the effect of κ (transmission coefficient)
on the plateau-limit obtained in electron-transfer processes and demonstrated the very large effect ($\kappa \sim 10^{-6}$)
for europium complexes.   Yet in studies by your group,
ours and others, the limiting plateaus for oxidative and
reductive geometry of polypyridyl complexes of ruthenium,
osmium, iridium, and chromium are very similar, even
though the "one-electron" orbitals involved in the reductive and oxidative processes are quite different. Can you
comment on the significance of this?

Prof. Balzani:
   According to our approach, the limiting plateau value reflects the diffusion-controlled rate constant when the reaction is adiabatic or nearly adiabatic, and reflects the transmission coefficient when the reaction is strongly nonadiabatic. Differences in plateau values related to the different one-electron orbitals involved can show up only for strongly nonadiabatic reactions, which is not the case of these mentioned in your question.

Dr. N. Sutin, Brookhaven National Laboratory:
   I would like to comment on the Marcus inverted region. It is important to note that the rate decrease for very exothermic reactions is predicted by classical as well as by quantum mechanical (energy gap) considerations. The question then is, why there are so many systems in which the inverted region is not seen? One explanation, which Prof. Balzani has alluded to, is the formation of excited-state products. Another explanation is that the reaction has undergone a change in mechanism; the rate decrease in the inverted region is predicted for oxidation-reduction reactions that proceed by an electron-transfer mechanism; it is not expected for oxidation-reduction reactions proceeding by an atom-transfer mechanism. Perhaps some of the reactions that have been discussed involve hydrogen atom transfer rather than electron transfer.

Prof. Balzani:
   I agree that some decrease in reaction rate with increasing exergonicity is also expected to occur according to quantum mechanical models, although at more negative ΔG values than predicted by the Marcus theory. I also agree that in some cases mechanisms other than electron transfer (e.g., H-transfer) can intervene to keep the plateau high. I'd like to point out, however, that for some of the systems discussed, neither formation of excited-state products nor H-transfer are likely to occur. For example, the back reactions involving $Cr(bpy)_3^{2+}$ are not likely to proceed via H-transfer because the bipyridyl ligands do not carry a significant negative charge.

CHAPTER 5

PHOTOINDUCED WATER SPLITTING IN HETEROGENEOUS SOLUTION[1]

Michael Grätzel

Institut de Chimie Physique
Ecole Polytechnique Fédérale de Lausanne
Lausanne, Switzerland

I. INTRODUCTION

The field of photochemical conversion of solar energy has become an exciting and rapidly growing area of research over the last few years (1-5). In addition to unimolecular isomerization reactions (6-8) light-driven redox reactions have also attracted wide attention. The thermodynamics of such a process are depicted in Figure 1. In the aqeuous solution are present a sensitizer (S), which functions as the electron donor, and an electron acceptor, sometimes referred to as an electron relay (R). No spontaneous electron transfer can occur between the ground states of the species. However, electrons will be transferred from the sensitizer to the relay once the former is excited by visible light.

$$S + R \overset{h\nu}{\rightarrow} S^+ + R^- \qquad [1]$$

During this process a major part of the photon energy can be converted into the chemical potential of the products, S^+ and R^-. A similar redox process may be conducted with sensitizers that act as electron acceptors. In this case the electron relay is oxidized, as shown in Figure 2.

Applications of such systems to production of fuels and chemicals by solar energy encounters three major problems summarized in Figure 2. First, sensitizer/relay couples have to be found which are suitable from the viewpoint of light absorption and redox potentials and which undergo no chemical

---

[1]Supported, in part, by the Swiss National Science Foundation, Ciba-Geigy and Engelhard Industries.

standard potential
electron transfer
S*+R
S⁺+R⁻
E°(R/R⁻)
e⁻
E°(H2O/H2)
light absorption
hν
e⁻
thermal back reaction
E°(O2/H2O)
E°(S⁺/S)
S+R

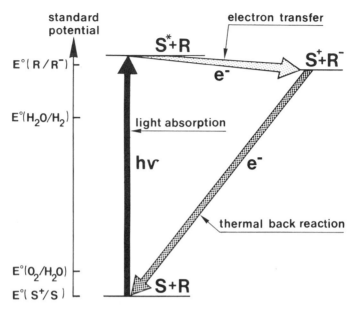

FIGURE 1. Schematic illustration of a light-induced redox reaction coupled with water cleavage.

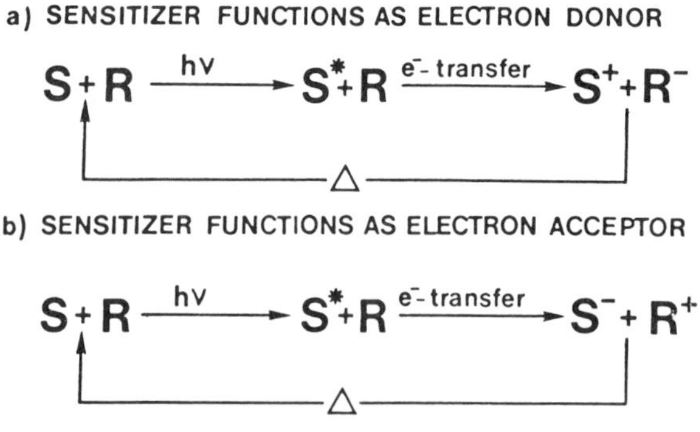

a) SENSITIZER FUNCTIONS AS ELECTRON DONOR
S+R
hν
S*+R
e⁻-transfer
S⁺+R⁻
△
b) SENSITIZER FUNCTIONS AS ELECTRON ACCEPTOR
S+R
hν
S*+R
e⁻-transfer
S⁻+R⁺
△

PROBLEMS TO BE SOLVED

1 FIND SUITABLE SENSITIZER-RELAY SYSTEMS

2 PREVENT THERMAL BACK REACTION

3 FUEL PRODUCTION FROM S^+, R^- OR S^-, R^+ REDOX PRODUCTS

FIGURE 2. Sensitizer-relay systems and major obstacles to be overcome to achieve photochemical solar energy conversion.

side-reactions in the oxidation state of interest. The sensitizer should have good light absorption properties with respect to the solar spectrum. Also the excited state should be long-lived and have high quantum yields for the processes of interest, and the electron-transfer reaction must occur with high efficiency, i.e., there must be good escape yields of the redox products from the solvent cage. The redox properties of the donor-acceptor relays obviously must be tuned to the fuel-producing transformation envisaged. If, for example, water cleavage by light is to be achieved, then the thermodynamic requirements are such that $E^o(D/D^+) > 1.23$ V (nhe) and $E^o(A/A^-) < 0$ V, under standard conditions.

Considerable progress has been made over the last few years in the design of sensitizer/relay couples suitable for photoinduced water decomposition (9-13). A number of systems have been explored which convert more than 90% of the threshold light energy required to excite the sensitizer into chemical potential. Also, several of these compounds satisfy the requirement that the reduced relay and oxidized sensitizer are thermodynamically capable of generating hydrogen and oxygen, respectively, from water as indicated in Eqs. [2] and [3].

$$R^- + H_2O \rightarrow 1/2\ H_2 + OH^- + R \qquad\qquad [2]$$

$$2S^+ + H_2O \rightarrow 1/2\ O_2 + 2H^+ + 2S \qquad\qquad [3]$$

Noteworthy examples of sensitizers are Ru(bipy)$_3^{2+}$ (14,15), porphyrin derivatives (16-20) such as:

ZnTMPyP^{4+}

(where R = N-methylpyridinium chloride), and acridine dyes such as proflavin (21-25).

The electron-relay compounds investigated include the viologens (26-28), Eu^{3+}, V^{3+} and their respective salycilate complexes (21), $Rh(bipy)_3^{3+}$ (29-31; see also, Chapter 6) and macrocyclic cobalt complexes (32). The latter reagents, after reduction by an excited sensitizer, are capable of evolving hydrogen from water but the presence of a catalyst is generally mandatory.

The second and third points indicated in Figure 2 are related to prevention of thermal back electron-transfer and coupling with the fuel-generating steps, Eqs. [2] and [3]. These processes will be the focus of this paper. We will discuss in particular the design of functional microheterogeneous systems consisting of cooperative units which are both suitable for the control of the various electron-transfer events and exhibit exceptionally high catalytic activity.

Figure 3 illustrates structural features of various colloidal assemblies that play a fundamental role in these processes. Inhibition of the back reaction has been achieved by employing solutions of micelles (32-44), microemulsions (45,46) and vesicles (47-50). These aggregates are distinguished by a charged lipid/water interface that may be exploited to exercise kinetic control over the electron-transfer events. Studies have also been carried out with other multiphase systems such as monolayers (51), lipid bilayers (52) and polyelectrolytes (53,54). In the following sections we will elucidate in more detail the role of micelles in producing charge-separation and storage effects.

II. STABILIZATION OF REDOX INTERMEDIATES BY MICELLAR SYSTEMS

Micelles are surfactant assemblies that form spontaneously in aqueous solution above a certain critical concentration (CMC) (55-57). They are of approximately spherical structure (radius 15-30 Å), the polar head groups being exposed to the aqueous bulk phase while the hydrocarbon tails protrude in the interior. Such an assembly is shown in Figure 4. An important feature of micellar solutions is their microheterogeneous character. The apolar interior of the aggregates is distinguished from the aqueous phase through its hydrocarbon-like character which facilitates the solubilization of hydrophobic species. Of crucial importance for light-energy conversion is the electrical double-layer formed around ionic micelles. This gives rise to surface potentials (Fig. 5) frequently exceeding values of 150 mV (58,59). This charged lipid-water interface provides a microscopic barrier for prevention of thermal back-transfer of the electrons. The role of the micellar double layer may thus be compared to that of the

PREVENTION OF BACK REACTION   CATALYSIS

MICELLES   MICROEMULSIONS   VESICLES

METAL (Oxide) SOLS

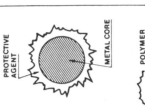

PROTECTIVE AGENT

METAL CORE

H₂O

~~~o SURFACTANT (SIMPLE FUNCTIONAL)

~~~o SURFACTANT
~~~× COSURFACTANT
— HYDROCARBON

~~~ SURFACTANT (SIMPLE FUNCTIONAL)

~~~ POLYMER SURFACTANT

Pt, Pd, Os, Rh, Ag
RuO_2, $NiCO_2O_4$

FIGURE 3. Structural features of colloidal assemblies employed in light-induced charge separation and redox catalysis.

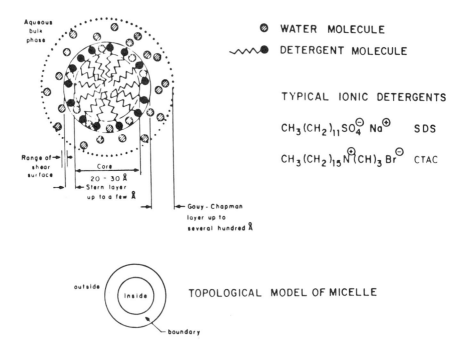

WATER MOLECULE

DETERGENT MOLECULE

TYPICAL IONIC DETERGENTS

$CH_3(CH_2)_{11}SO_4^{\ominus}$ $Na^{\oplus}$ SDS

$CH_3(CH_2)_{15}\overset{\oplus}{N}(CH_3)_3$ $Br^{\ominus}$ CTAC

Aqueous bulk phase

Range of shear surface

Core
20 - 30 Å
Stern layer up to a few Å

Gouy - Chapman layer up to several hundred Å

outside | Inside | boundary

TOPOLOGICAL MODEL OF MICELLE

FIGURE 4. Schematic illustration of a surfactant micelle.

depletion or accumulation layer present at the semiconductor/
electrolyte interface (see also, Chapters 9 and 10).   The
following examples illustrate this point.
    Consider a situation where both sensitizer and electron
relay are dissolved in the aqueous phase.   The micelle serves
here as a carrier for the reduced electron relay.   In the case
where the amphiphilic viologen derivative, N-tetradecyl,
N'-methyl viologen, is employed as an electron relay, the oxidized state exhibits a pronounced hydrophilic character, while
in the reduced state it has hydrophobic properties (60):

CH_3—$\overset{+}{N}$⬡⬡$\overset{+}{N}$—$[CH_2]_{13}$—CH_3 $C_{14}MV^{2+}$
(hydrophilic)

$\downarrow e^-$

CH_3—N·⬡⬡$\overset{+}{N}$—$[CH_2]_{13}$—CH_3 $C_{14}MV^{+}$
(hydrophobic)

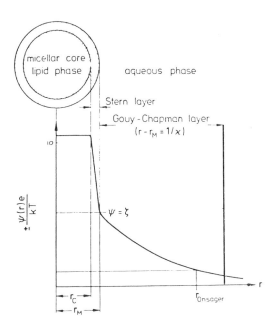

FIGURE 5. Potential-distance function in the vicinity of a spherical micelle.

$$[ZnTMPyP^{4+}]=5\times10^{-5}M, [C_{14}MV^{2+}]=10^{-3}M$$

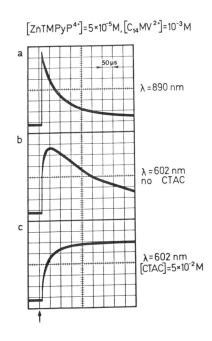

FIGURE 6. Light-induced reduction of $C_{14}MV^{2+}$ by the triplet state of $ZnTMPyP^{4+}$. (a) Triplet decay at 890 nm; (b) growth and decay of redox products in pure aqueous solution; and (c) stabilization of redox products by CTAC micelles.

In a solution containing $C_{14}MV^{2+}$ as an electron relay and the porphyrin $ZnTMPyP^{4+}$ as a sensitizer, excitation of the latter by visible light leads to formation of the porphyrin triplet state which subsequently transfers an electron to the viologen:

$$ZnTMPyP^{4+} + C_{14}MV^{2+} \underset{\Delta}{\overset{h\nu}{\rightleftarrows}} C_{14}MV^{+} + ZnTMPyP^{5+} \qquad [4]$$

We have studied the kinetics of this process by laser flash photolysis techniques. Figure 6 shows the temporal behavior of the absorbance of a solution of $ZnTMPyP^{4+}$ (5×10^{-5} M) and $C_{14}MV^{2+}$ (10^{-3} M). The absorbance at 890 nm is due to absorption by the porphyrin triplet state while that at 600 nm is characteristic of $C_{14}MV^{+}$ and $ZnTMPyP^{5+}$. Decay of the triplet state is accompanied by formation of redox products (Fig. 6b) which, however, undergo a rapid back reaction. In marked contrast to this behavior stand the results obtained in the presence of cationic micelles (cetyl trimethyl ammonium chloride, CTAC). As is apparent from Figure 6c, forward electron transfer in this case goes to completion without subsequent intervention of the back reaction. Detailed kinetic analysis shows that CTAC micelles retard the reverse electron transfer by at least a factor of 10^3.

This charge-separation effect induced by the micelles is explained schematically in Figure 7. Due to its hydrophilic character $C_{14}MV^{2+}$ is present mainly in the aqueous phase and does not associate with the CTAC aggregates. Forward electron transfer will therefore occur in water. In the reduced state the viologen relay acquires hydrophobic properties, which

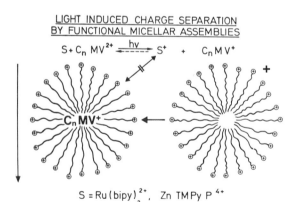

FIGURE 7. Principle of charge separation by cationic micelles in the photoinduced reduction of an amphiphilic viologen derivative.

leads to rapid solubilization in the CTAC assemblies. The oxidized porphyrin, on the other hand, is prevented from approaching the micelles by the positive surface charge. As the ζ-potential of CTAC is at least +100 mV (58,59), the probability of encounter with ZnTMPyP^{5+} is less than 2×10^{-9}. This explains the effective micellar inhibition of the back reaction.

III. HYDROGEN PRODUCTION

In the previous section examples were presented illustrating successful kinetic control of light-induced redox reactions by micellar systems. This achievement is, however, only a first step on the way to conversion of solar photons into chemical energy. A crucial problem to be solved is coupling of the photoredox events with catalytic steps leading to water decomposition. Here, we based our strategy on the concept of redox catalysis which was developed in 1938 by Wagner and Traud (61). Finely dispersed catalytic particles are added to the solution which serve as "microelectrodes" to afford water oxidation or reduction selectively. Consider first the hydrogen-evolution step:

$$2R^- + 2H_2O \rightarrow H_2 + 2OH^- + 2R \qquad [6]$$

In homogeneous solution this process is rendered difficult by the fact that it has to pass through the stage of a free H-atom whose free-energy is 2.1 eV above that of the H_2 molecule. In the presence of a suitable catalyst, formation of H-atoms is avoided, which reduces considerably the energy requirements for the water-reduction step, as shown in Figure 8.

The colloidal redox catalysts employed in our studies are functioning as local elements (61); oxidation of the reduced relay is coupled to hydrogen generation from water. The choice of the catalytic material may be based on the same considerations that apply for electrocatalytic reagents used on macroelectrodes; the exchange current densities for the anodic and cathodic electron transfer steps must be high. Colloidal platinum would then appear to be a suitable candidate to mediate reaction [6]. This fact had been recognized at the end of last century when numerous examples for the intervention of finely divided Pt in the process of water reduction by reagents such as Cr^{2+} and V^{2+} appeared in the German colloid literature (62).

A reaction of particular interest is the reduction of water by methyl viologen (Eq. [7]):

HYDROGEN GENERATION FROM WATER

PHOTO-REDOX REACTION

$$S + R \xrightarrow{\ h\nu\ } S^+ + R^-$$

TO BE COUPLED WITH FUEL GENERATING STEP

$$R^- + H_2O \rightarrow R + OH^- + \tfrac{1}{2}H_2$$

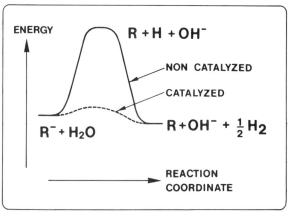

FIGURE 8.   Pathways of light-induced hydrogen evolution from water.

$$2MV^+ + 2H_2O \rightarrow H_2 + 2OH^- + 2MV^{2+} \qquad\qquad [7]$$

The fact that this process can be catalyzed by Pt dispersions was discovered by Green and Stickland in 1934 (63). More recently a number of photochemical systems have been developed (21,22,29-31,64-68) in which MV^{2+} is reduced in a light-driven electron-transfer reaction by a suitable sensitizer. The sensitizer cation undergoes a subsequent reaction with a third component which is irreversibly oxidized. Such sacrificial systems serve to optimize conditions for light-induced hydrogen evolution. An illustrative example is the case where $Ru(bipy)_3^{2+}$ serves as a sensitizer and EDTA as a sacrificial electron donor.

Figure 9 illustrates the influence of the radius of the Pt particles on photo-induced hydrogen evolution rates (69,70). These particles were stabilized by polyvinyl alcohol, and illuminations were carried out with a 450 W xenon lamp. One notices that a decrease of the particle radius from 500 to 100 Å leads to a drastic enhancement of the hydrogen-evolution rate to as high as 12 liter/day/liter of solution for the smallest particle size. In fact, in the last case, bubbling of hydrogen gas occuring under illumination of the solution can be readily seen. In Figure 9 the common feature is the Pt content of the solutions of different particle size, namely 3.5 mg Pt/25 cc of solution. In the lower left of this figure are also shown two additional points for 100 Å particle size but with low Pt levels;   0.35 mg Pt/25 cc (upper) and 0.25 mg Pt/25 cc (lower). Quantum yield measurements indicate a stoichiometric relation between the viologen reduced in the photoprocess and the amount of hydrogen produced. One might object to the high Pt levels (140 mg/liter) required to obtain these high yields.   Such concentrations are intolerable for practical systems.   However, reduction in particle size to a radius of only 15 Å leads to a 100-fold increase in the activity of the catalyst (71).

The preparation of these ultrafine and stable Pt-particles is indicated schematically on Figure 10. A solution of hexachloroplatinate is reduced by citrate according to the method of Turkevich et al. (72).   Excess citrate is then removed by stirring with an exchange resin.   After filtration an efficient protective agent such as Carbowax-20 M is added, or else the particles are deposited on a powdered support such as TiO_2 or $SrTiO_3$. The very small Pt particle sizes obtained in this manner have been verified by photon-correlation spectroscopy and scanning electron microscopy (SEM).   With such preparations a hydrogen output of ~12 liter/day/liter solution can be achieved with a Pt concentration of only ~1 mg/liter.

Another advantage of these finely divided platinum dispersions is that the solutions remain completely transparent even at high catalyst concentrations. This facilitates direct

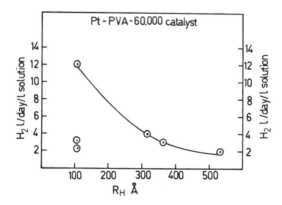

FIGURE 9. Correlation   between the observed yields of H_2 and the radius of the platinum catalyst.   The system contained $Ru(bipy)_3^{2+}$ (4×10^{-5} M), MV^{2+} (10^{-3} M), and EDTA (3×10^{-2}M) at pH 5.   The lower points at 110 Å contained 0.2 and 0.3 mg Pt per 25 cc of solution, respectively.

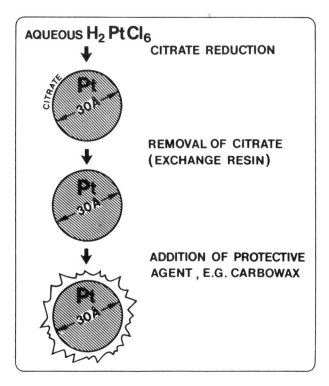

FIGURE 10. Procedure for the preparation of ultrafine platinum sols. Particle sizes were verified by photon-correlation spectroscopy and scanning electron microscopy.

study of the dynamics of the reaction of MV^+ with Pt particles using laser flash techniques. Figure 11 shows oscillgraphs illustrating the temporal behavior of the characteristic MV^+ absorbance at 602 nm in the absence and presence of the catalyst. The upward deflection of the signal after the laser pulse is due to formation of MV^+ via the photoredox process:

$$Ru(bipy)_3^{2+} + MV^{2+} \xrightarrow{h\nu} Ru(bipy)_3^{3+} + MV^+ \qquad [8]$$

In this case, no back reaction occurs since $Ru(bipy)_3^{3+}$ is rapidly reduced to the 2+ state by added EDTA. Whereas in the absence of catalyst the MV^+ absorption is stable (Fig. 11a), addition of colloidal Pt induces a decay of the signal, the rate of which increases sharply with Pt concentration (Figs. 11b and c). From a first-order plot of the absorbance decay curves, one obtains the rate constants which are plotted as a function of Pt concentration in Figure 11d.

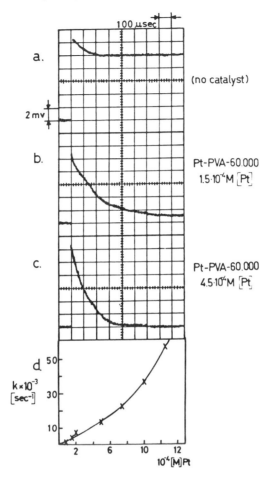

FIGURE 11. Effect of the Pt–PVA catalyst concentration on the behavior of MV^+ absorption at 602 nm. Representative oscilloscope traces are included in a, b and c. In d, the rate constants for MV^+ decay are plotted vs. platinum concentration in the Pt–PVA catalyst at pH 5.

The ascent of the curve is steeper than linear, indicating that the reaction order is greater than one with respect to Pt concentration. At a concentration of 10^{-4} M Pt in Pt–PVA-polymer, we observe a rate constant, k = 140 s^{-1}. At the highest concentration of catalyst (1.25×10^{-3} M in Pt), the rate constant was 5.7×10^4 s^{-1}. The lifetimes observed for MV^+ were shortened from about 7 ms to 15 µs when the concentration of catalyst varied by only a factor ~12. The high reaction rate of the PVA-protected Pt catalyst ($R_H \simeq 110$ Å) with the reduced viologen is exceeded even by the rates

obtained using Pt particles with a mean radius of 15 Å. In
this case the half-life of MV^+ in the presence of 10^{-4} M Pt
(20 mg/liter) is only ~40 μs. Taking into account that one
particle contains ~1200 Pt atoms, we derive a rate constant
for reaction [7] of ~2 × 10^{11} $M^{-1}s^{-1}$, indicating that this
process is essentially diffusion-controlled. These kinetic
results are of great importance for the design of a cyclic
water-decomposition system. It appears that by suitable
choice of the catalyst, conversion of A^- into A and simultaneous formation of hydrogen can be accomplished so rapidly
that it can compete efficiently with the thermal back reaction. (In the photostationary state achieved under sunlight
irradiation, the latter process, due to its bimolecular
nature, should require at least several milliseconds).

Figure 12 summarizes the mechanism of intervention of
these ultrafine particles in the hydrogen-formation process.
The particle is polarized cathodically through electron
transfer from the reduced relay species. Coupled with this
process is hydrogen ion discharge leading to the formation of
adsorbed hydrogen atoms. The latter subsequently combine to

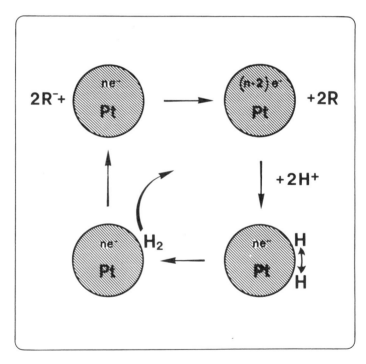

FIGURE 12.   Mechanism of intervention of ultrafine Pt
particles in hydrogen formation via the electron-pool effect.

yield H_2. The concept of coupling the two electron-transfer processes is supported by the fact that the rate of disappearance of R^- is enhanced in acid medium. In the case of $R = MV^{2+}$, this fact has been clearly established by using pulse radiolysis techniques (71).

The presence of a catalyst is not mandatory in all H_2-producing systems (29-31). In this context, it is worth mentioning a very interesting experiment by Gray et al. (73). These authors have shown that irradiation with visible light of a bridged dinuclear Rh^I leads directly to formation of hydrogen under strongly acidic conditions. Thus, the complex simultaneously assumes the roles of sensitizer and catalyst. This system is noncyclic, however, with H_2 yields stoichiometric with respect to the complex being photolyzed.

In the above examples, we have illustrated ways to couple a photoredox event with hydrogen production from water. This field is still in rapid development. New catalytic materials such as semiconductors (74) are presently being explored and alternative sensitizer/relay pairs are being identified. However, simple modification of a conventional electron relay, i.e., methyl viologen, can itself afford considerable improvement. Thus, the mere substitution of one methyl group on this compound by a tetradecyl chain gives excellent results when employed with a platinum sol protected by a cationic polysoap.

Photolysis experiments were carried out with solutions containing $C_{14}MV^{2+}$ (10^{-3} M), $Ru(bipy)_3^{2+}$ (10^{-4} M) and colloidal Pt (10^{-4} M, 30 Å radius) protected by 40 mg of cationic polysoap (71). Flash photolysis results are presented in Figure 13. The temporal behavior of the redox products $C_{14}MV^+$ and $Ru(bipy)_3^{3+}$ was monitored by following the absorbance of the solution at 602 and 470 nm. The initial rise of the signal at 602 nm is due to the formation of $C_{14}MV^+$ after laser excitation of $Ru(bipy)_3^{2+}$. The absorption decays sharply ($\tau \approx 60$ μs) back to the baseline indicating rapid consumption of $C_{14}MV^+$. Formation of $Ru(bipy)_3^{3+}$ is apparent from the bleaching of the 470 nm absorption. There is a fractional recovery of the negative signal to a plateau value from which no further changes are noted. This indicates that the major part of $Ru(bipy)_3^{2+}$ formed in the photoredox reaction is preserved and does not undergo back reaction with $C_{14}MV^+$.

The rapid disappearance of $C_{14}MV^+$ may be interpreted by a mechanism involving initial scavenging by the Pt particles through hydrophobic interaction with the protective agent, as shown in Figure 14. Subsequently, charge transfer and water reduction occur on the Pt-surface. Formation of hydrogen is readily observable under continuous irradiation. Charge separation is achieved by making use of electrostatic interactions; i.e., $Ru(bipy)_3^{3+}$ is strongly repelled from the surface of the positively charged aggregates. Hence, neither the

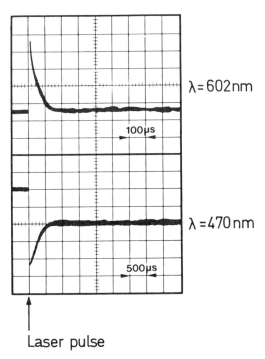

Laser pulse

FIGURE 13. Oscilloscope traces showing 530 nm laser flash photolysis of a solution containing $Ru(bipy)_3^{2+}$ (10^{-4} M) as the sensitizer, $C_{14}MV^{2+}$ (10^{-3} M), as the relay and colloidal Pt (10^{-4} M) protected by a cationic polysoap. Detection wavelengths were 602 and 470 nm. Solutions were freed from oxygen by flushing with argon.

reduced relay $C_{14}MV^+$ nor the Pt-particle itself can interact with $Ru(bipy)_3^{3+}$, which explains its extraordinarily high stability in such a system.

In the latter system, hydrogen formation and charge separation occur simultaneously. The overall reaction is the irreversible oxidation of the sensitizer to its cation, S^+, and simultaneous production of hydrogen from water (Fig. 15). The task remains to render this process cyclic, i.e., to convert S^+ back to S with simultaneous oxidation of water. To solve this problem, a second catalyst is added to the solution which promotes the generation of oxygen from the sensitizer cation S^+. The role of this redox catalyst is to preclude formation of free OH radicals, which are highly energetic and reactive intermediates in the homogeneous oxidation of water.

LIGHT INDUCED ELECTRON TRANSFER
COUPLED TO HYDROGEN EVOLUTION

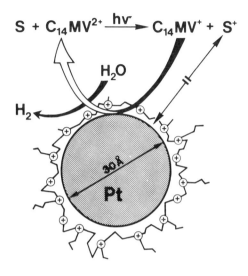

OVERALL REACTION :

$$S + H_2O \xrightarrow{h\nu} S^+ + \tfrac{1}{2}H_2 + OH^-$$

⊕∿⊕∿╲ CATIONIC POLYSOAP

$$C_{14}MV^{2+}: CH_3-N\langle\rangle-\langle\rangle N-(CH_2)_{13}-CH_3$$
$$2Cl^-$$

FIGURE 14. Combined charge separation and hydrogen evolution on Pt particles protected by a cationic polysoap.

IV. CATALYSIS OF OXYGEN PRODUCTION FROM WATER

Earlier work in our laboratory had led to the discovery that certain noble metal oxides can serve as mediators in the oxygen-generating reaction (75). Among these, RuO_2 has been most widely investigated (76–79; see also, Chapter 6). Meanwhile, a series of compounds such as Ce^{4+}, $Fe(bipy)_3^{3+}$ and $Ru(bipy)_3^{2+}$ have been identified which, in the presence of colloidal or macrodisperse RuO_2, are able to oxidize water

OXYGEN GENERATION FROM WATER

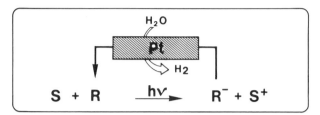

TO MAKE PROCESS CYCLIC S^+ HAS TO BE
RECONVERTED INTO S , FOR EXAMPLE

$$S^+ + \tfrac{1}{2}H_2O \;\rightarrow\; S + H^+ + \tfrac{1}{4}O_2$$

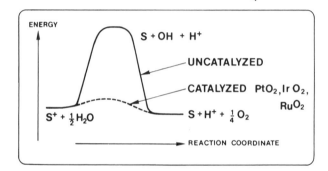

REACTIONS INVESTIGATED

$$Ce^{4+} + \tfrac{1}{2}H_2O \;\rightarrow\; Ce^{3+} + \tfrac{1}{4}O_2 + H^+$$

$$Fe(bipy)_3^{3+} + \tfrac{1}{2}H_2O \;\rightarrow\; Fe(bipy)_3^{2+} + H^+ + \tfrac{1}{4}O_2$$

$$Ru(bipy)_3^{3+} + \tfrac{1}{2}H_2O \;\rightarrow\; Ru(bipy)_3^{2+} + H^+ + \tfrac{1}{4}O_2$$

FIGURE 15.   Pathways of catalytic oxygen evolution from
water and the specific reactions investigated.

with a stoichiometric yield.   Other transition metal oxides
such as cobalt or nickel oxide also produce oxygen in the
presence of a suitable oxidizing agent, but the yield is less
than stoichiometric (80).

The mode of intervention of RuO_2 in the oxygen-generation
process may be best understood in terms of electrochemical
concepts.   The two coupled redox reactions occurring on the
RuO_2 particle are depicted schematically in Figure 16.   Holes
are first injected from the oxidized sensitizer into the particle;   here the holes   presumably represent   higher oxidation

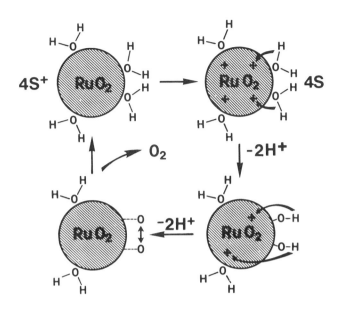

FIGURE 16. Mechanism of oxygen formation on hydrated RuO_2 particles.

states of Ru. Subsequent electron transfer and proton dissociation involve water adsorbed at the particle surface and
lead through the stage of bound OH• radicals to oxygen formation. Figure 17 shows schematic current-potential curves for
the two electron-transfer processes which occur on RuO_2 particles. The cathodic branch represents reduction of $Fe(bipy)_3^{3+}$
(or $Ru(bipy)_3^{3+}$) while the anodic branch corresponds to water
oxidation. A decrease in the particle potential results in an
increase of the rate of cathodic electron transfer while that
of the anodic reaction is retarded. Under steady-state conditions, the RuO_2 will assume a potential E_p given by the
intersection of the two current-voltage curves. At this
potential a current i_R will flow which defines the overall
reaction rate. It appears that the case of oxidizing agents
as $Fe(bipy)_3^{3+}$, $Ru(bipy)_3^{3+}$, Ce^{4+} or $ZnTMPyP^{5+}$ a driving force
of ~150 mV for water oxidation suffices to makes the process
occur rapidly and quantitatively.

V. CYCLIC WATER DECOMPOSITION

The case of water oxidation by the $Ru(bipy)_3^{3+}$ complex is
particularly interesting since this species is formed in the

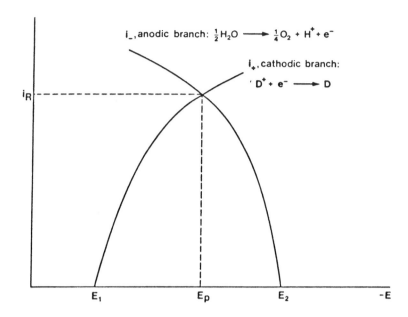

FIGURE 17. Current-voltage diagram for coupled redox processes occurring on RuO_2 particles. Symbols used are as follows: D^+ = $Fe(bipy)_3^{3+}$, $Ru(bipy)_3^{3+}$, Ce^{4+}; E_p = potential of particles; E_1 = $E^0 + 0.059$ $log\{[D^+]/[D]\}$; $E_2 = 1.23 - 0.059$ $log\{(P_{O_2})^{1/4} \times [H^+]\}$.

photoreaction that precedes hydrogen generation from the reduced relay. One recognizes here the possibility to close the cycle of water decomposition with visible light. In fact, irradiation of a solution containing colloidal Pt (protected by a copolymer of maleic anhydride and styrene) and macro-disperse RuO_2 as catalysts, apart from the $Ru(bipy)_3^{2+}$ sensitizer and the MV^{2+} electron relay, produces the two gases hydrogen and oxygen simultaneously (81). Successful operation of this system depends on the proper choice of the protective agent for the Pt sol. The copolymer of maleic anhydride and styrene employed in our experiments is suitable since it pro-vides functions with pronounced hydrophobicity. Of the redox products formed in the light reaction, $Ru(bipy)_3^{3+}$ is strongly hydrophilic while MV^+ is relatively hydrophobic. Hence, the latter will interact predominantly with the Pt particles giving rise to hydrogen formation. The $Ru(bipy)_3^{3+}$ left behind is prone to interact with the hydrophilic RuO_2 surface yielding oxygen from water. In this system the reaction rate of MV^+ with the Pt particles must also be high enough to com-pete with back electron transfer and the reduction of oxygen:

$$MV^+ + O_2 \rightarrow O_2^- + MV^{2+} \qquad\qquad [9]$$

The overall reaction scheme is illustrated in Figure 18. With the combination of catalysts employed initially, the quantum yield of water splitting was found to be relatively low, i.e., 1.5×10^{-3}. However, this value has since been improved drastically through the utilization of a bifunctional redox catalyst (82). Here, TiO_2 particles doped with RuO_2 were employed which serve at the same time as the support for an extremely fine Pt deposit. Surprisingly, with such dispersions we achieve as high as 20% of the quantum yield obtained in the sacrificial systems. It is significant that the H_2- and O_2-production rates do not decrease even over many hours of irradiation time. This is illustrated in Figure 19 which shows the amount of hydrogen produced by illuminating a solution containing 10^{-4} M $Ru(bipy)_3^{2+}$, 5×10^{-3} M MV^{2+} and the bifunctional catalyst. The hydrogen formation rate remains constant over at least 40 hours of irradiation time, oxygen being produced in stochiometric proportion. The average quantum yield of hydrogen formation in this time interval is ~5%. Thus the extent of cross reactions is less important with this bifunctional catalyst than with the earlier system.

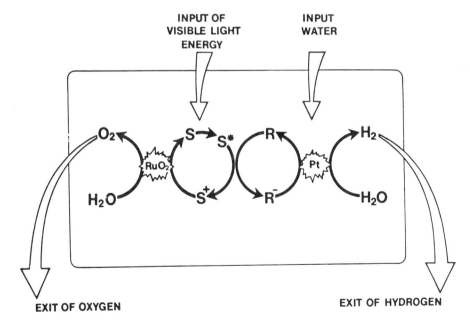

FIGURE 18. Scheme for cyclic photochemical water splitting in a coupled catalytic system.

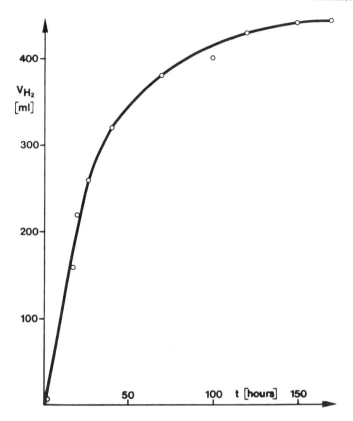

FIGURE 19. Sustained light-induced water splitting in a system containing the bifunctional redox catalyst $TiO_2/RuO_2/$ Pt. The ordinate is volume of H_2 produced per liter of solution and the abscissa is the irradiation time in hours. Light source was a 450 W xenon lamp with a 400 nm cutoff filter.

It is likely that adsorption of the reactants and/or participation of electronic states of the TiO_2 semiconductor in the redox events render the water-splitting process more efficient. Figure 20 illustrates a feasible mechanism for the intervention of the semiconducting redox catalyst. Excitation of the sensitizer $Ru(bipy)_3^{2+}$ is followed by electron transfer to the relay, in our case methyl viologen. The latter in turn injects an electron into the conduction band of the semiconductor. From there it is channelled to a Pt-site where H_2 evolution occurs. RuO_2, on the other hand, assists the back-conversion of S^+ to S with simultaneous oxygen formation from water. TiO_2 is a favorable material in that its conduction band is located close to the H_2/H^+ standard potential. Chemisorption of O_2 to this material may further assist water

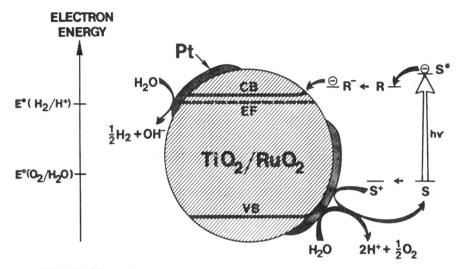

FIGURE 20. Schematic illustration of the intervention of a TiO_2-based catalyst in simultaneous H_2 and O_2 formation.

cleavage by light. These experiments of sensitized water decomposition on doped TiO_2 particles should be viewed in the context of photocatalytic decomposition of gaseous water over TiO_2 or $SrTiO_3$ surfaces (83–85). Interestingly, addition of an oxidation catalyst such as RuO_2 to TiO_2 (86) or NiO to $SrTiO_3$ (87) also results in a drastic increase of the quantum yield of water splitting. However, in all these cases the semiconductor was directly excited by UV irradiation.

An alternative approach to achieve light-induced cleavage of water is to separate hydrogen production from the oxygen-generating reaction. A device that achieves this goal is depicted schematically in Figure 21. The anode compartment contains only water and a RuO_2 electrode and is kept in the dark. It is coupled via an external circuit and a proton-conducting membrane to the cathode half-cell which is illuminated. The latter contains a sensitizer, an electron acceptor (R) and a platinum gauze electrode. Light quanta drive the electron transfer from S to R producing the species S^+ and R^-. The latter, in the presence of a suitable catalyst, will reduce water to hydrogen. Back conversion of the oxidized sensitizer to its original form occurs via the external circuit; electrons are furnished from water oxidation occurring at the RuO_2 electrode.

Finally, we have conducted a study on a photoredox system in which $Ru(bipy)_3^{2+}$ was employed as sensitizer and peroxodisulfate as the acceptor. The latter is irreversibly reduced to sulfate in the photoreaction. Photopotentials obtained

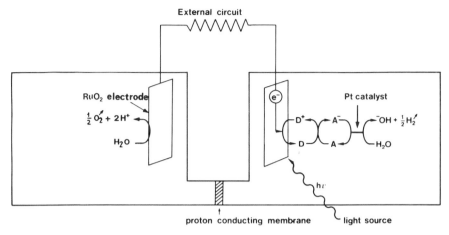

FIGURE 21. Schematic device for cyclic photolysis of water. H_2 and O_2 are evolved in two separate half-cells.

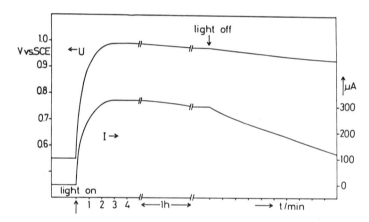

FIGURE 22. Time course of light-induced current and potential in a photoelectrochemical experiment. The illuminated half-cell contained $Ru(bipy)_3^{2+}$ (10^{-4} M) and $K_2S_2O_8$ (10^{-3} M) at pH 4.7 (acetate buffer). The dark half-cell contained a RuO_2 electrode in buffer solution. The photopotential was measured vs. sce. Light source was ambient room illumination

with such a system are sufficiently high to afford water oxidation, even in room light (Fig. 22). The observed photocurrents were on the order of several hundred $\mu A/cm^2$.

We are presently exploring a catalytic system in which the photoredox reaction is followed by reduction of water thus making the process cyclic with respect to the acceptor. Such a system can not only afford water cleavage by visible light, but has the additional advantage of producing H_2 and O_2 separately (88).

ACKNOWLEDGEMENT

I wish to express my gratitude to my collaborators whose enthusiasm and efforts contributed largely to this work.

REFERENCES

1. V. Balzani, L. Moggi, M. F. Manfrin, F. Bolletta and M. Gleria, Science 189, 852 (1975).
2. M. Calvin, Photochem. Photobiol. 23, 425 (1976).
3. G. Porter and M. D. Archer, Interdisc. Sci. Rev. 1, 119 (1976).
4. J. R. Bolton, Science 202, 705 (1978).
5. E. Schuhmacher, Chimia 32, 193 (1978).
6. W. H. F. Sasse, in "Solar Power and Fuels" (J. R. Bolton, ed.), Academic Press, New York, 1977, pp. 227-245.
7. R. E. Schwerzel and R. A. Nathan, U.S. Patent 4,004,571 (Jan. 25, 1977).
8. W. H. F. Sasse and T. Teitei, U.S. Patent 4,123,219 (Oct. 31, 1978).
9. V. Balzani, F. Boletta, M. T. Gandolfi and M. Maestri, Topics Current Chem. 75, 1 (1978).
10. G. Porter, Proc. R. Soc. (London) A362, 281 (1978).
11. M. Calvin, Accts. Chem. Res. 11, 369 (1978).
12. N. Sutin, J. Photochem. 10, 19 (1979).
13. D. Whitten, Accts. Chem. Res. 13, 83 (1980).
14. H. D. Gafney and A. W. Adamson, J. Am. Chem. Soc. 94, 8238 (1972).
15. C. R. Bock, T. J. Meyer and D. G. Whitten, J. Am. Chem. Soc. 96, 4710 (1974).
16. F. R. Hopf and D. G. Whitten, in "Porphyrins and Metallo-porphyrins", (K. M. Smith, ed.), Elsevier, Amsterdam, 1975, p. 667-700.
17. D. G. Whitten, Rev. Chem. Intermed. 2, 107 (1978).
18. I. I. Dilung and E. I. Kapinus, Russ. Chem. Rev. 47, 43 (1978).
19. D. Mauzerall, in "The Porphyrins" (D. Dolphin, ed.), Vol. V, Academic Press, New York, 1978, pp. 53-75.
20. K. Kalyanasundaram and M. Gratzel, Helv. Chim. Acta 63, 478 (1980).

21. B. V. Koryakin, T. S. Dzhabiev and A. E. Shilov, Dokl. Akad. Nauk SSSR 298, 620 (1977).
22. A. I. Krasna, Photochem. Photobiol. 29, 267 (1979).
23. J. J. Grimaldi, S. Boileau and J.-M. Lehn, Nature 265, 229 (1977).
24. J. S. Bellin, R. Alexander and R. D. Mahoney, Photochem. Photobiol., 17, 17 (1973).
25. K. Kalyanasundaram and M. Grätzel, J. Chem. Soc., Chem. Commun., 1137 (1979).
26. L. Michaelis and E. S. Hill, J. Gen. Physiol. 16, 859 (1933).
27. R. M. Elotson and R. L. Edsberg, Can. J. Chem. 33, 646 (1957).
28. J. E. Hornbaugh, J. E. Sandquist, R. H. Burris and W. H. Orme-Johnson, Biochemistry 15, 2633 (1976).
29. J.-M. Lehn and J.-P. Sauvage, Nouv. J. Chim. 1, 449 (1977).
30. M. Kirsch, J.-M. Lehn and J.-P. Sauvage, Helv. Chim. Acta 62, 1345 (1979).
31. G. M. Brown, S. F. Chan, C. Creutz, H. A. Schwarz and N. Sutin, J. Am. Chem. Soc. 101, 7638 (1979).
32. G. M. Brown, B. S. Brunschwig, C. Creutz, J. Endicott and N. Sutin, J. Am. Chem. Soc. 101, 1298 (1979).
33. M. Grätzel, in "Micellization and Micromulsions" (K. L. Mittal, ed.) Vol. 2, Plenum Press, New York, 1977, pp. 531-549.
34. M. Grätzel, Israel J. Chem. 18, 3 (1979).
35. C. Wolff and M. Grätzel, Chem. Phys. Lett. 52, 542 (1977).
36. Y. Waka, K. Hamamoto and N. Mataga, Chem. Phys. Lett. 53, 242 (1978).
37. B. Razem, M. Wong and J. K. Thomas, J. Am. Chem. Soc. 100, 1629 (1978).
38. S. A. Alkaitis, G. Beck and M. Grätzel, J. Am. Chem. Soc. 97, 5723 (1975).
39. S. A. Alkaitis and M. Grätzel, J. Am. Chem. Soc. 98, 3549 (1976).
40. M. Maestri, P. P. Infelta and M. Grätzel, J. Chem. Phys. 69, 1522 (1978).
41. Y. Moroi, A. M. Braun and M. Grätzel, J. Am. Chem. Soc. 101, 567 (1979).
42. Y. Moroi, P. P. Infelta and M. Grätzel, J. Am. Chem. Soc. 101, 573 (1979).
43. R. Humphry-Baker, M. Grätzel, P. Tundo and E. Pelizzetti, Angew. Chemie, Int. Ed. 18, 630 (1979).
44. R. Humphry-Baker, Y. Moroi, M. Grätzel, E. Pelizetti and P. Tundo, J. Am. Chem. Soc. 102, 3689 (1980)
45. J. Kiwi and M. Grätzel, J. Am. Chem. Soc. 100, 6314 (1978).

46. I. Willner, W. E. Ford, J. W. Otvos and M. Calvin, Nature 280, 823 (1979).
47. W. E. Ford, J. W. Otvos and M. Calvin, Proc. Natl. Acad. Sci. USA 76, 3590 (1979).
48. W. E. Ford, J. W. Otvos and M. Calvin, Nature 274, 507 (1978).
49. M. Calvin, Int. J. Energy Res. 3, 73 (1979).
50. P. P. Infelta, J. H. Fendler and M. Grätzel, J. Am. Chem. Soc. 102, 1479 (1980).
51. H. Kuhn, Pure and Appl. Chem. 51, 341 (1979), and references cited therein.
52. H. T. Tien, in "Topics in Photosynthesis - Photosynthesis in Relation to Model Systems" (J. Barber, ed.), Vol. 3 Elsevier/North-Holland, Amsterdam, 1979, pp. 116-173.
53. D. Meisel and M. S. Matheson, J. Am. Chem. Soc. 99, 6577 (1977).
54. C. D. Jonah, M. S. Matheson and D. Meisel, J. Phys. Chem. 83, 257 (1979).
55. C. Tanford, "The Hydrophobic Effect: Formation of Micelles and Biological Membranes", Wiley-Interscience, New York, 1973.
56. J. H. Fendler and E. J. Fendler, "Catalysis in Micellar and Macromolecular Systems", Academic Press, New York, 1975.
57. B. Lindman and H. Wennerström, Topics Current Chem., 87, 3 (1980).
58. M. S. Fernandez and P. Fromherz, J. Phys. Chem. 81, 1755 (1977).
59. S. McLaughlin, in "Current Topics in Membranes and Transport", Vol. 9, Academic Press, New York, 1977, pp. 71-144.
60. P. A. Brugger and M. Grätzel, J. Am. Chem. Soc. 102, 2461 (1980); P. A. Brugger, P. P. Infelta, A. M. Braun and M. Grätzel, J. Am. Chem. Soc. 103, 320 (1981)
61. C. Wagner and W. Traud, Z. Electrochem. 44, 397 (1938).
62. G. Bredig and R. Müller von Berneck, Ber. 31, 258 (1899), and references cited therein.
63. D. E. Green and L. H. Stickland, Biochem. J. 28, 898 (1934).
64. K. Kalyanasundaram, J. Kiwi and M. Grätzel, Helv. Chim. Acta 61, 2720 (1978).
65. A. Moradpour, E. Amouyal, P. Keller and H. Kagan, Nouv. J. Chim. 2, 547 (1978).
66. B. O. Durham, W. J. Dressick and T. J. Meyer, J. Chem. Soc., Chem. Commun., 381 (1979).
67. P. J. DeLaive, B. P. Sullivan, T. J. Meyer and D. G. Whitten, J. Am. Chem. Soc. 101, 4007 (1979).
68. T. Kawai, K. Tanimura and T. Sakada, Chem. Lett., 137 (1979).
69. J. Kiwi and M. Grätzel, Nature 282, 657 (1979).

70. J. Kiwi and M. Grätzel, J. Am. Chem. Soc. <u>101</u>, 7214 (1979).
71. P. A. Brugger, P. Cuendet and M. Grätzel, J. Am. Chem. Soc., in press (1981).
72. J. Turkevich, K. Aika, L. L. Ban, I. Okura and S. Namba, J. Res. Inst. Catalysis, Hokkaido Univ. <u>24</u>, 54 (1976).
73. K. R. Mann, N. S. Lewis, V. M. Miskowski, D. K. Erwin, G. S. Hammond and H. B. Gray, J. Am. Chem. Soc. <u>99</u>, 5525 (1977).
74. D. Dung, K. Kalyanasundaram and M. Grätzel, unpublished results.
75. J. Kiwi and M. Grätzel, Angew. Chemie, Int. Ed. <u>17</u>, 860 (1978).
76. J. Kiwi and M. Grätzel, Angew. Chemie, Int. Ed. <u>18</u>, 624 (1979).
77. J.-M. Lehn, J.-P. Sauvage and R. Ziessel, Nouv. J. Chim. <u>3</u>, 423 (1979).
78. J. Kiwi and M. Grätzel, Chimia <u>33</u>, 289 (1979).
79. K. Kalyanasundaram, O. Mićić, E. Promauro and M. Grätzel, Helv. Chim. Acta <u>62</u>, 2432 (1979).
80. V. Ya. Shafirovich, N. K. Khannov and V. V. Strelets, Nouv. J. Chim. <u>4</u>, 81 (1980).
81. K. Kalyanasundaram and M. Grätzel, Angew. Chem., Int. Ed. <u>18</u>, 701 (1979).
82. J. Kiwi, E. Bogarello, E. Pelizzetti, M. Visca and M. Grätzel, Angew. Chem., Int. Ed. <u>19</u>, 646 (1980).
83. G. N. Schrauzer and T. D. Guth, J. Am. Chem. Soc. <u>99</u>, 7189 (1977).
84. H. Van Damme and W. K. Hall, J. Am. Chem. Soc. <u>101</u>, 4373 (1979).
85. S. Sato and J. M. White, Chem. Phys. Lett. <u>72</u>, 83 (1980).
86. T. Kawai and T. Sakata, Chem. Phys. Lett. <u>72</u>, 87 (1980).
87. K. Komen, S. Naito, M. Soma, T. Onishi and K. Tamaru, J. Chem. Soc., Chem. Comm., 543 (1980).
88. M. Neumann-Spallart, K. Kalyanasundaram, C. K. Grätzel and M. Grätzel, Helv. Chim. Acta <u>63</u>, 1111 (1980). A similar system, where $Co(NH_3)_5Cl^{3+}$ was used as an acceptor and oxygen was evolved at a Pt electrode, has been reported recently (89).
89. D. P. Rillema, W. J. Dressick and T. J. Meyer, J. Chem. Soc., Chem. Comm., 241 (1980).

DISCUSSION

Dr. M. D. Archer, University of Cambridge:
Does the production of selectivity by placing a hydrophilic coating on RuO_2 and a hydrophobic coating on Pt
derive from the fact that A^- is not highly charged while
D^+ is, or from the fact that A^- is organic and D^+ is
inorganic?

Prof. Grätzel:
The selectivity of the hydrophilic coating on RuO_2 derives
from the extremely hydrophilic nature of Ru(III), which is
much more hydrophilic than Ru(II).

Dr. H. Tributch, C.N.R.S., Laboratoires de Bellevue:
In my opinion there is no fundamental difference between a
macroscopic photoelectrochemical or electrochemical system
or a microscopic one which is functioning in the form of
small particles or aggregation of particles in suspension.
One difference is that microscopic systems don't offer the
possibility to vary electrode potentials externally. However, a certain flexibility is available by selecting
appropriate supporting materials. Metallic platinum particles attached to TiO_2 particles can — for example,
through their very positive Fermi potential — extract
electrons from the TiO_2 semiconductor, thus providing a
more positive electrode potential for oxygen evolution.
All basic experiments could be performed with a macroscopic system and subsequently used for devising the
microscopic catalysts.

Prof. Grätzel:
I agree. In fact there is much opportunity to apply electrochemical techniques and principles to devise both
microscopic and macroscopic systems for photochemical
water splitting.

Dr. Tributch:
I was surprised to hear you mention PtO_2 together with
RuO_2 and IrO_2 as favorable catalysts for the evolution of
oxygen from water. Positively polarized platinum is generally considered as a bad catalyst for the evolution of
oxygen since it has a high overpotential for this reaction. This is generally attributed to the formation of
platinum oxides on the electrode surface. There is an
apparent contradiction.

Prof. Grätzel:

I'm very glad you raised that point. (Lengthy discussion followed).

Dr. A. Mackor, TNO Utrecht:

I would like to remark that you have used a doped TiO_2 powder for the preparation of your catalyst, while we have used a reduced TiO_2 powder for the same purpose, which essentially gives the same results. Our results are presented in poster VI-1.

Prof. Grätzel:

Did you use a Nb-doped TiO_2 for this purpose?

Dr. Mackor:

Not for this study, but I have used it in a related study of photoelectrolysis

CHAPTER 6

PHOTOINDUCED GENERATION OF HYDROGEN AND OXYGEN FROM WATER

Jean-Marie Lehn

Institut Le Bel, Université Louis Pasteur,
Strasbourg
and Collège de France,
Paris, France

I. INTRODUCTION

The search for photochemical systems capable of dissociating water into hydrogen and oxygen (Eq. [1]) is of interest

$$2H_2O \xrightarrow{h\nu} 2H_2 + O_2 \qquad\qquad [1]$$

from at least three major points of view: **chemical** -- it may
lead to the realization of abiotic systems capable of performing "artificial photosynthesis"; in the process much new
basic knowledge will be acquired, especially because the coop-
eration of various areas of research from organic synthesis to
solid-state physics is required; **biological** -- it should help
our understanding of the water-splitting process which occurs
in natural photosynthesis and in designing biomimetic systems;
economical -- it represents a potential source of fuel which,
as part of the energy package of the future, may play an
important role in meeting the energy needs of the world.
    Feedback   between   abiotic   and   biomimetic   approaches
provides opportunities for mutual fertilization.   Artificial
photosynthetic systems are not subject to the constraints
imposed upon natural photosynthesis to be compatible with the
process of life.   Abiotic systems may be designed to operate
at higher temperatures than photosynthetic organisms, thus
making use of the heat generated by solar radiation to increase the efficiencies and/or reaction rates, rather than
dissipating it by water evaporation.   Purely synthetic systems
may also employ light-absorbing materials of higher chemical
stability than the natural pigment, chlorophyll.

PHOTOCHEMICAL CONVERSION AND
STORAGE OF SOLAR ENERGY
161

It is not my intention here to discuss the large amount of interesting work performed on a problem that spans such a vast area; references (1-17) give a selection of some recent reviews covering various domains. Instead, I shall limit my presentation to our own work in the field.   Other aspects are discussed elsewhere in this Volume.

Our interest in the photodissociation of water and in related photochemical processes (e.g., CO_2 or N_2 fixation) evolved from three areas of our work: the synthesis of macropolycyclic ligands which form inclusion complexes with two metal cations, i.e., dinuclear cryptates (18-20) which may be able to act as catalytic sites for the exchange of several electrons (21); the development of carriers and processes for the transport of cations, anions and electrons through artificial membranes (22); and the coupling of electron transport with light excitation (23) and the cotransport of other substrates (24).

Di- or polynuclear cryptates appeared to be of potential interest for catalyzing the two-electron reduction, Eq. [2], and the four-electron oxidation, Eq. [3], of water:

$$2H^+ + 2\ e^- \rightarrow H_2 \qquad\qquad E_7 = -0.41\ V \qquad\qquad [2]$$

$$2H_2O \rightarrow O_2 + 4H^+ + 4e^- \qquad\qquad E_7 = +0.82\ V \qquad\qquad [3]$$

where E_7 is the redox potential at pH 7. On the other hand, membrane systems to transport electrons and protons may be

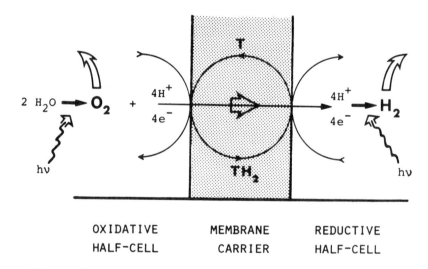

$$2\ H_2O \rightarrow O_2 \ + \quad \frac{4H^+}{4e^-} \qquad T \qquad \frac{4H^+}{4e^-} \rightarrow H_2$$

$$TH_2$$

OXIDATIVE	MEMBRANE	REDUCTIVE
HALF-CELL	CARRIER	HALF-CELL

FIGURE 1.   Schematic diagram of a water-splitting photochemical cell composed of an oxidative half-cell, a reductive half-cell and a membrane permeable to electrons and protons.

required for the development of a water-splitting cell. Such membranes may be used to separate the oxidative and reductive half-cells, thus minimizing short-circuits between reduced and oxidized species participating in Eqs. [2] and [3] (Fig. 1; see also Fig. 1 in ref. 25). Furthermore, generation of H_2 and O_2 in separate compartments would facilitate recovery of the gases and thereby enhance the practical utilization of such a system (see also, Chapters 1 and 5). Although membrane systems present several attractive features, single-cell systems may also be feasible if suitable catalysts can be found. In principle the latter should be simpler to operate but will require separation of the cogenerated gases.

The potentials for the various redox processes of water are depicted in Figure 2. Reactions [2] and [3] are thermodynamically the most favorable ones, and may therefore be effected photochemically over the broadest range of wavelengths of the solar spectrum; since $\Delta G = 1.23$ eV per electron for reaction [1] via processes [2] and [3], light of $\lambda <$ 1000 nm would be effective, corresponding to a threshold energy of 119 kJ/Nhν, so that about 45% of the energy in the solar spectrum could, in principle, be stored (5). However, when conversion-efficiency factors are taken into account, the thresholds are shifted to shorter wavelengths: $\lambda \leqslant 611$ nm and $\lambda \leqslant 877$ nm for one hν/electron and two hν/electron, respectively (see refs. 3,5,12, and Chapter 11). Figure 3 is a schematic representation of the light-intensity distribution in the solar spectrum, showing the threshold energies and wavelengths for dissociation of water by one-, two-, and four-electron processes (see also, Fig. 2).

Light of any wavelength shorter than the threshold should be capable of inducing the reaction. However, all photon energy in excess will be transformed into heat, thus higher efficiencies of solar energy conversion should be achieved in multiprocess systems in which several reactions having thresholds distributed over the entire solar spectrum would operate simultaneously. Water photodissociation is not the only reaction to be considered, and other reactions for producing solar fuels deserve attention (5,10).

Although originating from our work on metal cation cryptates and electron transport-membranes, our interest in photochemical processes for water dissociation and other energy-storage reactions (like CO_2-reduction) evolved towards quite different areas of investigation. The approach was to proceed stepwise: (a) First, we set up systems capable of generating separately H_2 or O_2 via reactions [2] and [3]; these processes require photoactive materials possessing suitable redox properties for water reduction or oxidation, and catalysts capable of mediating multielectronic processes. (b) We then searched for a single process which would produce

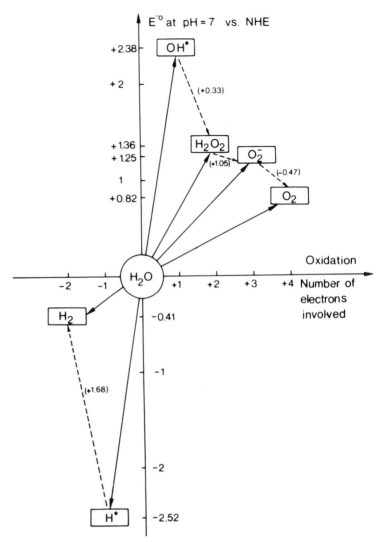

FIGURE 2. Redox potentials for the various redox reactions of the water molecule (at pH 7, vs. nhe).

H_2 and O_2 simultaneously. (c) The last requirement is that the spectral response of the system be as wide as possible, which limits the selection of otherwise suitable photoactive materials. Achieving these three goals implies finding solutions to the thermodynamic and mechanistic (a), kinetic (b) and spectral (c) problems for conversion and storage of light energy by water photolysis. In the following sections, I will summarize our work along these lines together with related

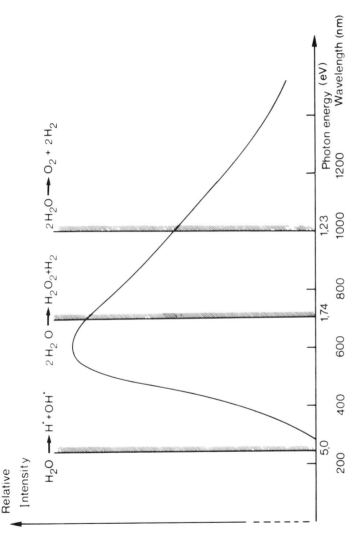

FIGURE 3. Schematic representation of solar spectral distribution and the theoretical threshold energies and wavelengths for dissociation of water by one-, two-, and four-electron processes. Taking efficiency factors into account shifts the thresholds to shorter wavelengths (3,5; see also Chapter 11, this Volume).

studies reported in the recent literature. The extensive work on photoelectrochemical decomposition of water will not be discussed here. (See Chapters 9 and 10, this Volume).

II. PHOTOINDUCED HYDROGEN GENERATION FROM WATER

A. Photochemical Hydrogen-Generation Scheme

Conversion of solar energy via photochemical redox processes (Fig. 4) relies on electron-transfer reactions between the excited state of a light-absorbing species and another substance which is either reduced or oxidized (26-28). If spontaneous recombination can be overcome, the resulting charge separation may drive the formation of energy-storage products that are stable under the reaction conditions.

Photogeneration of hydrogen as a storage compound, requires production of a reduced species R^- at a redox potential for $R/R^- \leqslant -0.41$ V and a catalyst, C_{red}, to facilitate the two-electron reduction of water, Eq. [2]. This R^- species may be formed by electron transfer from the excited state of a suitable photosensitizer, such that the redox potential of the excited state, PS^+/PS^*, is more negative than that of the R/R^- couple. In such a scheme, substance R acts as a relay between PS and the catalyst, C_{red}, providing a means for intermediate storage of electrons.

Finally, since the first goal is to set up a hydrogen-generation system whose components possess the required thermodynamic properties, complications due to kinetically fast recombination processes (e.g., between PS^+ and R^-) should be avoided at this stage. This may be realized by using an electron donor D which allows fast back conversion of PS^+ into PS and is consumed in the process by a fast, irreversible decomposition of the oxidized D^+ species thus formed.

The complete system (Fig. 5) thus comprises a photosensitive species PS, a relay species R, an electron donor D, and a catalyst C_{red}: i.e., $PS/R/D/C_{red}$. It operates via the following steps:

● light absorption by PS:

$$PS + h\nu \rightarrow PS^* \qquad\qquad\qquad\qquad [4]$$

● electron transfer to R by oxidative quenching:

$$PS^* + R \rightarrow PS^+ + R^- \qquad\qquad\qquad [5]$$

● rapid back-reduction of PS^+ by D, with subsequent rapid decomposition of D^+:

$$PS^+ + D \rightarrow PS + D^+ \qquad\qquad\qquad [6]$$

$$D^+ \rightarrow \text{decomposition} \qquad\qquad\qquad [7]$$

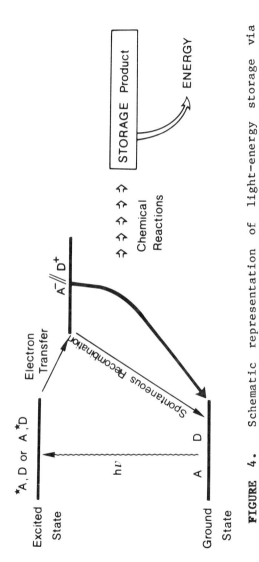

FIGURE 4. Schematic representation of light-energy storage via photochemical redox processes. The light-absorbing substance may be an electron acceptor A or an electron donor D.

o hydrogen production by reduction of water by R^- (or by a transformation product thereof) in the presence of the catalyst C_{red}:

$$2R^- + 2H^+ \xrightarrow{C_{red}} 2R + H_2 \qquad [8]$$

In the oxidative quenching scheme represented in Figure 5, PS and R follow catalytic cycles, while D and H^+ are consumed. Of course, this sequence represents only one possibility; reaction [5] could also be replaced by a reductive quenching process to yield the reduced species PS^-, which then undergoes further transformations leading to H_2 production. A more detailed discussion has been given elsewhere (25).

B. Components of Photochemical Hydrogen-Generating Systems

The choice of the respective components in a system of the type represented in Figure 5 must meet a number of requirements with respect to excited-state properties, redox potentials, and reaction kinetics. The extensively studied $Ru(bipy)_3^{2+}$ complex has been shown to undergo facile light-induced electron-transfer reactions and to possess appropriate redox properties for reducing or oxidizing water, either from the excited $Ru(II)^*$ or from its reduced $Ru(I)$ or oxidized

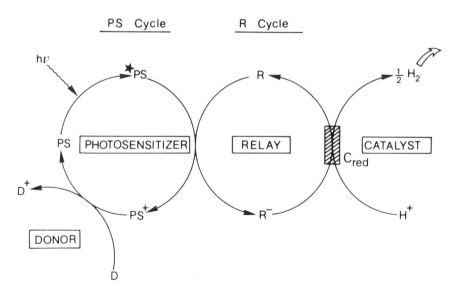

FIGURE 5. Schematic representation of hydrogen generation by a relay-mediated, photoinduced water reduction process.

Ru(III) ground states, respectively (26-28; see also, Chapter 4, this Volume). It is thus a suitable candidate for the sensitizer since it also strongly absorbs visible light.

The R/R^- redox potential of the relay must lie between those of H^+/H_2 and PS^+/PS^*. We used the rhodium complex, $Rh(bipy)_3^{3+}$, as a relay species. A major reason for choosing a Rh(III) complex was the possibility of exchanging two electrons by interconversion of Rh(III) and Rh(I). This property is of interest in view of the two-electron requirement for water reduction, Eq. [2]. It has been known for a long time that heterogeneous catalysts like colloidal platinum or Adams' catalyst are efficient redox catalysts (29-31). Finally, the photooxidation and subsequent rapid decomposition of tertiary amines (such as EDTA or triethanolamine, TEOA) make them suitable candidates as sacrificial electron donors (25,32). In our studies, we made use of the system: $Ru(bipy)_3^{2+}/Rh(bipy)_3^{3+}/$ TEOA/Pt (25,33). The redox properties of the various states of these components are represented in Figure 6.

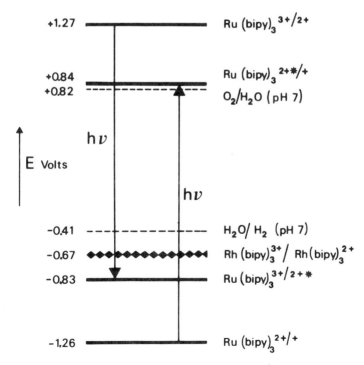

FIGURE 6. Redox potentials of $Ru(bipy)_3^{2+}$ and $Rh(bipy)_3^{3+}$ complexes and water vs. nhe at pH 7. Both the oxidizing and reducing properties of the complex are enhanced by the energy of the excited state, 2.10 eV (see also ref. 25).

Generation of reduced methyl viologen cation MV^+ by visible photolysis of aqueous solutions containing MV^{2+}, proflavine (PF) as photosensitizer, and EDTA as donor provides the basis for another suitable system (34). These components have been employed previously in light-induced electron-transport experiments (23). Furthermore, the MV^+ species has a sufficiently low potential ($MV^{2+}/MV^+ = -0.44$ V) to reduce water in the presence of a platinum catalyst (35). Thus systems using MV^{2+} as a relay species and either $Ru(bipy)_3^{2+}$ or PF as the sensitizer have been shown to yield hydrogen under irradiation by visible light of aqueous solutions containing a Pt catalyst and EDTA as donor (36-42).

A number of systems have been investigated which use different components, like metalloporphyrins as the sensitizer (43-47; see also, Chapters 1 and 5), other viologens (48,49) or the (Fe_4S_4) iron-sulfur cluster (50) as the relay, or hydrogenase as the catalyst (37,44,51). Studies have also been performed on systems in which one component is omitted, e.g., the MV^{2+} relay (52) or the Pt catalyst (53). A photoelectrochemical variant has also been reported (54).

The efficiency of the heterogeneous redox catalyst is of particular importance for making hydrogen evolution competitive with other redox reactions in the system. The influence of the nature of the protective colloid and of the particle size of the Pt catalysts has been investigated (55), although another study has indicated that in the $Ru(bipy)_3^{2+}/MV^{2+}/EDTA$ system there is no appreciable effect of particle size on hydrogen-evolution rates (56). Colloidal silver, gold and platinum also catalyze hydrogen formation from organic radicals produced by ionizing radiation and other electron-transfer processes (57-59).

Hydrogen evolution has been observed in systems involving the reductive quenching of $Ru(bipy)_3^{2+*}$ by Eu(II) or ascorbate (using a macrocyclic Co(II) complex as relay) (60) or by triethylamine (61). Reductive quenching of excited Cr(III)-polypyridine complexes by EDTA in the presence of Pt also yields hydrogen (62). Other hydrogen-generating experiments involving photolysis of coordination compounds have been reported (25), and systems involving di- and tetranuclear Rh(I) complexes have been investigated recently (63,64). Also, oxidative quenching of $Ru(bipy)_3^{2+*}$ by Ag^+ yields Ag^0 as the reduced species (65).

C. Hydrogen Generation with the $Ru(bipy)_3^{2+}/Rh(bipy)_3^{3+}/TEOA/Pt$ System

Irradiation with visible light of an aqueous solution (pH 7) containing these components leads to efficient generation of hydrogen gas with high catalytic turnover numbers with respect to the metal complexes (25,33). D_2 is obtained when the reaction is conducted in D_2O, indicating that water is the proton source. After the initial discovery of this hydrogen-generating process (33), a study of its mechanism was undertaken (25). Another mechanistic investigation of the same system has also been carried out (66), and the nature of the interconversion of the different rhodium species involved in the process has been analyzed (67).

The mechanism of hydrogen generation by this system has been found to follow the general lines shown in Figure 5. The scheme depicted in Figure 7 contains four components: two homogeneous catalytic cycles involving ruthenium and rhodium, respectively; a heterogeneous catalytic reduction step on platinum yielding hydrogen from water; and an oxidation reaction of the TEOA donor. The ruthenium cycle is of the oxidative quenching type and comprises the following steps:

- absorption of light

$$Ru(bipy)_3{}^{2+} \xrightarrow{h\nu} Ru(bipy)_3{}^{2+*} \qquad [9]$$

- oxidative quenching

$$Ru(bipy)_3{}^{2+*} + Ru(bipy)_3{}^{3+} \rightarrow Ru(bipy)_3{}^{3+} + Rh(bipy)_3{}^{2+} \qquad [10]$$

- back conversion

$$Ru(bipy)_3{}^{3+} + TEOA \rightarrow Ru(bipy)_3{}^{2+} + TEOA^{+} \qquad [11]$$

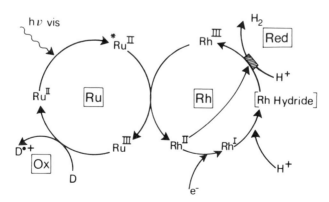

FIGURE 7. Schematic representation of hydrogen generation by the $Ru(bipy)_3{}^{2+}/Rh(bipy)_3{}^{3+}/TEOA/Pt$ system. The bipyridyl ligands have been omitted for clarity. D is the electron donor, TEOA. Two possible pathways are indicated for the rhodium cycle (see text).

Rapid decomposition of TEOA$^+$ into diethanolamine and hydroxy-acetaldehyde makes reaction [11] irreversible and thus allows cycling of the ruthenium complex.

The rhodium component (Fig. 7) involves a complicated series of species and transformations which depend on the pH at which the reaction is performed (25,66,67). At basic pH, formation and accumulation of the pink Rh(bipy)$_2$$^+$ species occur, which protonates to form rhodium hydrides at acidic pH. It has been suggested that hydrogen generation may occur via a "Rh(I)-Rh(hyrides)" pathway involving protonation and facilitation on the heterogeneous Pt catalyst (25). The presumed Rh(bipy)$_2$$^+$ entity was found to yield, around neutral pH, a mixture of species from which thermal evolution of H$_2$ appears to proceed only slowly (67).

Flash photolysis experiments (66) also been carried out to study the formation and fate of Rh(bipy)$_3$$^{2+}$ (Eq. [10]) which, in the presence of Pt catalyst, may lead directly to H$_2$ without going through an intermediate Rh(I) species:

$$Rh(bipy)_3^{2+} + H^+ \xrightarrow{Pt} Rh(bipy)_3^{3+} + \tfrac{1}{2} H_2 \qquad [12]$$

These kinetic investigations favor the "Rh(II)" pathway, reaction [12], as the predominant hydrogen-generating pathway under steady-state illumination (66).

The nature and transformations of the rhodium-bipyridyl species depend markedly on the actual conditions (pH, temperature, light intensity, excess ligand, etc.), and a number of them still remain to be characterized unambiguously. Further studies of the chemistry of this system would therefore be of interest both from the photochemical and coordination chemistry points of view. The possibility that interconversion of Rh(III) and Rh(I) may offer a two-electron pathway for water reduction is of great interest, as mentioned above. In this respect, it has been found that the doubly reduced Ru(bipy)$_3$^{o} species yields hydrogen from water stoichiometrically, whereas very little or no H$_2$ is obtained from Ru(bipy)$_3$$^+$ (68).

The overall process of hydrogen formation for reactions [9] through [12] is:

$$R_3N + H_2O \xrightarrow{h\nu} R_2NH + HOCH_2CHO + H_2 \qquad [13]$$

where R = (-CH$_2$CH$_2$OH). Whether or not net energy storage occurs depends on the exact combustion energies of the organic compounds; these are unfortunately not known with sufficient accuracy at present.

The quantum yields of hydrogen production are 0.11 for the Ru(bipy)$_3$$^{2+}$/Rh(bipy)$_3$$^{3+}$/TEOA/Pt system (66) and 0.13 for the

$Ru(bipy)_3^{2+}/MV^{2+}/EDTA/Pt$ system (55). However, the stability of the latter system is limited by a process that consumes methyl viologen via a platinum-dependent hydrogenation reaction (39). The high catalytic turnover numbers obtained with the rhodium system are indicative of its long term stability; further protection may result from the addition of excess free pyridine ligand (25).

D. Modified Hydrogen—Generating Systems

A number of systems based on the Ru(II)/Rh(III)/TEOA/Pt system have been investigated which involve other sensitizers (e.g., proflavine, $Os(bipy)_3^{2+}$), other relays (cobalt and copper bipyridyl complexes, other rhodium complexes, MV^{2+}) and other donor species. Efficient light-induced evolution of hydrogen has been obtained in several cases (25). Systems lacking an added photosensitizer may also yield hydrogen; for example, irradiation of $Rh(bipy)_3^{3+}/TEOA/Pt$ with a UV lamp or the sun (25,69) and visible irradiation of Ir(III) bipyridyl complexes in the presence of TEOA and Pt (25). In these reactions the excited states of the Rh and Ir complexes undergo reductive quenching by the electron donor (25,69,70), and the Rh(II) and Ir(II) complexes thus formed may yield hydrogen in the presence of the Pt catalyst or may undergo further transformation. Of interest also is the observation that hydrogen is formed by irradiation of $Ru(bipy)_3^{2+}/Rh(bipy)_3^{3+}/EDTA$ with with visible light at pH 5.2 in the absence of a Pt catalyst. This may be an indication in favor of the two-electron "Rh(I)" type pathway discussed above (25).

The search for highly efficient redox catalysts is of particular importance, since to realize a complete water-photolysis system, formation of H_2 and O_2 must be competitive with the various recombination reactions. Studies of the efficiencies of heterogeneous Pt catalysts of various types have already been mentioned (55,56). We have prepared several types of catalytic materials and compared their activities in standard hydrogen-generation experiments involving the system: $Ru(bipy)_3^{2+}/MV^{2+}/EDTA/catalyst$ at pH 4.8 (71). In addition to unsupported colloidal Pt obtained from K_2PtCl_4 and Pt/polymer species, deposition of Pt, Ru or Ir on solid supports like zeolites or the semiconductor, $SrTiO_3$, was also investigated (Fig. 8). In these experiments, the most efficient catalysts were unsupported colloidal Pt and either Pt or Rh deposited on $SrTiO_3$ powder. The latter observation is of interest with respect to further developments involving semiconductors, as discussed below.

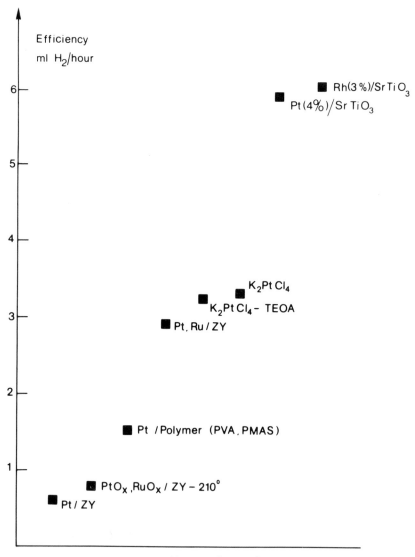

Nature of the Catalyst

FIGURE 8. Comparative efficiencies of various redox catalysts for photogeneration of hydrogen. Systems of the type: $Ru(bipy)_3^{2+}/MV^{2+}/EDTA/catalyst$ were studied at pH 4.8 under irradiation with visible light; ZY-210° = zeolite support prepared at 210°C; PVA = polyvinylalcohol polymer and PMAS = hydrolyzed maleic anhydride-styrene copolymer supports; $SrTiO_3$ support (1 hour irradiation; 30 ml solution).

The photochemical hydrogen-generating systems discussed previously represent the reductive component of a water-photolysis process. Their development is an important step towards the realization of a complete system. On the other hand, they might allow production of hydrogen from cheap donors available in large quantities, such as organic wastes or biomass. Photogeneration of $Rh(bipy)_2^+$ using glucose as donor (25) and of MV^+ with various organic molecules (37) and cellulose (72) as donors represent steps in this direction. The next stage towards cyclic water photolysis was to find ways to generate oxygen from water photochemically, thus realizing the oxidative component of the complete system.

III. PHOTOINDUCED OXYGEN GENERATION FROM WATER

A. Photochemical Oxygen-Generation Scheme

Photogeneration of oxygen from water requires the photo-induced production of an oxidized species of redox potential >0.82 V and a redox catalyst C_{ox} to facilitate the four-electron oxidation of water, Eq. [3]. Oxidative quenching of the excited state PS* of a sensitizer by an electron acceptor A leads to photogeneration of PS^+:

$$PS^* + A \rightarrow PS^+ + A^-$$ [14]

Recombination of the charge-separated species PS^+ and A^- may be avoided or minimized if A^- rapidly undergoes further irreversible transformations, as in Eq. [15]:

$$A^- \rightarrow decomposition$$ [15]

Reaction of the one-electron oxidant PS^+ with water may yield oxygen in presence of a suitable redox catalyst capable of mediating the water-oxidation reaction:

$$4PS^+ + H_2O \xrightarrow{C_{ox}} 4PS + 4H^+ + O_2$$ [16]

This general scheme is represented in Figure 9, in which the sensitizer PS follows a catalytic cycle while the acceptor A and water are consumed. The redox potentials of the couples involved must obey the relations: O_2/H_2O (+0.82 V) $<$ PS^+/PS and $PS^+/PS^* < A/A^-$. The two main problems in realizing such a $PS/A/C_{ox}$ system for oxygen generation are to find a suitable oxidant and catalyst.

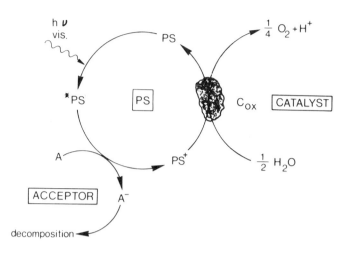

FIGURE 9. Schematic representation of oxygen generation by photoinduced water oxidation.

B. Components of Photochemical Oxygen-Generating Systems

High oxidation states of metal cation complexes may have redox potentials suitable for water oxidation (>0.82 V). Indeed, addition of alkali hydroxide to solutions of ML_3^{3+} (where M = Fe, Ru or Os and L = bipyridine or phenanthroline) causes reduction to the M^{2+} state (73). Detailed mechanistic and oxygen-production studies for the thermal reductions of $Fe(bipy)_3^{3+}$ (74,75) and $Ru(bipy)_3^{3+}$ (76) by water at various pH values have been performed. These reactions involve addition of hydroxide to the bipyridine ligand in the complex.

It has been shown that irradiation of $Ru(bipy)_3^{2+}$ in the presence of a cobalt(III) complex as electron acceptor yields $Ru(bipy)_3^{3+}$ (77), whose redox potential is sufficient to oxidize water in neutral or even acidic solutions. Thus, as in the hydrogen-generation system, $Ru(bipy)_3^{2+}$ is again a suitable photosensitizer. The redox properties of the various states of these compounds are shown in Figure 10.

In search of a suitable redox catalyst C_{ox}, our interest focused initially on ruthenium dioxide materials, since it was known that RuO_x anodes show high electrocatalytic activity (i.e., low overvoltage) for oxygen evolution in electrolysis of water (78-80). Studies of oxygen evolution at a number of metal oxide electrodes have also been reported (81-86). Furthermore, it has been shown that PtO_2 and IrO_2 accelerate the reduction of Ce^{4+} and $Ru(bipy)_3^{3+}$ by water (87). These observations have been extended to RuO_2 (88) and to the reduction of $Fe(bipy)_3^{3+}$ in the presence of powdered or colloidal

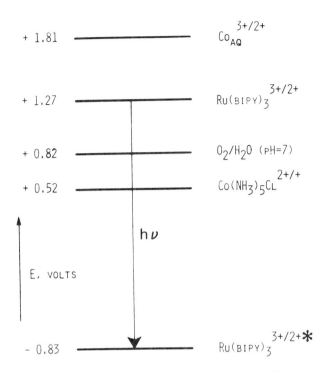

FIGURE 10. Redox potentials of $Ru(bipy)_3^{2+}$, $Co(NH_3)_5^{2+}$, $Co(aq)^{2+}$ and water vs. nhe at pH 7.

RuO_2 (89). In agreement with our expectations based on the electrolysis results, we found that nonstoichiometric RuO_x powders markedly accelerate thermal reduction of $Ru(bipy)_3^{3+}$ by water with evolution of oxygen (90). It has also been reported that hydrolyzed cations like Fe(II) or Co(II) catalyze oxidation of water and reduction of $M(bipy)_3^{3+}$ complexes, where M = Fe, Ru or Os (91-93).

C. Oxygen Generation with the $Ru(bipy)_3^{2+}/Co(NH_3)_5Cl^{2+}/RuO_x$ System

Combination of RuO_x catalysts with the photochemical reaction that generates $Ru(bipy)_3^{3+}$ (77) leads to efficient evolution of oxygen on irradiation with visibile light of aqueous solutions of $Ru(bipy)_3^{2+}/Co(NH_3)_5Cl^{2+}/RuO_x$ (90,94). Using $H_2^{18}O$ yields mainly $^{18}O_2$. The system is chemically stable and shows high catalytic turnover numbers on both $Ru(bipy)_3^{2+}$ and RuO_x (90).

The mechanism of oxygen generation by this system follows the pathway depicted in Figure 11. It comprises the following steps:

o oxidative quenching of the $Ru(bipy)_3^{2+*}$ excited state with a quantum yield of 6.3% (77,95):

$$Ru(bipy)_3^{2+*} + Co(NH_3)_5Cl^{2+} \rightarrow Ru(bipy)_3^{3+} + Co(NH_3)_5Cl^{+}$$
[17]

o thermal back conversion by oxidation of water on the RuO_x catalyst:

$$Ru(bipy)_3^{3+} + \tfrac{1}{2} H_2O \xrightarrow{RuO_x} Ru(bipy)_3^{2+} + H^+ + \tfrac{1}{4} O_2$$
[18]

o rapid decomposition of the reduced cobalt complex to the aquo-Co(II) cation:

$$Co(NH_3)_5Cl^{+} + 5H^+ \rightarrow Co(aq)^{2+} + 5NH_4^{+} + Cl^-$$ [19]

The ruthenium complex undergoes a catalytic cycle, while the Co(III) complex and water are consumed. The overall process amounts to Eq. [20]:

$$Co(NH_3)_5Cl^{2+} + 4H^+ + \tfrac{1}{2} H_2O \longrightarrow$$

$$Co(aq)^{2+} + 5NH_4^{+} + Cl^- + \tfrac{1}{4} O_2$$ [20]

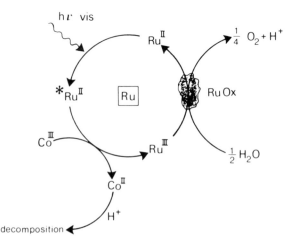

FIGURE 11. Photochemical oxygen generation in the $Ru(bipy)_3^{2+}/Co(NH_3)_5Cl^{2+}/RuO_x$ system.

Marked dependence of the rates on pH has been found, with a maximum at about pH 4.5 (89,96).

The exact mechanism of the oxidation step (Eq. [18]) is not known at present. One may consider that the Ru(III) complex charges the catalyst by injecting holes and oxidizing surface-bound hydroxyl groups. Analogous mechanisms have been proposed for electrolysis of water using RuO_x anodes (79).

D. Modified Oxygen–Generating Systems

Various modifications of the $Ru(bipy)_3^{2+}/Co(NH_3)_5Cl^{2+}/RuO_x$ system have been investigated. Other electron acceptors may be used, such as different Co(III) complexes and $TiCl_3$ (94). As in the case of hydrogen generation, the efficiency of the redox catalyst is of great importance. Pure, stoichiometric RuO_2 is ineffective, whereas the so-called "hydrated" powder variety is active. For this reason the catalyst is better designated as RuO_x. Polymer-supported RuO_x has been investigated (89,94), and a study of catalysts obtained by depositing metal oxides on solid supports (like γ-alumina, silica or Y zeolites) has revealed a number of interesting features (96). The most efficient materials are (RuO_x + IrO_x) materials deposited on Y zeolites at low temperature (200°C). This is especially true when the ratio of O_2-production rates to metal content is considered; the activity of this mixed catalyst is more than ten times higher than that of unsupported RuO_x. Zeolite-supported metal oxides of the first transition series, like manganese oxide (which is of special interest because of the presence of manganese ions in photosystem II of green plants; see Chapters 1 and 2), are also active, although much less so than supported Ru and Ir oxides (96). Catalysis by hydrolyzed Fe(II) and Co(II) cations has been studied (93). It should be mentioned that generation of oxygen in photoelectrochemical systems has made use of $Ru(bipy)_3^{3+}$ as the oxidant in the presence of an acceptor species (97–101). Photoreduction of manganese complexes in aqueous solution has also been reported (7,102,103).

IV. PHOTOCHEMICAL GENERATION OF HYDROGEN AND OXYGEN FROM WATER

A. Schemes for Simultaneous Generation of Hydrogen and Oxygen by Water Photolysis

The two types of systems described above can carry out separate photogeneration of either H_2 or O_2 with consumption

of an electron donor or acceptor, respectively, whose role is
to compete with recombination reactions. As such, these
processes represent the reductive and oxidative components of
a complete water-splitting system; they have allowed us to
discover suitable catalysts and to fullfill the thermodynamic
and mechanistic requirements for water reduction and oxidation
in a continuous catalytic process under irradiation with
visible light. In order to be able to photogenerate H_2 and O_2
simultaneously, conditions must be found in which the recombination reactions between charge-separated species are minimized in the absence of trapping materials, i.e., the kinetic
requirements must also be fullfilled.

Two main schemes may be considered, each of which presents
several variants: single-cell systems ("SCS") in which the
redox catalysts found to be efficient for separate evolution
of H_2 and O_2 are combined in a single aqueous phase; and
membrane-separated systems ("MSS") (see Fig. 1) in which the
two half-cells are separated by a membrane which is permeable
to electrons and protons and contain the respective components
required for separate production of H_2 and O_2.

In an electrochemical variant of the MSS, electrons are
led through an external circuit via electrodes and protons
permeate through the membrane (see also, Fig. 1 in ref. 89).
Schematic representations of an SCS and an MSS are shown in

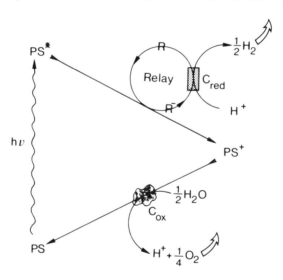

Redox catalysts

FIGURE 12. Schematic representation of a single-cell
system (SCS) for photochemical water splitting.

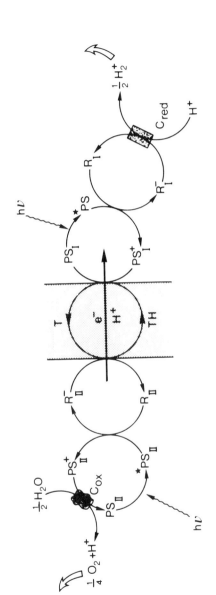

FIGURE 13. Schematic representation of a membrane separated system (MSS). Biphotonic water photolysis occurs in the coupled half-cells containing $PS_I/R_I/C_{red}$ in the reductive half-cell and $PS_{II}/R_{II}/C_{ox}$ in the oxidative half-cell; electron carrier T is in the membrane.

Figures 12 and 13, respectively. Other variants involve micellar systems (Chapter 5), microemulsions (6,104,105), and monolayer assemblies (106).

Using the redox catalysts developed previously, systems of both types may be set up. Realizing a SCS, requires that the water-reduction and -oxidation reactions compete successfully with recombination processes; this depends largely on the efficiency of the redox catalysts. The triple heterogeneity of the SCS (a liquid phase consisting of the aqueous solution, two solid or colloidal phases containing the catalysts, and a gas phase comprised of the H_2 and O_2 products) may help solve the kinetic problem. Studies directed towards the development of systems of the SCS or MSS type are being pursued. Simultaneous generation of H_2 and O_2 in an SCS set-up has been reported recently (94).

Another approach is to render the system irreversible by rapid physical separation (e.g., precipitation) of one of the components formed after charge separation. In this way a reduced solid material might accumulate, whereas oxygen would be evolved by the oxidized counterpart on contact with the C_{ox} catalyst. Preliminary experiments indicate that tungsten salts might be suitable materials (107). Accumulation of the reduced material as a solid would also be of practical interest with respect to ease of separation and storage.

Homogeneous redox catalysts should be considered in addition to the heterogeneous catalysts discussed previously. In Section I it was noted that dinuclear cryptates, polynuclear metal-cation complexes and metal clusters in general may be potential catalysts for the exchange of several electrons, as required for water dissociation according to Eqs. [2] and [3]. Such species are expected to possess many oxidation states and are especially worth investigating. Iron-sulfur clusters (50) and polyoxometalates (e.g., polytungstates and polymolybdates) (107) might also be suitable candidates. On the other hand, a dinuclear bis-Cu(II) cryptate (structure[12]-[12]) has been shown to function as a two-electron acceptor species (19-21), exchanging two electrons in

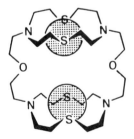

Structure [12]-[12]

a single wave by electrochemical reduction or oxidation (21).
A bis-Rh(III) complex might be able to exchange four electrons
by Rh(III)-Rh(I) interconversion.

With such ideas in mind, we have also synthesized a
special quaterpyridine, pQP, which is formed by two independent bipyridine subunits maintained in a twisted conformation
around the central linkage by two methyl groups in the ortho
and ortho' positions with respect to that bond (107). It was
expected that hindrance to planarity would prevent this "perpendicular" quaterpyridine from binding only a single metal
cation by using three pyridine rings in a terpyridine fashion
and leaving the fourth ring free. The twisted geometry:

should favor the formation of dinuclear complexes, the pQP
ligand behaving like two bipyridyl units. This is indeed the
case, and several dinuclear complexes with Mn(II), Co(II),
Ni(II), Cu(I), Ru(II) have been isolated and characterized;
complexes containing either two cations and one pQP, or two
cations and two pQP ligands have been obtained (107). Their
properties, especially with respect to electron exchange and
redox behavior, are being studied. Finally, in view of the
efficiency shown by the C_{red} and C_{ox} catalysts prepared on
solid supports (71,96), we became interested in the possibility of using the solid itself as the photoactive component for
deposition of the catalytic species; we were thus led to investigate semiconductors as supporting materials.

B.   Semiconductors as Photoactive Supports

Irradiation of a semiconductor (SC) with light of energy
equal to or higher than the bandgap leads to an electron-hole
pair $e^- - h^+$ by promotion of an electron from the valence band
to the conduction band:

$$SC \xrightarrow{h\nu} e^- + h^+ \qquad\qquad [21]$$

The electron and hole thus generated can be used to drive
chemical reactions in a non-spontaneous ($\Delta G > 0$) direction

provided: (a) that the bandgap is larger than the energy
required for the reaction;   (b) that the redox potentials of
the e^- and h^+ species are sufficient to induce the relevant
reduction and oxidation processes, respectively; and (c) that
the rates of these redox reactions are fast enough to compete
with electron-hole recombination.   These energetic relationships are illustrated in Figure 14.

Photocatalysis on semiconductors is of great interest for
energy conversion,   and numerous   photoelectrochemical systems
have been studied (for recent reviews see refs. 4,8,14-17,
108-113, and Chapter 10, this Volume).   In particular, photoassisted decomposition of water at semiconductor electrodes
(Fig. 14) has been the subject of active research in recent
years,   since the initial report on water photoelectrolysis at
TiO_2 electrodes by Fujishima and Honda in 1971 (114).   Figure
15 indicates the energetic relationships between the properties of several semiconductors and the redox potentials of the
water-reduction and -oxidation reactions, Eqs. [2] and [3].

Photoassisted   electrolysis   of   water   at   zero   applied
potential   has   also   been   demonstrated   using   monocrystalline
$SrTiO_3$ or $KTaO_3$ photoelectrodes   (115-118).   Photolysis   of
chemisorbed   water   on TiO_2 has   been   reported   (119),   but   its
catalytic   nature   has   been   questioned   (120).   Recent   papers
have reported water photolysis on illumination of a suspension
of $SrTiO_3$ powder in distilled water, with evolution rates of
about 0.7 µmole (~16 µl) of H_2 and 0.35 µmole (~8 µl) of O_2 in

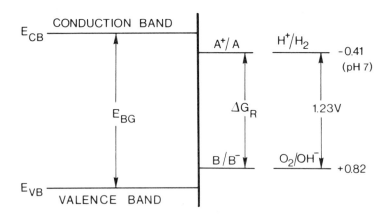

FIGURE 14.   Energetic relationships required for driving
the   reaction: $A^+ + B^- \rightarrow A + B$ by   irradiation   of   a   semiconductor. E_{CB} = conduction   band, E_{VB} = valence   band, E_{BG} =
bandgap energy of the semiconductor; A^+/A and B/B^- represent
the redox potentials and ΔG_R is the reaction free-energy.   The
case of water reduction and oxidation is given as an example.

12 hours (121), as well as photocatalytic H_2 production from water on metal-free $SrTiO_3$ crystals in alkali hydroxide solutions (122). Formation of H_2 and O_2 has been observed on irradiation of "photochemical diodes", which consist of n-type TiO_2/p-type GaP particles suspended in water (123).

Heterogeneous photosynthetic processes on semiconductors present major kinetic difficulties: the formation of stable products has to be faster than both electron-hole recombination and the reactions of reduced and oxidized species produced on the particle with each other. The search for highly efficient redox catalysts capable of making product formation competitive with these short-circuit processes is therefore of crucial importance for the development of photocatalytic systems. In particular, the presence of an efficient catalyst for water reduction (C_{red}) and/or water oxidation (C_{ox}) might sufficiently accelerate H_2- and/or O_2-formation from the electron-hole pair photogenerated in the semiconductor support to compete with recombination and back reactions:

$$2e^- + 2H^+ \xrightarrow{\text{C}_{red}} H_2 \qquad [22]$$

$$4h^+ + 2H_2O \xrightarrow{\text{C}_{ox}} O_2 + 4H^+ \qquad [23]$$

For our initial studies, $SrTiO_3$ appeared to be an appropriate material, since: (a) it is stable under the reaction conditions; (b) its bandgap energy (~3.2 eV) is larger than the energy required for water dissociation (>1.23 eV); and (c) the electron-hole pair formed on excitation has the required redox properties (i.e., flat-band potentials: $e^- \sim -0.6$ V and $h^+ \sim +2.6$ V vs. nhe at pH 7) to reduce and oxidize water, respectively, via reactions [2] and [3] (108-113) (Fig. 15). The light required for excitation lies in the UV range (absorption increases below the threshold at ~385 nm and levels off below ~330 nm), so that such a material responds to only a small fraction of the energy of the sun (~3%). However, it has been shown that optical-to-chemical energy conversion occurs with high efficiency (~30%) in photoelectrochemical cells using $SrTiO_3$ electrodes irradiated with monochromatic UV light (~330 nm) (116-124).

In view of their ability to reduce overpotentials in water electrolysis (78-86) and their use in photochemical systems (described above), deposition of ruthenium, rhodium and iridium as catalytic species on $SrTiO_3$ was investigated (125). Photodeposition of platinum metal on TiO_2 powder has been described recently (110,126). The decomposition of water (1N H_2SO_4) by UV light on platinized titanium dioxide has been claimed (127). H_2 and O_2 are reported (124) to be formed by irradiation of platinized monocrystalline $SrTiO_3$ and $KTaO_3$ electrodes in aqueous base. Deposition of small amounts of

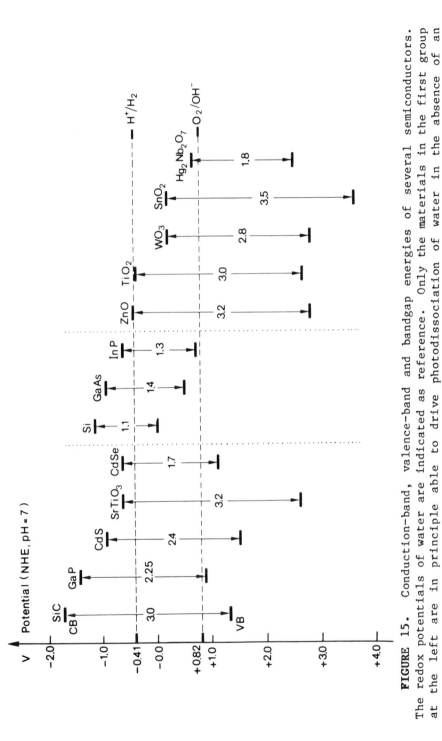

FIGURE 15. Conduction-band, valence-band and bandgap energies of several semiconductors. The redox potentials of water are indicated as reference. Only the materials in the first group at the left are in principle able to drive photodissociation of water in the absence of an applied external potential. (Values adapted from Table 1, ref. 111).

transition metals on p–GaP has been reported to catalyze H_2 evolution from water (128,129), and recent papers have described the photocatalytic decomposition of liquid water on platinized TiO_2 powder (130) and of water vapor on mixtures of TiO_2 and RuO_2 powders (131) and on $NiO–SrTiO_3$ powder (132).

C. Photogeneration of Hydrogen and Oxygen from Water on Metal–Loaded Strontium Titanate

A number of metal-loaded stontium titanate materials have been prepared by photodeposition from a solution of the salt on the suspended semiconductor powder (125). Irradiation of these suspensions in water with UV light ($\lambda > 300$ nm) leads to evolution of H_2 and O_2 at rates that depend on the nature of the deposit, but in all cases with efficiencies very much higher than for undoped $SrTiO_3$. By far the most active catalyst is $Rh/SrTiO_3$, which for instance yields 10 ml of gas ($H_2 + 1/2 \ O_2$) when 1.2 mg of material (corresponding to 2.2×10^{-8} moles of deposited rhodium) is suspended in 5.5 ml water and irradiated for 64 hours (125). This corresponds to turnover numbers of 90 and 2.7×10^4 on $SrTiO_3$ and Rh, respectively.

Figure 16 represents the relative efficiencies of the various $M/SrTiO_3$ materials assayed to date. One notes that on progressing from metal-free $SrTiO_3$ to $Rh/SrTiO_3$ the efficiency increases by a factor of about 10^4, without any attempt to optimize the catalytic material. A study of temperature and pressure effects indicates that the photocatalysis might also involve water vapor. The catalyst is entirely conserved, as far as can be judged from present results, and no decrease in activity was observed after 30 hours of irradiation.

The nature of the metallic deposit and the mechanism of catalysis are not known at present. The deposit may catalyze one or both of the H_2- and O_2-producing reactions, Eqs. [2] and [3], rendering evolution of at least one of the gases fast enough to compete with recombination (Fig. 17). Another possibility is that the deposited metals decrease the surface-recombination rates of electrons and holes; such an effect has been observed for ruthenium ions chemisorbed on n–GaAs but the deposits may be of a quite different nature (133).

Due to its large bandgap (3.2 eV), excitation of $SrTiO_3$ requires UV light. Sensitization of TiO_2 electrodes with Cr-dopant impurities has been shown to extend the spectral response to the red, improving by 10% the solar conversion efficiency and allowing photoelectrolytic generation of hydrogen and oxygen using visible light (111,134). Indeed, preliminary experiments on additional doping of the $Rh/SrTiO_3$ catalyst with Cr or Fe gave encouraging results regarding the

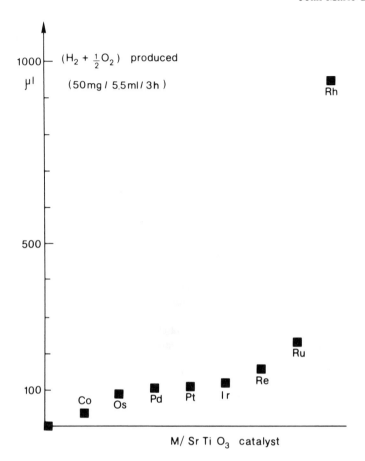

FIGURE 16. Relative efficiencies of M/SrTiO3 catalysts for water photolysis. The ordinate is the volume of gas (H_2 + 1/2 O_2) generated by irradiation of about 50 mg of M/SrTiO3 suspended in 5.5 cc of water for 3 hours with light of $\lambda > 300$ nm from a 1 kW lamp. The samples were kept under vacuum but were not thermostatted (125).

possibility of shifting the spectral response of the water-photolysis process towards the visible region of the spectrum.

The simplicity of such heterogeneous water photolysis makes it a very attractive photosynthetic process. Of course, much work remains to be done on optimization of the catalytic materials, on the nature and mechanism of the action of the catalytic deposit, and on extension of the spectral response into the visible by bulk- or surface-doping by dye sensitization (135-139) or by using semiconductors with smaller band gaps.

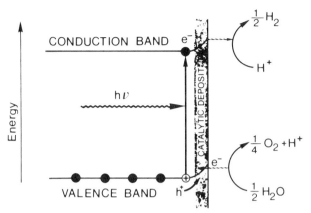

FIGURE 17.   Schematic representation of water photolysis on a semiconductor loaded with a catalytic deposit.

Another scheme may be imagined for shifting the photon-energy requirements for H_2 and O_2 generation into the visible region. Metal-loaded $SrTiO_3$ materials have been found to be very efficient catalysts for separate photoinduced generation of hydrogen (by $Pt/SrTiO_3$ or $Rh(3\%)/SrTiO_3$, Fig. 8) or oxygen (RuO_x or $IrO_x/SrTiO_3$), when used in the photosensitized systems discussed above (71,107). Such catalysts might, therefore, allow simultaneous generation of both hydrogen and oxygen from water under visible light irradiation of a PS/R/SC(Cat) system containing: a photosensitizer which absorbs in the visible range (e.g., $Ru(bipy)_3^{2+}$), a relay species of suitable reduction potential (like a rhodium complex or a viologen), and heterogeneous catalyst(s) (co)deposited on a semiconductor of appropriate flat-band potential ($E(R/R^-) < -0.41$ V). In this case, the light-absorbing component is the PS and not the semiconductor, which serves merely as mediator by electron injection into its conduction band. This mechanism might play a role in the Pt or $Rh/SrTiO_3$ experiments shown earlier in Figure 8. Figure 18 shows a representation of the operation of such a system. The reverse process, involving reduction of a relay species by electron ejection from a semiconductor, may also be envisaged. Such schemes have been considered in various systems containing semiconductor/dye combinations (see, for example, refs. 110,136-142 and literature cited therein).

Judiciously doped semiconductors may also be used for establishing photosynthetic or photocatalytic processes involving molecules other than water, e.g., CO_2, N_2 or X_2/XH couples (X = halogen). In fact, interesting results on the reduction of CO_2 using $Ru/SrTiO_3$ catalyst have already been

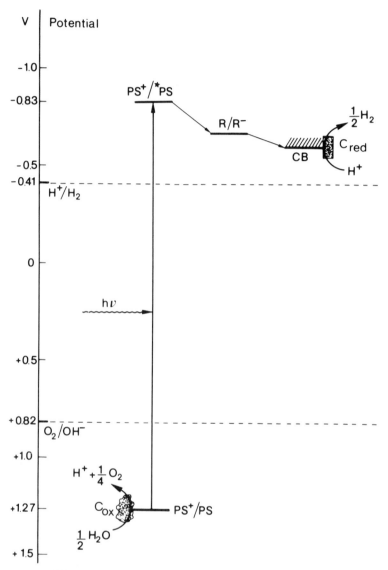

FIGURE 18. Schematic representation of a possible scheme
for simultaneous photogeneration of H_2 and O_2. The components $PS/R/C_{red}$ and C_{ox} are either deposited separately or co-
deposited on a semiconductor. The relative positions required
for the various redox potentials (pH 7) are as indicated. In
the case illustrated, the following materials are taken as an
example: $PS = Ru(bipy)_3^{2+}$, $R = Rh(bipy)_3^{3+}$ and $SC = SrTiO_3$.
Direct injection of an electron from the excited photosensitizer PS^* into the conduction band CB without intervention of
a relay species R may also be considered.

obtained (107). Doping single-crystal electrodes in a similar way may also be of interest for facilitating photoelectrochemical processes.

Finally, the development of homogeneous catalysts for multielectronic processes (water reduction and oxidation; reduction of oxygen, nitrogen or carbon dioxide by two-, four-, six- or even eight-electron processes) remains an open field. Polynuclear metal complexes and clusters are obvious candidates, as already pointed out. Possibly, the metal-loaded solid catalysts represent a crude state of future catalysts that might be obtained by combining the features of polynuclear complexes with those of solid-state materials used as supports.

V. CONCLUSION

Artificial photosynthesis is in its prime age. Because of the fundamental knowledge to be gained, the breadth of fields that it spans and the economic impact it may have, research in this area should enjoy a bright future indeed. In a rather short time, significant progress has been made in realizing systems capable of producing hydrogen and/or oxygen from water under irradiation by visible light. Notable improvements may be expected in the future, and are required for large scale use. Other light-driven reactions are also of interest. Whether and when sunlight-driven processes providing solar fuels will become of practical significance depends not only on chemistry, but on political, economic and environmental considerations as well.

ACKNOWLEDGMENTS

I wish to express my gratitude to my collaborators: J.-P. Sauvage, R. Ziessel and M. Kirch, who performed the work described in this Chapter. Financial support was provided by PIRDES, CNRS.

REFERENCES

1. R. J. Marcus, Science 123, 399 (1956).
2. G. Stein, Israel J. Chem. 14, 213 (1975).
3. V. Balzani, L. Moggi, M. F. Manfrin, F. Bolletta and M. Gleria, Science 189, 852 (1975).

4. M. D. Archer, J. Appl. Electrochem. $\underline{5}$, 17 (1975).
5. J. R. Bolton, J. Solid State Chem. $\underline{22}$, 3 (1977); Science $\underline{202}$, 705 (1978).
6. M. Calvin, Accts. Chem. Res. $\underline{11}$, 369 (1978).
7. G. Porter, Proc. R. Soc. (London) $\underline{A362}$, 281 (1978).
8. H. Gerischer, Ber. Bunsenges Phys. Chem. $\underline{80}$, 1046 (1976).
9. E. Schumacher, Chimia $\underline{32}$, 193 (1978).
10. G. Calzaferri, Chimia $\underline{32}$, 241 (1978).
11. T. Ohta and T. N. Veziroglu, Int. J. Hydrogen Energy $\underline{1}$, 255 (1976); P. Hagenmüller, La Recherche $\underline{8}$, 756 (1977).
12. A. Moradpour, L'actualité chimique, Feb. 1980, p. 7.
13. J. O'M. Bockris, Pure Appl. Chem. $\underline{47}$, 25 (1976).
14. J. R. Bolton (ed.),"Solar Power and Fuels", Academic Press, New York, 1977.
15. M. S. Wrighton, Chem. & Eng. News, Sept. 3, 1979, p. 30.
16. R. R. Hautala, R. B. King and C. Kutal, (eds.), "Solar Energy: Conversion and Storage", Humana Press, Clifton, New Jersey, 1979.
17. T. Ohta (ed.), "Solar-Hydrogen Energy Systems", Pergamon Press, Oxford, 1979.
18. J.-M. Lehn, Accts. Chem. Res. $\underline{11}$, 49 (1978).
19. J.-M. Lehn, S. H. Pine, E. I. Watanabe and A. K. Willard, J. Am. Chem. Soc. $\underline{99}$, 6766 (1977); A. H. Alberts, R. Annunziata and J.-M. Lehn, ibid., 99, 8502 (1977).
20. J.-M. Lehn, Pure Appl. Chem. $\underline{52}$, 2441 (1980).
21. J. P. Gisselbrecht, M. Gross, A. H. Alberts and J.-M. Lehn, Inorg. Chem. $\underline{19}$, 1386 (1980).
22. J.-M. Lehn, Pure Appl. Chem. $\underline{51}$, 979 (1979).
23. J. J. Grimaldi, S. Boileau and J.-M. Lehn, Nature $\underline{265}$, 229 (1977).
24. J. J. Grimaldi and J.-M. Lehn, J. Am. Chem. Soc. $\underline{101}$, 1333 (1979).
25. M. Kirch, J.-M. Lehn and J.-P. Sauvage, Helv. Chim. Acta $\underline{62}$, 1345 (1979), and references cited therein.
26. T. J. Meyer, Israel J. Chem. $\underline{15}$, 200 (1976/77); Accts. Chem. Res. $\underline{11}$, 94 (1978).
27. V. Balzani, F. Bolletta, M. T. Gandolfi and M. Maestri, Topics Current Chem. $\underline{75}$, 1 (1978).
28. N. Sutin, J. Photochem. $\underline{10}$, 19 (1979).
29. A. Picini and L. Marino, Z. Anorg. Allg. Chem. $\underline{32}$, 55 (1902).
30. D. E. Green and L. H. Stickland, Biochem. J. $\underline{28}$, 898 (1934).
31. L. D. Rampino and F. F. Nord, J. Am. Chem. Soc. $\underline{63}$, 2745 (1941); L. D. Rampino, K. E. Kavanaugh and F. F. Nord, Proc. Natl. Acad. Sci. USA $\underline{29}$, 246 (1943); W. P. Dunworth and F. F. Nord, Adv. Catalysis $\underline{6}$, 125 (1954).

32. S. G. Cohen, A. Parola and G. H. Parsons, Jr., Chem. Rev. 73, 141 (1973); S. G. Cohen and R. J. Baumgarten, J. Am. Chem. Soc. 87, 2996 (1965).
33. J.-M. Lehn and J.-P. Sauvage, Nouv. J. Chim. 1, 449 (1977).
34. F. Millich and G. Oster, J. Am. Chem. Soc. 81, 1357 (1959); J. S. Bellin, R. Alexander and R. D. Mahoney, Photochem. Photobiol. 17, 17 (1963).
35. L. Michaelis and E. S. Hill, J. Gen. Physiol. 16, 859 (1933).
36. B. V. Koriakin, T. S. Dzhabiev and A. E. Shilov, Dokl. Akad. Nauk S.S.S.R. 233, 620 (1977).
37. A. I. Krasna, in "Biological Solar Energy Conversion", (A. San Pietro and A. Mitsui, eds.), Academic Press, New York, 1977, pp. 53-60; Photochem. Photobiol. 29, 267 (1979); ibid., 31, 75 (1980); Enzyme Microb. Technol. 1, 165 (1979); T. Pow and A. I. Krasna, Archives Biochem. Biophys. 194, 413 (1979); M. M. Rosen and A. I. Krasna, Photochem. Photobiol. 31, 259 (1980).
38. A. Moradpour, E. Amouyal, P. Keller and H. Kagan, Nouv. J. Chim. 2, 547 (1978).
39. P. Keller, A. Moradpour, E. Amouyal and H. B. Kagan, Nouv. J. Chim. 4, 377 (1980).
40. K. Kalyanasundaram, J. Kiwi and M. Grätzel, Helv. Chim. Acta 61, 2720 (1978).
41. I. Okura and N. Kim-Thuan, J. Mol. Catalysis 5, 311 (1979).
42. O. Johansen, A. Launikonis, A. W.-H. Mau and W. H. F. Sasse, personal communication, (to be submitted).
43. I. Okura and N. Kim-Thuan, J. Molec. Catalysis 6, 227 (1979).
44. I. Okura and N. Kim-Thuan, J. Chem. Soc., Chem. Comm., 84 (1980); J. Chem. Research (S), 344 (1979).
45. K. Kalyanasundaram and M. Grätzel, Helv. Chim. Acta. 63, 478 (1980).
46. G. McLendon and D. S. Miller, J. Chem. Soc., Chem. Comm., 533 (1980).
47. Y. Harel and J. Manassen, J. Am. Chem. Soc. 99, 5817 (1977); Y. Harel, J. Manassen and H. Levanon, Photochem. Photobiol. 23, 337 (1976).
48. P. Keller, A. Moradpour, E. Amouyal and H. Kagan, J. Molec. Catalysis 7, 539 (1980).
49. E. Amouyal, B. Zidler, P. Keller and A. Moradpour, personal communication.
50. Y. Okuni and O. Yonemitsu, Chem. Lett., 959 (1980).
51. I. Okura, S. Nakamura, N. Kim-Thuan and K. I. Nakamura, J. Molec. Catalysis 6, 261 (1979).
52. K. Kalyanasundaram and M. Grätzel, J. Chem. Soc., Chem. Comm., 1137 (1979).

53. M. Gohn and N. Getoff, Z. Naturforsch. 34a, 1135 (1975).
54. B. Durham, W. J. Dressick and T. J. Meyer, J. Chem. Soc., Chem. Comm., 381 (1977).
55. J. Kiwi and M. Grätzel, Nature 282, 657 (1979); J. Am. Chem. Soc. 101, 7214 (1979).
56. P. Keller and A. Moradpour, J. Am. Chem. Soc. 102, 7193 (1980).
57. A Henglein, Angew. Chem. Int. Ed. 18, 418 (1979); J. Phys. Chem. 83, 2858 (1979), Ber. Bunsenges Phys. Chem. 84, 253 (1980).
58. D. Meisel, J. Am. Chem. Soc. 101, 6133 (1979); K. Kopple, D. Meyerstein and D. Meisel, J. Phys. Chem. 84, 870 (1980).
59. C. K. Grätzel and M. Grätzel, J. Am. Chem. Soc. 101, 7741 (1979).
60. G. M. Brown, B. S. Brunschwig, C. Creutz, J. F. Endicott and N. Sutin, J. Am. Chem. Soc. 101, 1298 (1979).
61. P. J. DeLaive, B. P. Sullivan, T. J. Meyer and D. G. Whitten, J. Am. Chem. Soc. 101, 4007 (1979).
62. R. Ballardini, A. Juris, G. Varani and V. Balzani, Nouv. J. Chim. 4, 563 (1980).
63. K. R. Mann, N. S. Lewis, V. M. Miskowski, D. K. Erwin, G. S. Hammond and H. B. Gray, J. Am. Chem. Soc. 99, 5525 (1977).
64. K. R. Mann and H. B. Gray, in "Inorganic Compounds with Unusual Properties", (R. B. King, ed.), Advances in Chemistry Series, No. 173, American Chemical Society, Washington, D.C., 1979, pp. 225-235; I. S. Sigal, K. R. Mann and H. B. Gray, J. Am. Chem. Soc. 102, 7252 (1980).
65. T. K. Foreman, C. Giannotti and D. G. Whitten, J. Am. Chem. Soc. 102, 1170 (1980).
66. G. M. Brown, S.-F. Chan, C. Creutz, H. A. Schwarz and N. Sutin, J. Am. Chem. Soc. 101, 7638 (1979); S.-F. Chan, M. Chou, C. Creutz, T. Matzubara and N. Sutin, ibid., 103, 369 (1981).
67. Q. G. Mulazzani, S. Emmi, M. Z. Hoffmann and M. Venturi, J. Am. Chem. Soc. 103, 3362 (1981).
68. H. D. Abruña, A. Y. Teng, G. J. Samuels and T. J. Meyer, J. Am. Chem. Soc. 101, 6745 (1979).
69. K. Kalyanasundaram, Nouv. J. Chim. 3, 511 (1979).
70. R. Ballardini, G. Varani and V. Balzani, J. Am. Chem. Soc. 102, 1719 (1980).
71. J.-M. Lehn, J.-P. Sauvage and R. Ziessel, reported at the "Atelier sur la Photochimie des Composés de Coordination", Paris, 19-20 June, 1980; Nouv. J. Chim. 5, 291 (1981).
72. M. Kaneko, J. Motoyoshi and A. Yamada, Nature 285, 468 (1980).

73. F. Blau, Monatsh. 19, 647 (1898); W. W. Brandt, F. P. Dwyer and E. C. Gyarfas, Chem. Rev. 54, 959 (1955); M. Anbar and J. Pecht, Trans. Faraday Soc. 4, 744 (1968).
74. G. Nord and O. Wernberg, J. Chem. Soc., Dalton, 866 (1972); ibid., 845 (1975).
75. V. Ya. Shafirovich, A. P. Moravskii, T. S. Dzhabiev and A. E. Shilov, Kinetika i Kataliz 18, 509 (1977).
76. C. Creutz and N. Sutin, Proc. Natl. Acad. Sci. USA 72, 2858 (1975).
77. H. D. Gafney and A. W. Adamson, J. Am. Chem. Soc. 94, 8238 (1972).
78. S. Trasatti and G. Buzzanca, J. Electroanal. Chem. 29, app. 1 (1971).
79. L. D. Burke, O. J. Murphy, J. F. O'Neill and S. Venkatesan, J. Chem. Soc., Faraday I, 73, 1659 (1977); L. D. Burke, O. J. Murphy and J. F. O'Neill, J. Electroanal. Chem. 81, 391 (1977).
80. D. R. Rolison, K. Kuo, M. Umana, D. Brundage and R. W. Murray, J. Electrochem. Soc. 126, 407 (1979).
81. M. H. Miles and M. A. Thomason, J. Electrochem. Soc. 123, 1459 (1976).
82. A. C. C. Tseung and S. Jasem, Electrochim. Acta 22, 31 (1977).
83. S. Gottesfeld and S. Srinivasan, J. Electroanal. Chem. 86, 89 (1978).
84. Y. Matsumoto and E. Sato, Electrochim. Acta. 24, 241 (1979).
85. L. D. Burke and E. J. M. O'Sullivan, J. Electroanal. Chem. 97, 123 (1979).
86. E. J. Frazer and R. Woods, J. Electroanal. Chem. 102, 127 (1979).
87. J. Kiwi and M. Grätzel, Angew. Chem. Int. Ed. 17, 860 (1978).
88. J. Kiwi and M. Grätzel, Chimia 33, 289 (1979).
89. K. Kalyanasundaram, O. Mićić, E. Pramauro and M. Grätzel, Helv. Chim. Acta. 62, 2432 (1979).
90. J.-M. Lehn, J.-P. Sauvage and R. Ziessel, Nouv. J. Chim. 3, 423 (1979).
91. V. V. Strelets, O. N. Efinnov and V. Ya. Shafirovich, Kinetika i Kataliz 18, 646 (1977).
92. V. Ya. Shafirovich and V. V. Strelets, Nouv. J. Chim. 2, 199 (1978).
93. V. Ya. Shafirovich, N. K. Khannanov and V. V. Strelets, Nouv. J. Chim. 4, 81 (1980).
94. K. Kalyanasundaram and M. Grätzel, Angew. Chem. Int. Ed. 701 (1979).
95. G. Navon and N. Sutin, Inorg. Chem. 13, 2159 (1974).
96. J.-M. Lehn, J.-P. Sauvage and R. Ziessel, Nouv. J. Chim. 4, 355 (1980).

97. C. T. Lin and N. Sutin, J. Phys. Chem. 80, 97 (1976).
98. M. Gleria and R. Memming, Z. Phys. Chem. N.F. 98, 903
     (1976); ibid.,101, 171 (1976); R. Memming, F. Schröppel
     and U. Bringmann, J. Electroanal. Chem. 100, 307 (1979).
99. S. O. Kobayashi, N. Furuta and O. Simamura, Chem. Lett.
     503 (1976).
100. D. P. Rillema, W. J. Dressick and T. J. Meyer, J. Chem.
     Soc., Chem. Comm., 247 (1980).
101. M. Neumann-Spallart, K. Kalyanasundaram, C. Grätzel and
     M. Grätzel, Helv. Chim. Acta. 63, 1111 (1980).
102. A. Harriman and G. Porter, J. Chem. Soc., Faraday II,
     75, 1543 (1979).
103. Y. Otsuji, K. Sawada, I. Orishita, Y. Taniguchi and
     K. Mizuno, Chem. Lett., 983 (1977).
104. W. E. Ford, J. W. Otvos and M. Calvin, Nature 274, 507
     (1978); I. Willner, W. E. Ford, J. W. Otvos and M.
     Calvin, ibid., 280, 823 (1979).
105. S. A. Alkaitis and M. Grätzel, J. Am. Chem. Soc. 98,
     3549 (1976); J. Kiwi and M. Grätzel, ibid., 100, 6314
     (1978).
106. H. Kuhn, J. Photochem. 10, 111 (1979); D. G. Whitten,
     Angew. Chem. Int. Ed. 18, 440 (1979).
107. J.-M. Lehn, J.-P. Sauvage and R. Ziessel, unpublished
     results.
108. J. O'M. Bockris and K. Uosaki, in "Solid State Chemistry
     of Energy Conversion and Storage", (J. B. Goodenough and
     M. S. Whittingham, eds.), Advances in Chemistry Series,
     No. 163, American Chemical Society, Washington, D.C.,
     1977, pp. 33-70.
109. K. Rajeshwar, P. Singh and J. Dubow, Electrochim. Acta.
     23, 1117 (1978).
110. A. J. Bard, J. Photochem. 10, 59 (1979); Science 207,
     139 (1980).
111. H. P. Maruska and A. K. Ghosh, Solar Energy 20, 443
     (1978), and references cited therein.
112. A. J. Nozik, Annu. Rev. Phys. Chem. 29, 189 (1978).
113. M. S. Wrighton, Accts. Chem. Res. 12, 303 (1979).
114. A. Fujishima and K. Honda, Bull. Chem. Soc. Japan 44,
     1148 (1971); Nature 238, 37 (1972).
115. T. Watanabe, A. Fujishima and K. Honda, Bull. Chem. Soc.
     Japan 49, 355 (1976).
116. M. S. Wrighton, A. B. Ellis, P. T. Wolczanski, D. L.
     Morse, H. B. Abrahamson and D. S. Ginley, J. Am. Chem.
     Soc. 98, 2774 (1976); A. B. Ellis, S. K. Kaiser and
     M. S. Wrighton, J. Phys. Chem. 80, 1325 (1976).
117. J. G. Mavroides, J. A. Kafalas and D. F. Kolesar, Appl.
     Phys. Lett. 28, 241 (1976).
118. R. D. Nasby and R. K. Quinn, Mater. Res. Bull. 11, 985
     (1976).

119. G. N. Schrauzer and T. D. Guth, J. Am. Chem. Soc. 99, 7189 (1977).
120. H. Van Damme and W. K. Hall, J. Am. Chem. Soc. 101, 4373 (1979).
121. H. Yoneyama, M. Koizumi and H. Tamura, Bull. Chem. Soc. Japan 52, 3449 (1979).
122. F. T. Wagner and G. A. Somorjai, Nature 285, 559 (1980).
123. A. J. Nozik, Appl. Phys. Lett. 30, 567 (1977).
124. M. S. Wrighton, P. T. Wolczanski and A. B. Ellis, J. Solid State Chem. 22, 17 (1977).
125. J.-M. Lehn, J.-P. Sauvage and R. Ziessel, Nouv. J. Chim. 4, 623 (1980).
126. B. Kraeutler and A. J. Bard., J. Am. Chem. Soc. 100, 4317 (1978).
127. A. V. Bulatov and M. L. Khidekel, Izvest. Akad. Nauk SSSR, Ser. Khim., 1902 (1976).
128. Y. Kakato, K. Abe and H. Tsubomura, Ber. Bunsenges. Phys. Chem. 80, 1002 (1976).
129. Y. Nakato, S. Tonomura and H. Tsubomura, Ber. Bunsenges. Phys. Chem. 80, 1289 (1976).
130. S. Sato and J. White, Chem. Phys. Lett. 72, 83 (1980).
131. T. Kawai and T. Sakata, Chem. Phys. Lett. 72, 87 (1980).
132. K. Domen, S. Naito, M. Soma, T. Onishi and K. Tamaru, J. Chem. Soc., Chem. Comm., 543 (1980).
133. R. J. Nelson, J. S. Williams, H. J. Leany, B. Miller, H. C. Casey, Jr., B. A. Parkinson and A. Heller, Appl. Phys. Lett. 36, 76 (1980).
134. A. K. Ghosh and H. P. Maruska, J. Electrochem. Soc. 124, 1516 (1977).
135. K. D. Snell and A. G. Keenan, Chem. Soc. Rev. 8, 259 (1979).
136. H. Gerischer and F. Willig, Topics Current Chem. 61, 31 (1976).
137. C. D. Jaeger, F.-R. F. Fan and A. J. Bard, J. Am. Chem. Soc. 102, 2592 (1980).
138. A. A. Krasnovskii and G. P. Brin, Dokl. Akad. Nauk SSSR 213, 1431 (1973).
139. T. Takizawa, T. Watanabe and K. Honda, J. Phys. Chem. 84, 51 (1980); A. Fujishima, I. Iwase and K. Honda, J. Am. Chem. Soc. 98, 1625 (1976).
140. F.-R. F. Fan, B. Reichman and A. J. Bard, J. Am. Chem. Soc. 102, 1488 (1980).

DISCUSSION

Dr. A. Mackor, TNO, Utrecht:
With respect to the questions raised by Prof. Lehn in the last part of his lecture on water photolysis using $SrTiO_3$, I think that we have some answers, as you can see in our Contributed Paper VIII-9. We show that you can sensitize $SrTiO_3$ single crystals by a thin film of $Ru(bipy)_3^{2+}$. Also we have improved the spectral response so that we are able to obtain photocurrents with visible light using bulk and surface doping with chromium. We have thereby extended the work of Maruska and Ghosh on $TiO_2:Cr_2O_3$. Secondly, I would like to comment on the stability of $SrTiO_3$ in acid and neutral solutions, where we find extensive photocorrosion using single crystals; I would expect this to take place under your neutral conditions too.

Prof. Lehn:
$SrTiO_3$ is known to be a stable material (see refs. 109, 112). Determination of Sr^{2+} by flame emission spectrometry after completion of our experiments did not indicate more free Sr^{2+} than expected on the basis of the solubility product of $SrTiO_3$ in water at neutral pH.

Dr. W. H. F. Sasse, CSIRO, Melbourne:
I would like to comment that we find that hydrogenation produces compound(s) which retard the evolution of hydrogen on Pt catalysts.

Prof. M. Grätzel, Ecole Polytechnique Fédérale, Lausanne:
I would like to respond to Dr. Sasse. Indeed H_2 reacts with reduced methyl viologen, not with the oxidized form. This reaction is suppressed at lower pH and by hydrophobicly substituted viologens. Secondly, I want to point out that in combined catalytic systems, it is necessary to have specific catalysts. High activity of each catalyst is only one prerequisite to achieve water decomposition. One must do a laser experiment, showing that the reaction goes on a microsecond time-scale and is specific with respect to the reduced relay. It is not as simple as putting two catalysts together.

Prof. Lehn:
There are two answers to that. With respect to the reaction of H_2 with methyl viologen, I think it has been known for a long time that MV^+ hydrogenates very readily but this can be inhibited by catalyzing the competing reaction, namely H_2 evolution. With respect to your second

point of putting two catalysts together, I completely
agree that one must know exactly the rate and specificity
of each catalyst. All I am saying is that the most active
catalysts we have for each component (i.e., H_2 and O_2
evolution) do not work together in the combined system.

Dr. A. Alberts, TNO, Utrecht:
Do you have any evidence whether $Co(NH_3)_5Cl^{2+}$ produces O_2
in the presence of RuO_2 but in the absence of $Ru(bipy)_3^{2+}$?

Prof. Lehn:
A typical control experiment in which the system:
$Co(NH_3)_5Cl^{+2}/RuO_x/(AcO^-/AcOH)$ buffer at pH 4.05 is irradiated with visible light (150 W; $\lambda > 400$ nm) for 20 hours
did not yield detectable (< 5 μl) amounts of oxygen. In
similar conditions with $Ru(bipy)_3^{2+}$ present, 2.2 ml O_2
were produced after 5 hours irradiation. On the other
hand, aquo-Co(III) is thermodynamically able to oxidize
water and may be expected to do so in the presence of RuO_x
without irradiation. We cannot exclude that when only
small amounts of O_2 are formed it may be dissolved in the
buffer and remain undetected in our VPC analysis. In
fact, since O_2 is about twice as soluble as N_2 in water,
solutions (or buffers) which have been left standing in
contact with air must be degassed before use.

Prof. Grätzel:
I would like to comment on what Dr. Alberts said concerning O_2 evolution from the reaction of $Ru(bipy)_3^{2+}$ with
$Co(NH_3)_5Cl^{2+}$. Dr. Infelta in our laboratory has established that O_2 evolution occurs even in the absence of
RuO_2 catalyst. The yield is 15-20%, stoichiometric with
respect to the amount of the cobalt complex. Apparently,
hydrolyzed Co^{2+} species can catalyze O_2 formation after
being produced in the photoredox process.

Dr. J. S. Connolly, Solar Energy Research Institute:
What are the net solar efficiencies of the processes studied by you and also by Prof. Grätzel? Also, what are the
optimum achievable efficiencies in these systems?

Prof. Lehn:
Norman Sutin and his collaborators have found the quantum
yield for hydrogen generation to be about 11% for the
$Ru(bipy)_3^{2+}/Ru(bipy)_3^{3+}/TEOA/Pt$ system. The quantum yield
for the formation of $Ru(bipy)_3^{3+}$ by irradiation of a
$Ru(bipy)_3^{2+}/Co(NH_3)_5Cl^{2+}$ solution was found to be 6.3% by
Gafney and Adamson. We have not yet tried to determine

the solar efficiencies for O_2 or $(H_2 + O_2)$ evolution in our systems.

Prof. Grätzel:
We now know—and I believe both Balzani and Sutin have examined the problem as well—that the initial reduction (quantum) yield is a maximum of 30% which, of course, severely restricts the solar efficiencies that can be realized. However, with porphyrins you can get to to 85%. In the case of our cyclic ruthenium system, we observed a quantum efficiency of about 3-5%.

Prof. J. R. Bolton, University of Western Ontario:
The thermodynamic potential for water splitting is 1.23 V. To this must be added the thermodynamic loss of ~0.37 V which means that a photon energy of at least 1.60 eV must be used for a four-photon system. This corresponds to an effective bandgap wavelength (λ_g) of ~775 nm. The optimum wavelength for conversion of solar energy to chemical energy is 840 nm for AM1.2 sunlight. Any "friction" in the system (i.e., overpotentials, back-reactions, etc.) will increase the loss and hence decrease the threshold wavelength further below the optimum. An eight-photon system will, on the other hand, allow for some friction and still allow the threshold to be kept near the optimum. Of course, an eight-photon system is more complex and it may be that the simpler four-photon systems of Grätzel and Lehn would be preferred, even with a lower possible yield. Nevertheless, it is interesting that photosynthesis, which drives a reaction with $E^o = 1.24$ V, i.e., almost the same as for water photolysis, utilizes an eight-photon system.

Prof. R. T. Ross, Ohio State University:
The efficiency of photoelectrolysis using four photons per O_2 molecule evolved can be as high as 25% with an over-voltage of 0.2 V. A lower overvoltage will permit a higher efficiency (up to 29%).

Dr. J. Gobrecht, Solar Energy Research Institute:
Will the proposed photochemical systems be more efficient than a conventional electrolysis cell coupled to a solid-state solar cell?

Prof. Lehn:
It may well be, especially since our system has potential cost advantages, at least over present day solid-state devices.

CHAPTER 7

PHOTOGALVANIC CELLS AND EFFECTS

Mary D. Archer

Department of Physical Chemistry
University of Cambridge
Cambridge, U.K.

M. Isabel C. Ferreira[1]

Ciencias Exactas e Tecnologia
Universidade do Minho
Braga, Portugal

I. INTRODUCTION

Photoelectrochemical devices for direct conversion of solar energy to electrical energy can be divided into three main categories: (a) those in which absorption of ultra-band-gap light by a semiconducting electrode or particle results in minority carrier injection into solution thereby driving a desirable Faradaic process at a thermodynamic underpotential; (b) those in which a homogeneous photochemical reaction, nearly always an electron transfer, yields electroactive products which can diffuse to, and react at, conventional electrodes; and (c) those in which light absorption in a pigmented micelle, membrane or coating yields charge separation in or across an insulating support medium.

We consider here only the second category of devices, which are generally termed photogalvanic cells. A photogalvanic effect may be defined as a change in the current-potential characteristic of an electrode/solution half-cell produced by photolysis of the solution. Such an effect is to be expected in any photochemical reaction which alters the

[1]Supported by NATO Research Grant No. 1742.

concentration of any component of the solution to which the electrode is responsive, and it can be created both by exergonic and by endergonic photochemical reactions. However, since the electrochemical cell reaction in a power-producing photogalvanic cell must be spontaneous and must represent net conversion of radiant energy to electrical energy, it follows that only an endergonic photochemical reaction may be made the basis of a photogalvanic cell. Furthermore, if the cell is to operate indefinitely without consuming materials, the cell reaction must recycle photochemical products to reactants.

Nearly all photogalvanic cells are based upon photochemical electron-transfer reactions, since these are relatively free from irreversible side reactions and yield products to which indifferent electrodes (i.e., electrodes which are simply a source of, or sink for, electrons) may respond.

For simplicity, let us suppose this reaction to be a one-electron transfer, as shown in Figure 1, although more complicated sequences may occur in practice. The photoredox process is represented by Eq.[1]:

$$A + Z \xrightarrow{h\nu} B + Y \qquad (\Delta G > 0) \tag{1}$$

and may be initiated by absorption of light by A, Z or a sensitizer. A/B and Y/Z are redox couples:

$$A + e^- \rightleftarrows B \tag{2}$$

$$Y + e^- \rightleftarrows Z \tag{3}$$

and the standard redox potential E^0 of A/B must be less than that of Y/Z. The permissible electrode reactions are represented by Eqs. [2] and [3], although only certain combinations

FIGURE 1. Schematic diagram of a typical photogalvanic cell.

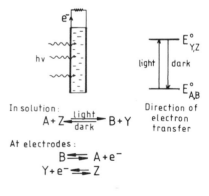

of these can produce power. The cell reaction should ideally be the reverse of [1]:

$$B + Y \xrightarrow{k} A + Z \qquad (\Delta G < 0) \qquad [4]$$

and should occur only at the electrodes. However, this is a spontaneous process, and it can and generally does occur also in homogeneous solution before B and Y can diffuse to, and react at, the electrodes; this homogeneous back reaction may greatly reduce the cell efficiency.

If the back reaction occurs sufficiently slowly to allow B and Y to be separated, or so slowly that they could coexist in appreciable concentrations in solution (the thermal back reaction between Fe^{3+} and I^- provides an example of this), then the photogalvanic cell would have useful storage capacity; i.e., radiant energy could be converted to chemical energy on recharge and chemical energy to electrical energy on discharge. Such devices have been described in the literature (1), although they are in fact often based upon an open cycle of chemical events in which an electron donor or acceptor is irreversibly consumed; this has sometimes been overlooked in data interpretation (2). These cells are sacrificial in the same sense, illustrated in Figure 2, as are some of the more recently developed noble-metal dispersions which mediate light-induced hydrogen and oxygen evolution from water, and where action originates in a light-driven electron-transfer reaction of the same type as that which occurs in a photogalvanic cell. Figure 2 compares the sacrificial use of EDTA in Eisenberg's old photogalvanic cell (3):

$$Pt|Proflavine,EDTA||K_2SnF_6,K_4SnF_6|Pt \qquad [5]$$

and in hydrogen-evolving systems, mediated by colloidal platinum, of the type described by Lehn, Grätzel and others (ref. 4; see also, Chapters 5 and 6, this Volume). The sacrifice of an electron donor (or acceptor) is unacceptable in both instances, and in both cases cyclic systems have been devised in which materials (other than water) are not consumed. These are compared in Figure 3.

The aim of the conventional photogalvanic cell is generation of electricity, and that of the dispersed system is water decomposition. The mass-transfer regime in the two cases is very different. A quantitative comparison of the extent to which the homogeneous back reaction can be suppressed in the two systems has yet to be made. It must, however, be admitted that spherical diffusion, across an effective diffusion layer of thickness on the order of the radius of the particle, to large numbers of small spherical microelectrodes appears to be

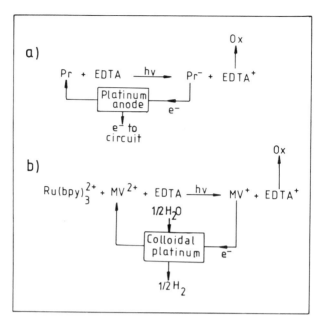

FIGURE 2. A comparison between sacrificial photoredox systems in (a) conventional photogalvanic cells and (b) micro-electrode systems. The electron donor, EDTA, is sacrificed in both cases, forming products, Ox, irreversibly; Pr = profla-vine, MV^{2+} = methyl viologen.

more promising in this respect than the semi-infinite linear diffusion to planar surfaces characteristic of the conventional photogalvanic cell.

II. CRITERIA FOR THE SUCCESSFUL PHOTOGALVANIC CELL

The characteristics required of an efficient photogalvanic cell have been examined in some detail (5,6) and will be summarized only briefly here.

Cell efficiency, expressed as the ratio of power output at the maximum power point to incident radiant power, depends upon several factors: viz., solution photochemistry and other homogeneous kinetics, mass transport and electrode kinetics. A thin layer ($\sim 10^{-2}$ cm) cell is preferable to a thicker one since the distance over which species must diffuse across the cell and the ohmic loss are minimized. In such a thin cell there will be little convection unless appreciable tempera-ture gradients exist. Also, a background electrolyte may be

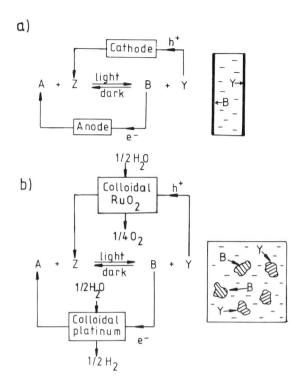

FIGURE 3. A comparison between non-sacrificial systems: (a) conventional photogalvanic cells and (b) microelectrode systems.

may be required to bring the specific conductance to a sufficient value, although if at least one ionic component is present at a concentration of ~0.1 M ($\kappa \simeq 10^{-2}$ ohm^{-1} cm^{-1}) this would not be necessary; in our analysis, we have assumed Y and Z to be ionic and present in excess. One electrode in the photogalvanic cell must be translucent to admit light, all of which should be absorbed near the front face of the cell.

The electrode kinetics of A/B and Y/Z play a crucial role in determining the maximum power available from a photogalvanic cell. However far to the right-hand side of Eq. [1] the photostationary state in the irradiated solution lies, the photopotential and the power developed will be zero if both electrodes are reversible to both couples, because the mixed potential adopted by the 'illuminated' electrode will be identical with the thermodynamically determined potential adopted by the 'dark' electrode (7). Some selectivity in electrode behavior is essential, and potentially useful cells

fall into one of two categories, as illustrated in Figure 4.
In the concentration cell (Fig. 4a) the electrodes are identical and both are reversible to one couple (A/B) and irreversible to the other (Y/Z). This cell yields power on irradiation by virtue of the differing concentration ratios of B to A
in the vicinities of the 'dark' and 'illuminated' electrodes.
However, the maximum power point is very close to the short-
circuit condition. One cannot draw significant currents from
a concentration cell and at the same time maintain very different concentrations at the two electrodes. We have estimated that the maximum power-conversion efficiency of a concentration cell is less than 0.2% (8).

A much better performance is in principle obtainable from
the differential cell illustrated in Figure 4b, in which the
illuminated electrode is still reversible to A/B and irreversible to Y/Z, but the dark electrode is either reversible to
Y/Z and irreversible to A/B or reversible to both couples. If
all the light is absorbed close to the illuminated electrode
these two possibilities produce identical performances.
Provided that certain criteria are met regarding homogeneous
and electrode kinetics, differential cells are capable of
delivering power at a voltage of $\Delta E^O \simeq |E^O(A/B) - E^O(Y/Z)|$
with a current collection efficiency of unity. The achievable
efficiency depends on the value assumed for ΔE^O and for E_g,
the threshold photon energy for reaction [1]. Assuming opti-
mistically that $\Delta E^O = 1.1$ V and $E_g = 1.8$ eV, we estimate that
solar conversion efficiencies of up to 18% are possible, provided that the following requirements are met:

(i) The light is absorbed by an unbleached solution in a
distance short compared with the reaction length of B, and
this length is itself smaller than the distance between the
electrodes.[2] A very soluble (~0.1 M) dye with a high extinction coefficient is required.

(ii) The quantum yield for charge transfer must be high.

(iii) The homogeneous back reaction, Eq. [4], must be
rather slow (k[Y] < 40 s^{-1}).

(iv) The standard rate constant for interconversion of A
and B at the illuminated selective electrode must be at least
0.1 cm s^{-1}, an almost impossibly high value. Likewise, the
standard rate constant for interconversion of Y and Z at this
electrode must be less than 10^{-12} cm s^{-1}, an almost impossibly
low value. However, it is possible that a dye-derivatized
electrode might have the required reactivity and selectivity.

---

[2]The reaction length of B is given by $(D/k[Y])^{1/2}$ where k
is the rate constant of reaction [4] and D is the diffusion
coefficient of B.

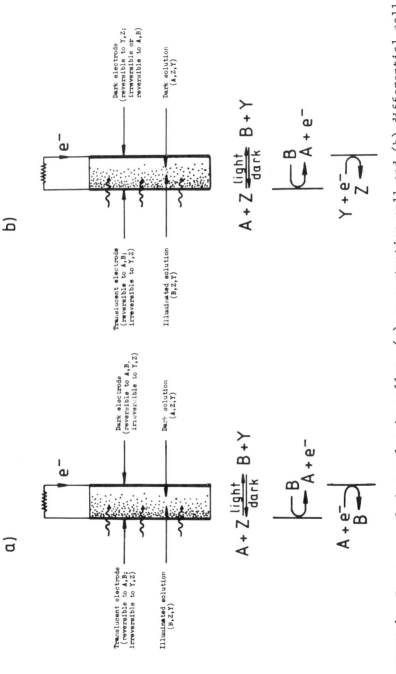

FIGURE 4. Two types of photogalvanic cells: (a) concentration cell and (b) differential cell.

(v) Finally, the standard rate constant for interconversion of Y and Z at the dark electrode must be at
least 5×10^{-3} cm s^{-1}. Also, Y and Z should be present in
excess to preclude an appreciable diffusion overpotential at
the dark electrode. The current–potential and current–voltage
characteristics in a cell that conforms to these criteria are
sketched in Figure 5.

A comment is in order regarding requirement (i) that all
the light be absorbed close to the illuminated electrode.
This arises from the requirement that the unstable B species
created near the electrode diffuse to it, rather than away to
the bulk solution leading to eventual destruction by homogeneous back reaction with Y. However, dye-solubility limitations

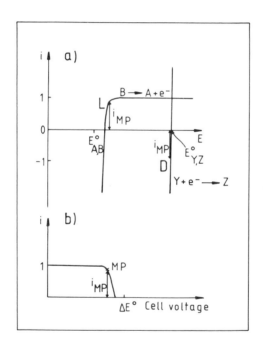

FIGURE 5. Characteristics of an ideal differential photogalvanic cell. In this example, the Y/Z couple is present in
considerable excess, so the dark electrode (D) is virtually
unpolarized. The degree of conversion of A to B near the
illuminated electrode (L) is small (i.e., the solution is not
bleached). ΔE^{o} has been taken as 1 V. The normalized current
density, i, is the actual current density divided by the flux
of incident photons of $E > E_{g}$. MP denotes the maximum power
point. Note how extremely irreversible the reduction of Y
must be at the illuminated electrode for the cell characteristic to be as shown in (b).

are such that it is more feasible to employ what Lichtin and co-workers have termed a totally illuminated-thin layer (TI-TL) cell in which the irradiance is virtually constant across the width of the cell. Several such cells would have to be stacked to absorb all the incident sunlight, thus leading to possible reflection losses. An efficient TI-TL cell would require two selective electrodes rather than the single one required for a differentially illuminated thin-layer (DI-TL) cell. This is because B and Y are generated throughout the cell volume, and will be destroyed on the 'wrong' electrode if selectivity does not operate. Moreover, the low dye concentration employed in a TI-TL cell increases the likelihood of bleaching with consequent reduction in current (5).

III.   RECENT WORK ON PHOTOGALVANIC CELLS

In this section relevant work published after mid-1977 is discussed, making reference where possible to the criteria listed above. Some of our own previously unpublished work is included.

A.   Homogeneous Processes and Composition of the Photostationary State

Given the mechanistic and kinetic data for all the thermal and photochemical reactions undergone in a photogalvanic cell, the composition of the photostationary state, and its dependence on irradiance and other parameters, can be calculated. Knowledge of this composition is required for prediction or interpretation of both photocurrents and photovoltages.

Calzaferri and Grüniger (9) have considered the position of the photostationary state in two systems: those in which one photon, at most, can produce one redox equivalent, such as the iron-thionine system, and those in which one photon can produce two redox equivalents, such as the Fe^{2+}-I_2 system; the latter is clearly capable of producing twice as much current as the former. Systems in which disproportionations such as that of semithionine occur have been considered by Albery et al. (10), who have shown that it is possible to deduce the reaction mechanisms operative in an irradiated iron-thionine solution from the composition of the photostationary state; below $[Fe^{3+}]$ = 10^{-3} M, the major route for semithionine decay is disproportionation, whereas above this concentration the undesirable back reaction with Fe^{3+} becomes dominant in acid aqueous sulfate media (11).

Because the disproportionation route is favored by high semithionine concentrations, the quantum yield of leuco- thionine is a function of irradiance as well as of solution composition, and thus it is not surprising that quantum yields of leucothionine reported by different groups are not in agreement. Schafer and Schmidt (12,13) studied ligand and solvent effects in the photoreactions of thionine and methyl- ene blue with Fe^{2+} and found that quantum yields of leucodye production in aqueous solution varied considerably with the nature of the ligand (NO_3^-, Cl^-, SO_4^{2-}, HPO_4^{2-}, acetate, phthalate). The highest yield they reported was 0.32 for thionine in 0.1 M $FeSO_4$ containing 0.05 M H_2SO_4 irradiated at 3.3×10^{15} quanta cm^{-2} s^{-1}. Higher yields are obtained in some mixed aqueous-organic systems, although the increase achievable for leucothionine is comparatively modest, a quantum yield of 0.35 being the maximum reported in a sulfate medium containing >40 vol % tetrahydrofuran. These effects are interpreted as due to complexation of Fe^{2+} ions by the anionic ligands, an effect which is enhanced by the presence of an organic solvent. The enhancement of the yield of the leucodye parallels the enhancement of the overall rate of quenching of the triplet by Fe^{2+}; a plausible interpretation is that electron-transfer quenching is facilitated by an anion bridge between the (positively charged) dye and Fe^{2+}. Albery et al. (14) reported lower quantum yields in a solution of 0.05 M H_2SO_4 containing 0.01 M Fe^{2+} and 6.0×10^{-6} M thio- nine; at an irradiance of 7.2×10^{15} quanta cm^{-2} s^{-1}, the quantum yield of leucothionine was found to be 0.18.

Complexation is also involved in the back reaction. Fe^{2+} and leucothionine complex rapidly and reversibly, so that pseudo first-order rate constants for leucothionine reoxida- tion tend slowly towards an asymptotic limit (of 0.45 s^{-1} in 0.1 M H_2SO_4, with 50 vol % acetonitrile). Values in the pres- ence of chloride ion are higher, which is undesirable; anions may be involved in these complexes as well (15,16).

Quenching of thionine fluorescence by Fe^{2+} has recently been reexamined (17). This quenching is a well-attested phe- nomenon (18) and, in the case of methylene blue, occurs mainly via electron transfer which generates the semireduced dye. Thus fluorescence quenching is not, in terms of the photo- galvanic cell, a loss-producing reaction. However, it has been reported (19) to be accompanied (for an excitation wave- length of 300 nm) by production of solvated electrons and of hydrogen atoms which attack the dye with a quantum yield of 1.4×10^{-5}. Such a side reaction would degrade a working photogalvanic cell in a matter of days, and would have to be avoided by filtering out light below 400 nm, where the second absorption band of the dye, which is reported (19) to be re- sponsible for this effect, commences.

B.   Studies on Electrode Selectivity

Electrode selectivity is now recognized as playing an important role in photogalvanic cells. Previously, it was encountered more by accident than design, as the case history of the platinum electrode illustrates.

Both thionine/leucothionine and Fe^{3+}/Fe^{2+} have rather fast electrode kinetics at a clean platinum electrode, and if this reversibility were maintained in a mixed system, the photogalvanic effect would be predicted to be zero. However, over a wide potential range, thionine adsorbs strongly on platinum pre-cleaned by a reductive technique (20). Cyclic voltammetry experiments, illustrated in Figure 6, as well as chronocoulometric and ring-disc studies, show approximately Langmuir behavior, and monolayer coverage is reached at bulk concentrations of thionine of $\sim 10^{-4}$ M. This adsorbed thionine layer considerably diminishes the rate of the Fe^{3+}/Fe^{2+} reaction at platinum, as illustrated in Figure 7, although the thionine/ leucothionine couple remains reversible.

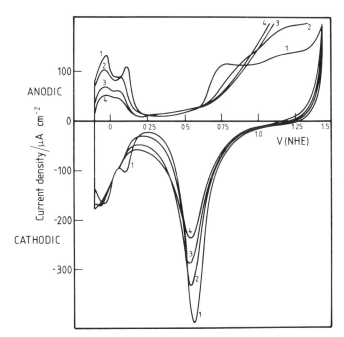

FIGURE 6. Cyclic voltammograms of spherical platinum microelectrodes. Solution contains 0.1 M K_2SO_4 (pH 2.5). Sweep rate 0.1 V s^{-1}, temperature 21°C. Thionine concentrations: (1) none, (2) 5.3×10^{-6} M, (3) 9.8×10^{-6} M, (4) 1.65×10^{-5} M. The thionine was adsorbed for 5 minutes and the curves were recorded after ~ 10 sweeps.

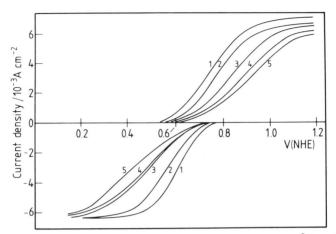

FIGURE 7. Current-potential curves for Fe^{3+}/Fe^{2+} at a reduced platinum rotating disk electrode in the presence of thionine. Upper curves, 10^{-2} M Fe $^{3+}$ and lower curves 10^{-2} M Fe^{2+}; rotation speed 50 Hz, room temperature. Thionine concentrations (all × 10^{-5} M): (1) 0, (2) 1.0, (3) 2.5, (4) 5.0, (5) 10.0 in 0.1 M K_2SO_4 (pH 2.5).

Thus the mixed potential exhibited by a platinum electrode in an irradiated iron-thionine solution lies closer to the Nernst potential of the thionine/leucothionine couple than of the Fe^{3+}/Fe^{2+} couple, although the selectivity is far from perfect (Fig. 8). The degree of bleaching, f, in this figure defines the composition of the photostationary state (pss):

$$f = [leuco]_{pss}/([thionine]_{pss} + [leuco]_{pss}) \qquad [6]$$

As an electrode, $n-SnO_2$ is also rather selective for thionine, although all redox processes tend to be slower at this quasi-metallic surface as compared with platinum. Photopotential and photocurrent data (21) both indicate that $n-SnO_2$ is more selective than platinum, which may be due not only to dye adsorption but also to the semiconductor band structure (22-24). Increasing the fraction of acetonitrile in the solvent somewhat diminishes the selectivity of $n-SnO_2$, possibly by decreasing the extent of thionine adsorption (22).

Greater selectivity can be engendered on platinum, gold and SnO_2, and possibly on other electrode materials as well, by coating the electrode with an anodically deposited thionine multilayer (6,25,26). It is not known how this coating is bound to the electrode surface; however, up to twenty monolayers can be durably attached, producing a modified electrode at which many inorganic redox reactions, including that of

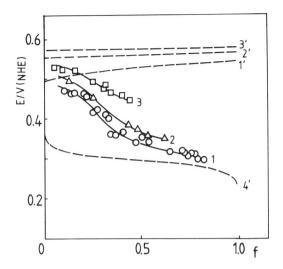

FIGURE 8. Open-circuit potential as a function of the extent of bleaching. The abscissa f is defined by Eq. [6]. Composition of the (dark) solution: thionine (3×10^{-5} M) + FeSO$_4$ (10^{-2} M) + the following Fe^{3+} concentrations: (1) 10^{-5} M (2) 1.1×10^{-4} M (3) 2.1×10^{-4} M. The dotted lines (1'-3') are the Nernst potentials for the Fe^{3+}/Fe^{2+} couple at the Fe^{3+} concentrations corresponding to (1-3); curve 4' is the Nernst potential for the thionine/leucothionine couple at the experimental pH (2.5).

Fe^{3+}/Fe^{2+}, are extremely sluggish, as illustrated in Figure 9, while the thionine/leucothionine process remains almost reversible for both bound and dissolved species. Thus the coated electrode should, and does, show better selectivity than the corresponding uncoated electrode. A comparison of dark- and photo-voltammetric curves obtained under identical conditions at uncoated and coated platinum is shown in Figure 10; oxidation of Fe^{2+} is almost entirely repressed on the coated electrode but thionine reduction is affected very little. The open-circuit potential of the coated electrode in the irradiated solution coincides quite closely with the calculated Nernst potential for the thionine/leucothionine couple in the photostationary state. Thus a coated electrode will function as the anode in a photogalvanic cell at a slightly lower potential than will an uncoated one, which is advantageous. The coated electrode will also produce a slightly higher photocurrent, as indicated in Figure 10, since reduction of Fe^{3+} is virtually absent throughout its working potential range.

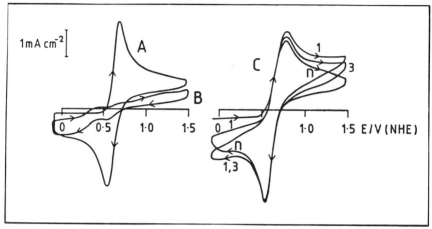

FIGURE 9. Cyclic voltammograms obtained with various types of Pt electrodes. All solutions contained 5×10^{-3} M $FeSO_4$ in 0.1 M K_2SO_4 at pH 2.5. (A) Reductively cleaned Pt, (B) Pt precoated with a thionine multilayer by the procedure described in ref. (20), and (C) anodized Pt. Sweep rate 0.1 V s^{-1}, temperature 25°C.

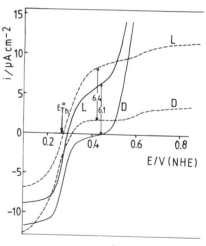

FIGURE 10. Current–potential curves for bare and thionine-coated Pt electrodes. Curve D represents dark and L illuminated solutions at stationary, bare (————) and thionine-coated (– – – –) Pt. The dark solution contained 3×10^{-5} M thionine, 10^{-2} M $FeSO_4$ and 0.1 M Na_2SO_4 at pH 2.5; f, the extent of bleaching (Eq. [5]), is 0.85 in the illuminated solution. The potential E_{Th} indicated by ↓ is the Nernst potential calculated for the thionine/leucothionine couple at this f and pH.

C. Studies of Iron–Thionine Half–Cell and Cell Performance

Cell output is a function of a rather large number of parameters. Although several studies relating to particular aspects have been carried out, it seems fair at present to say that no one has completed a systematic optimization. The difference, ΔE^o, between the redox potentials of the Fe^{3+}/Fe^{2+} and thionine/leucothionine couples is a function of pH. The former is virtually independent of pH below 3, where Fe^{3+} starts to hydrolyze, but the latter shifts cathodically by 90 mV per unit of pH increase. The observed cathodic shift in the onset of photocurrent at a platinum electrode (Fig. 11) follows this shift extremely closely, thus providing evidence that leucothionine, rather than semithionine, is the major electroactive species. At pH 1, ΔE^o is so small that a well-developed plateau in the photocurrent due to leucothionine oxidation is not observed. The optimal pH for an all-aqueous system is probably in the region of 2.5–3.5. Above pH 3.5, the potentials decline but the reason for this has not been established (27).

Half-cell and cell efficiency may be functions of the irradiance, I, for the following reasons:

(i) The concentration of leucothionine in the photosta-tionary state may not be proportional to I; other factors being equal, the functional dependence obtained depends upon the dominant reaction mechanism (10).

(ii) The solution in the vicinity of the illuminated electrode may become bleached at high I, leading to a drop in efficiency.

(iii) The reaction length of leucothionine will decrease with increasing I since Fe^{3+} in the photostationary state will

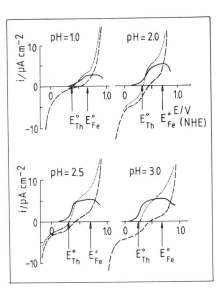

FIGURE 11. Current-potential behavior as function of pH at a stationary, bare platinum electrode. The solutions contained thionine (3×10^{-5} M) + $FeSO_4$ (0.01 M) in 0.1 M K_2SO_4 at room temperature. The dashed lines represent currents observed in the dark and the dotted lines the currents observed on irradiation with 400–800 nm light; the solid curves show the photocurrents. The arrows show the standard redox potentials of thionine and Fe^{3+} at the specified pH.

increase. This is likely to lead to a drop in collection
efficiency, but the effect will be negligible if there is a
substantial concentration of Fe^{3+} already present in the dark.

(iv) The open-circuit cell voltage is logarithmically
related to solution composition, and the voltage at maximum
power may therefore be a function of I.

As these effects operate in opposing directions, it seems
likely that cell efficiency could, in principle, be optimized
with respect to irradiance. However, for most solar applications the latter is God-given, so it may not be possible to
operate at such an optimum. Published data on output as a
function of irradiance are in agreement that the limiting
anodic photocurrents vary linearly with I (28,29), although we
find a somewhat nonlinear dependence (Fig. 12). The photocurrent must saturate eventually as I increases, since the
degree of conversion of thionine to leucothionine in the
photostationary state obviously cannot exceed 100%. At irrad-
iances characteristic of unfocussed sunlight, we conclude that
cell efficiency should be a weak function of irradiance, tend-
ing to drop as the irradiance decreases.

Addition to a photogalvanic solution of ions such as
fluoride, citrate and 2-aminopropionate, which complex more
strongly with Fe^{3+} than with Fe^{2+}, will exert an influence on
cell output for the following reasons:

(i) The standard potential of the Fe^{3+}/Fe^{2+} couple will
be shifted toward more negative values. This will decrease
the output of an ideal differential cell by decreasing its
open-circuit voltage, and it will also lower the voltage of an
imperfectly selective concentration cell.

(ii) The standard rate constant of the Fe^{3+}/Fe^{2+} redox
reaction at the electrodes will be altered. An increased
value may be helpful at the dark electrode of a differential
cell, but will be unhelpful in all other cases.

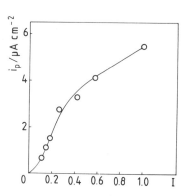

FIGURE 12. Photocurrent density
at a stationary semitransparent bare
Pt electrode as a function of relative light intensity. The composition of the solution and other conditions are as listed for Figure 11.

(iii) Homogeneous kinetics will be affected. A slowing of the reactions between leucothionine or semithionine and Fe^{3+} is both likely and desirable, since if it occurs, the concentration of leucothionine in the photostationary state and its reaction length will be increased, and hence photocurrents will be increased.

A recently published patent specification (30) shows that the output of iron-thionine concentration cells with platinum electrodes is considerably enhanced (by a factor of up to 500) by the addition of complexants for Fe^{3+}. The relative roles of the three effects summarized above were not examined, but both cell-voltages and cell-currents were very considerably enhanced, particularly in systems containing added Fe^{3+}, so effects (ii) and (iii) must outweigh (i). The best cell output under solar intensities is still very low; at an irradiance of 18.2 mW cm^{-2} (for wavelengths between 510 and 610 nm), an intensity which is slightly higher than that of bright sunlight, the output was reported to be 0.54 µW cm^{-2}, corresponding to a conversion efficiency in this spectral range of 3×10^{-3}%.

Figure 13 shows the effect of a modest addition of fluoride on dark- and photo-voltammograms at bare and thionine-coated platinum. The photocurrent wave obtained from leucothionine oxidation is more clearly defined on the coated electrode than on the bare electrode, and the anodic photocurrent at E < 0.25 V, due to the diminution of thionine in the illuminated solution, is also more clearly observable. These effects are due to the virtually complete suppression of Fe^{3+} reduction at the coated electrode. The addition of fluoride does not affect the form of the photocurrent greatly (except at E > 0.5 V on a bare electrode, at which Fe^{2+} oxidation is apparently hindered by fluoride), but has simply scaled up the values about two-fold. This we ascribe to the increased reaction length of leucothionine that results from the slowing of the back reaction with Fe^{3+}.

It thus appears that addition of a complexant for Fe^{3+} should be helpful. However, effect (i), described above, must be borne in mind. The formation constant for FeF$_3$ is approximately 10^{12} M^{-3}, and the formal redox potential of the FeF$_3$/Fe(aq)$^{2+}$ couple is thus only 0.06 V (nhe). Such a low value makes the dark reduction of thionine by Fe^{3+} in the presence of high fluoride concentrations spontaneous and renders the system incapable of photogalvanic power conversion.

Shigehara et al. (31,32) have studied thin layer differential cells of the type:

SnO$_2$|polymer-bound thionine|Pt   [7]

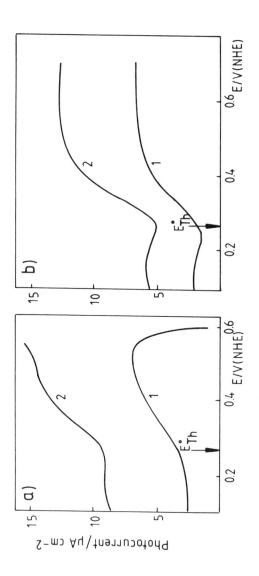

FIGURE 13. Photocurrent-potential curves at: (a) bare Pt, and (b) thionine-coated Pt electrodes. The dark solutions contained thionine (3×10^{-5} M) + $FeSO_4$ (0.01 M) in 0.1 M Na_2SO_4 at pH 2.5 and room temperature. In the illuminated solutions, $f = 0.85$; (1) no added fluoride, (2) 4.2×10^{-4} M NaF. The potential indicated by the arrows (E_{Th}) is the Nernst potential calculated for the thionine/leucothionine couple at this f and pH.

The intent is to slow the back reaction by electrostatic repulsion between the cationic polymer and Fe^{2+}. Thionine concentrations can be considerably higher in a polymer-bound or gel system, but even at the highest concentration (10^{-3} M) employed, the output is less than in an all-aqueous system.

Kamat et al. (33) have studied two-compartment cells in which the dark compartment contains a higher ratio of $[Fe^{3+}]$ to $[Fe^{2+}]$ than does the illuminated compartment. As one would predict from raising the cathode potential and current, the output is increased. However, these cells are no longer properly regenerative but will exhibit a dark voltage and will discharge until the Fe^{3+}/Fe^{2+} concentration ratio is the same in the two compartments.

IV.   CAN THE IRON-THIONINE CELL BE IMPROVED?

The sunlight-engineering efficiency of 0.03% reported by Lichtin and coworkers (34) for a single thin-layer iron-thionine cell with 1:1 (by volume) water-acetonitrile solvent remains the highest on record. None of the experiments described in this paper would lead directly to cells of improved performance because they have been carried out with thionine concentrations so low ($<10^{-4}$ M) that the absorbance length of the solution is much greater than the reaction length of leucothionine, and thus the current-collection efficiency is very poor. Thionine concentrations on the order of 10^{-4} M lead to limiting leucothionine oxidation currents on the order of 10 µA cm^{-2}. Even if the reaction length could be significantly increased, such low thionine concentrations would still give rise to unacceptably low photocurrents. The value reported by Lichtin's group (34) was for a thionine concentration of 10^{-3} M, which produced a short-circuit photocurrent of ~200 µA cm^{-2}, an open-circuit voltage of 0.15 V and a fill factor of 0.33.

The question to be asked is whether present capabilities would allow us to construct a more efficient cell. Possible improvements can be considered under two headings: improved voltage efficiency and improved current collection efficiency. Voltage efficiency could be somewhat improved by using a thionine-coated anode. However, a major improvement can result only from finding a selective Fe^{3+}/Fe^{2+} electrode and working with higher Fe^{3+} concentrations in the dark. Derivatization of the cathode with a ligand that raises the redox potential of Fe^{3+}, such as o-phenanthroline, would be attractive if it not only produced greater selectivity but also raised ΔE^{o}. The latter could also be raised by working in more alkaline media, provided Fe^{2+} were complexed to avoid

precipitation.   It is difficult to envisage ΔE^o values and
open-circuit photovoltages in excess of 0.5 V resulting from
all such measures.   The fill factor, which in present generation cells is lowered significantly by the cathode, would also
be improved, perhaps two-fold, by these measures.

Current-collection efficiency can be improved towards the
value of unity only by raising the dye concentration considerably, to about 0.1 M (5).   This improvement would be contingent upon finding conditions in which dye self-quenching
did not lead to a severe fall-off in the charge-transfer
quantum yield and upon minimizing the rate constant of the
back reaction.   Self-quenching is a ubiquitous phenomenon in
concentrated, homogeneous media.   For instance, it has been
suggested in the case of chlorophyll that the dye molecules
must be kept apart by more than 1 nm if the phenomenon is to
be avoided (35).   Incorporation of bulkier substituent groups
might assist this separation, and this has also been proposed
as a way of lowering the rate of the back reaction (36);
however, the forward electron-transfer rate might also be
reduced by the presence of such bulky substituents.   The back-
reaction rate can be lowered by decreasing the temperature,
and recent results suggest that this might improve cell performance (37).   However, this is not a practical course to
take in a working photogalvanic cell.

An improvement of ten-fold in the current collection efficiency is, in our view, an optimistic projection.   With the
projected improvements in the voltage efficiency and fill
factor, a cell with ~2.5% efficiency would be achieved.   This
is almost certainly below the value required for cost-
effective operation.

V.   STUDIES ON OTHER SYSTEMS

Baumann et al. (38) have published an interesting study of
the iron-iodine system, in which the overall solution
chemistry is:

$$2Fe^{2+} + I_3 \underset{dark}{\overset{light}{\rightleftarrows}} 2Fe^{3+} + 3I^- \qquad [8]$$

Although ΔE^o is rather low (formally 0.236 V), selective electrodes for each couple (n-SnO_2 for Fe^{3+}/Fe^{2+} and glassy carbon
for I_3^-/I^-) exist, and substantial photovoltages (up to 0.1 V)
are observed on irradiation.   The back reaction is so slow
that the photostationary state takes several hours to decay.
Short-circuit photocurrents on the order of 400 $\mu A\ cm^{-2}$
result from the much higher product concentrations in the

photostationary state obtained in these solutions. Observed monochromatic conversion efficiencies at 447 nm are on the order of $5 \times 10^{-3}\%$; if the estimated (38) solar conversion efficiency of 0.3% is correct, then this represents the best photogalvanic system investigated to date.

The $Ru(bipy)_3^{2+}/Fe^{3+}$ system (39) provides a contrast in that the ΔE^o is fairly high (0.47 V) and the rate of the back reaction between $Ru(bipy)_3^{3+}$ and Fe^{2+} is fast (k $\simeq 5 \times 10^6$ $M^{-1}s^{-1}$). The current-collection efficiency is low ($\sim 6 \times 10^{-4}$) and is determined by the reaction length of $Ru(bipy)_3^{3+}$. ΔE^o for the process:

$$6H^+ + Ru(bipy)_3^{2+} + Co(C_2O_4)_3^{3-} \xrightarrow{\text{light}}$$ [9]

$$Ru(bipy)_3^{3+} + Co^{2+} + 3H_2C_2O_4$$

is even higher (~ 0.72 V), and the observed photopotentials follow the Ru couple because of the irreversibility of the Co electrode kinetics (40). However, this system is unsuitable for use in a photogalvanic cell because decomposition of $Co(C_2O_4)_3^{4-}$ is probably irreversible and oxalic acid is oxidized by $Ru(bipy)_3^{3+}$.

Rhodamine B is subject to photochemical reduction by hydroquinone and gives rise to anodic photocurrents at $n-SnO_2$ electrodes, at which the quinone/hydroquinone couple is fairly irreversible (41). The maximum power output reported for a differentially illuminated cell containing an $n-SnO_2$ anode and a gold cathode irradiated by white light (150 W) is 1.3×10^{-8} W cm^{-2} (42). The power-conversion efficiency of the Au|rhodamine B,Fe(III)|Au photogalvanic cell is also very low (2.4×10^{-6} % at 528 nm) (43). Rapid back reactions or irreversible side reactions of the oxidized and reduced dye are presumably responsible for these low efficiencies. The same is true of photogalvanic systems based on triphenylmethane dyes (44,45). A dual-purpose cell for simultaneous production of hydrogen and electricity in which the excited state of $Ru(bipy)_3^{2+}$ is oxidatively quenched by methyl viologen is also based upon irreversible consumption of $Ru(bipy)_3^{3+}$ by a sacrificial donor to prevent back electron transfer from reduced methyl viologen (ref. 46; see also, Chapters 5 and 6, this Volume).

VI. THE OUTLOOK FOR PHOTOGALVANIC CELLS

Despite several years of research on photogalvanic cells and the increased understanding that has resulted, the over 100-fold gap between achievement and requirement remains a daunting one. The criteria for efficient photogalvanic cells are in many ways analogous to those for photovoltaic cells, their natural competitors. In a photovoltaic cell, light should be absorbed in or close to the space-charge region, and charge-carrier lifetimes must be long enough for holes and electrons to migrate out of that region to the electrodes, which are loss-less ohmic contacts. In a photogalvanic cell, light should be absorbed within the reaction length of the product B and its lifetime must be long enough to allow B to diffuse to the illuminated electrode. However, the diffusion coefficients of ions and molecules in solution are on the order of 10^{-6} to 10^{-5} cm^2 s^{-1}, whereas those of charge carriers in semiconductors are typically 1 to 100 cm^2 s^{-1}. Consequently, the constraint on acceptable back-reaction rates for B and Y in a photogalvanic cell (Eq. [1]) is much more irksome than that on bulk hole-electron recombination rates in a semiconductor. Moreover, while ohmic contacts to a semiconductor are readily achieved, the electrochemical analog, which is reversible behavior of the appropriate redox couple, is not. The required electrode selectivity, arising from the presence of B and Y throughout the cell volume, is a further troublesome requirement in a photogalvanic cell. In contrast, the process of selection in a photovoltaic cell is carried out by the junction region.

Two suggestions for improving the performance of photogalvanic cells deserve mention. Daul et al. (47) have pointed out that the transient response is often much larger than the steady-state response because depletion layers take time to form, and have suggested that an ac-operated cell, in which each electrode is illuminated in turn with an alternation rate of ~200 Hz, would have improved performance. Setting aside the problem of the added equipment and parasitic power required to actuate such light-chopping, the fact remains that this device would be suitable for use only with a concentration cell, which is intrinsically inefficient.

Tien et al. (48,49) have investigated what they have termed a "photogalvanovoltaic" cell, in which the cathode of a conventional iron-thionine cell is replaced by glassy carbon coated with adsorbed Mg(II) tetraphenylporphin. The latter behaves like a p-type semiconductor and adds a photovoltage of nearly 200 mV to the cell response. This suggests the possibility of combining a regenerative photoelectrochemical cell, containing one or two conventional semiconductor electrodes, immersed in a photogalvanic solution, thereby creating a tandem device. For the iron-thionine system, as illustrated in Figure 14, the energetic requirements for usefully additive effects are rather exacting. The conduction-band edge of the p-semiconductor should lie above the iron redox level but below the thionine redox level, while the valence-band edge of

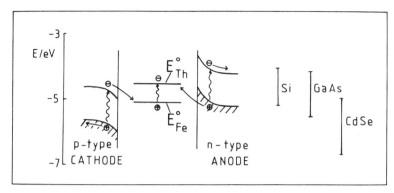

FIGURE 14. A hypothetical photogalvanovoltaic cell using the iron-thionine system and various n-type semiconductors (see text).

the n-type semiconductor should lie implausibly high—above the iron level although below the thionine level. If a suitable p-type cathode could be combined with a thionine-coated selective anode in a cell in which both the electrodes and the solution were irradiated, improved performance should result.

Neither conventional photogalvanic cells nor photogalvanovoltaic cells can be efficient if the criteria listed earlier are not met; in conclusion, we give our opinion of the likelihood of doing so. Investigations of electrode selectivity and methods of achieving it are in their infancy, which allows us to hope that improved electrodes will be found. Synthesis of photogalvanic dyes for which photochemical electron transfer occurs with high efficiency in a concentrated solution remains an unsolved problem. However, even if a cell of several percent efficiency were to be developed, degradation resulting from oxygen ingress, photochemical side-reactions and electrode aging remains a disturbing probability. Perhaps the most damning aspect is that a successful photogalvanic cell will not meet an unsatisfied need, but will be in competition with solid-state photovoltaic cells, which transduce solar to electrical energy with proven reliability and performance (Chapter 9). All in all, it must be admitted that photogalvanic cells, like Thursday's child, have far to go if they are to develop into cost-effective devices for conversion of solar energy.

ACKNOWLEDGEMENT

We thank Professor W. J. Albery for helpful discussions.

REFERENCES

1. Some examples of photogalvanic cells of this type are discussed in M. D. Archer, Specialist Periodical Report on Photochemistry, The Chemical Society 6, 739 (1975); M. D. Archer, J. Appl. Electrochem. 5, 17 (1975).

2. H. Tsubomura, Y. Shimoura and S. Fujiwara, J. Phys. Chem. 83, 2103 (1979).

3. M. Eisenberg and H. P. Silverman, Electrochim. Acta 5, 1 (1961).

4. J. Kiwi and M. Grätzel, Nature 281, 657 (1979), and references cited therein.

5. W. J. Albery and M. D. Archer, Nature 270, 399, (1977), and references cited therein.

6. W. J. Albery and A. W. Foulds, J. Photochem. 10, 41 (1979).

7. W. J. Albery and M. D. Archer, Electrochim. Acta 21, 1155 (1976).

8. W. J. Albery and M. D. Archer, J. Electroanal. Chem. 86, 1 (1978).

9. G. Calzaferri and H.-R. Grüniger, Z. Naturforsch. 32A, 1036 (1977).

10. W. J. Albery, W. R. Bowen and M. D. Archer, J. Photochem. 11, 15 (1979).

11. W. J. Albery, W. R. Bowen, M. D. Archer and M. I. Ferreira, J. Photochem. 11, 27 (1979).

12. H. Schafer and W. Schmidt, Z. physik. Chem. (Leipzig) 260, 817 (1979); ibid., 260, 890 (1979).

13. H. Schafer, R. Stahn and W. Schmidt, Z. physik. Chem. (Leipzig) 260, 862 (1979).

14. W. J. Albery, P. N. Bartlett, W. R. Bowen, F. S. Fisher and A. W. Foulds, J. Electroanal. Chem. 107, 23 (1980).

15. M. Z. Hoffman and N. N. Lichtin, in "Solar Energy: Chemical Conversion and Storage" (R. R. Hautala, R. B. King and C. Kutal, eds.), Humana Press, Clifton, N.J., 1979, pp. 153-187.

16. T. L. Osif, N. N. Lichtin and M. Z. Hoffman, J. Phys. Chem. 82, 1778 (1978).

17. S. N. Guha, P. N. Moorthy and K. N. Ras, Mol. Photochem. 9, 183 (1979).

18. M. D. Archer, M. I. C. Ferreira, G. Porter and C. J. Tredwell, Nouv. J. Chim. 1, 9 (1977).

19. S. Solar and N. Getoff, Int. J. Hydrogen Energy 4, 403 (1979).

20. M. D. Archer, M. I. C. Ferreira, W. J. Albery and A. R. Hillman, J. Electroanal. Chem. 111, 295 (1980).

21. Y. Suda, Y. Shimoura, T. Sakata and H. Tsubomura, J. Phys. Chem. 82, 268 (1978).

22. D. E. Hall, P. D. Wildes and N. N. Lichtin, J. Electrochem. Soc. 125, 1365 (1978).
23. P. D. Wildes and N. N. Lichtin, J. Am. Chem. Soc. 100, 6568 (1978).
24. N. N. Lichtin, P. D. Wildes, T. L. Osif and D. E. Hall, in "Inorganic Compounds with Unusual Properties", (R. B. King, ed.) Advances in Chemistry Series, No. 173, American Chemical Society, Washington, D.C., 1979, pp. 296-306.
25. W. J. Albery, A. W. Foulds, K. J. Hall, A. R. Hillman, R. G. Egdell and A. F. Orchard, Nature 282, 793 (1979).
26. W. J. Albery, A. W. Foulds, K. J. Hall and A. R. Hillman, J. Electrochem. Soc. 127, 654 (1980).
27. H. T. Tien and J. M. Mountz, J. Electrochem. Soc. 125, 885 (1978).
28. M. Wyart-Remy, A. Kirsch-De Mesmaeker and J. Nasielski, Nouv. J. Chim. 3, 304 (1979).
29. T. Sakata, Y. Suda, J. Tanaka and H. Tsubomura, J. Phys. Chem. 81, 537 (1977).
30. J. Cornelia, M. Brokken-Zijp and F. J. Reinders, British Patent No. 1,557,627; filed May 27, 1977, complete specifications published Dec. 12, 1979.
31. K. Shigehara, M. Nishimura and E. Tsuchida, Electrochim. Acta 23, 855 (1978).
32. K. Shigehara, H. Sano and E. Tsuchida, Macromol. Chem. 179, 1531 (1978).
33. P. V. Kamat, M. D. Karkhanavala and P. N. Moorthy, Ind. J. Chem. 18A, 210 (1979).
34. D. E. Hall, W. D. K. Clark, J. A. Eckert, N. N. Lichtin and P. D. Wildes, Bull. Amer. Ceram. Soc. 56, 408 (1977).
35. G. Porter, Proc. R. Soc. (London) A362, 281 (1978).
36. D. G. Whitten, Accts. Chem. Res. 13, 83 (1980).
37. P. V. Kamat, M. D. Karkhanavala and P. N. Moorthy, Ind. J. Chem. 18A, 206 (1979); Solar Energy 20, 173 (1978).
38. J. Baumann, H.-R. Grüniger and G. Calzaferri, Z. physik, Chem. 118, 11 (1979).
39. W. J. Albery, W. R. Bowen, F. S. Fisher and A. D. Turner, J. Electroanal. Chem. 107, 11 (1980).
40. B. Durham and T. J. Meyer, J. Am. Chem. Soc. 100, 6286 (1978).
41. J. Nasielski, A. Kirsch-De Mesmaeker and P. Leempoel, Electrochim. Acta 23, 605 (1978).
42. J. Nasielski, A. Kirsch-De Mesmaeker and P. Leempoel, Nouv. J. Chim. 2, 497 (1978).
43. T. I. Quickenden and G. K. Yim, Solar Energy 19, 283 (1977).
44. A. Kirsch-De Mesmaeker, J. Kanicki, P. Leempoel and J. Nasielski, Bull. Soc. Chim. Belg. 87, 849 (1978).
45. A. Kirsch-De Mesmaeker, P. Leempoel and J. Nasielski, Nouv. J. Chim. 3, 239 (1979).

46. B. Durham, W. J. Dressick and T. J. Meyer, J. Chem. Soc.,
    Chem. Comm., 381 (1979).
47. C. Daul, O. Haas and A. von Zelewsky, J. Electroanal.
    Chem. 107, 49 (1980).
48. J. M. Mountz and H. T. Tien, Solar Energy 21, 291 (1978).
49. H. T. Tien, J. Higgins and J. M. Mountz, in ref. (15),
    pp. 203-236.

DISCUSSION

Dr. E. Berman, ARCO Solar, Inc.:
I would like to comment on the efficiencies required for
practical devices that produce electricity from radiant
energy, assuming that the device itself costs nothing.
These requirements are site- and application-specific.
"Worst cases" require 18% efficiency, "best cases" 2.5%.

Dr. E. J. J. Groenen, Shell Laboratory, Amsterdam:
You made a remark concerning the advantage of having a
hydrophobic anode. In our study of carbon electrodes it
appeared that a "carbon-paste" electrode is the most
selective and it is also the most hydrophobic one.

Dr. Archer:
The comment I made was that the hydrophobic coating
Prof. Grätzel uses on his Pt particles suggested to me
that a simple hydrophobic coating on any electrode might
produce considerable selectivity for the hydrophobic substance as compared with the hydrophilic one, e.g.,
Fe(III). I am very interested to hear that this is indeed
the case; carbon-paste electrodes of course contain
oils. They are, however, not very transparent.

Dr. R. Memming, Philips Forschungslaboratorium, Hamburg:
According to the calculations of theoretical conversion
efficiencies you and Dr. Albery made, I got the impression
that the efficiencies depend strongly on the boundary conditions. Can you give us some reasonable theoretical
value which one could expect?

Dr. Archer:
Provided the criteria for the characteristic lengths $(X_e,$
$X_k,$ etc.) and those regarding the electrode kinetics are
met in the system, then the efficiency is not a function
of the boundary conditions but simply of what one assumes
about the ΔE^o and I_o (effective photon-flux density) that
could be compatible with these criteria being fulfilled.

In our Nature paper we assumed with unrealistic optimism
that all photons of energy greater than 1.1 eV could be
absorbed from sunlight in a system in which ΔE^o was also
1.1 eV. This produces an efficiency of 18%. However,
more realistic assumptions lower this figure. If, for
example, one assumes that all photons of $E > 2$ eV are
absorbed and that ΔE^o is 0.5 V, then the efficiency comes
down to 7%. Altering the boundary conditions is equivalent to altering the electrode kinetics. Any deviation
from the requirement of perfect selectivity and electrochemical reversibility for the A/B couple at the illuminated electrode, and reversibility for the Y/Z couple at
the dark electrode will further reduce the efficiency.

Prof. T. G. Spiro, Princeton University
What are the prospects for developing an electrode coating
e.g., with thionine, which is thick enough to absorb most
of the light and still permit effective electron transfer?

Dr. Archer:
The prospects for this seem to me to be good. However,
whether self-quenching of the excited states produced by
light absorption in the dye layer would severly diminish
the quantum yield of charge transfer is another matter.
As I mentioned, this tends to happen in concentrated homogeneous solutions. However, self-quenching is avoided in
the chlorophyll antenna system in green plants, and presumably could be avoided in a suitably constructed dye-
aggregate system.

Prof. G. Calzaferri, University of Bern:
I have the feeling that photoredox reactions in which a
reversible dissociation step or a reversible precipita-
tion-dissolution step is included have not yet been investigated very thoroughly. In such systems one should expect to overcome the problem of the thermal back reaction.

Dr. Archer:
I agree that removal of one of the products of the photochemical reaction could be an effective means of stopping
the back reaction. If a solid were formed, there would be
the additional attraction of high energy-storage density
in the product. Some early examples of this type involving Ag/AgCl electrodes exist in the literature. However,
phase separation is too slow to compete with the back reaction in these systems. Intermediate steps in the overall reaction, such as disproportionation of semithionine,
also tend to slow the effective back-reaction rate considerably although they also reduce the open-circuit voltage.

Dr. N. C. Fawcett, University of Southern Mississippi:
I have a comment regarding your statement that you felt
the ring-sulfur was participating somehow in the laying
down of thionine on a platinum electrode at positive
potentials. We have tried to do the same thing with the
oxygen analog of thionine, oxonine, and find that this dye
will not form a modified electrode on which the oxonine
remains electroactive. That is certainly circumstantial
evidence for the participation of sulfur in thionine.

Dr. Archer:
It appears that thionine itself is the only substance yet
discovered that can be laid down on platinum in the
described manner; no thionine derivative yet investigated
will do it. I suppose the reason could be steric effects.

Dr. P. V. Kamat, Boston University:
I would like to add to the comment made earlier with regard to employing a thionine-coated electrode to utilize
the incident flux efficiently. My opinion in this regard
is that one can successfully use the self-quenching reaction to generate semireduced and semioxidized species in
the electrode itself. One of the two species so generated
could produce a photogalvanic effect. The electron defi-
ciency in the electrode could then be compensated by the
redox system (electrolyte phase). Such a cell might work,
similar to a semiconductor-redox photochemical cell, with
electron flow from the electrode to the external circuit
followed by electron supply from the electrolyte to the
electrode.

Dr. Archer:
I agree that such a useful self-quenching could, in principle, occur. However, there is no evidence that I know
of that it actually does happen in the thionine coating.

CHAPTER 8

CHEMICAL ASPECTS OF PHOTOVOLTAIC CELLS[1]

Sigurd Wagner[2]

Photovoltaic Research Branch
Solar Energy Research Institute[3]
Golden, Colorado
U.S.A.

I. INTRODUCTION

Today's single-crystal silicon solar cell is a mature product. It is based on processes of photoexcitation and charge separation in space the physics of which are well understood. However, economical application of photovoltaic modules on a large scale is contingent on substantial cost reductions. Conventional semiconductor fabrication technology needs to be supplanted by processes characteristic of industrial chemistry. This transformation depends on the solution of a number of chemical problems which range from solid-state chemistry to chemical engineering.

Solar cells are semiconductor diodes that convert light directly to electric power (1). Photovoltaic converters, based on wafers of single-crystal silicon, have been in use on spacecraft for over twenty years. The incipient shortage of inexpensive conventional energy supplies has brought new emphasis to photovoltaic conversion. Technology-development

[1] Also presented at the Annual Meeting of the Bunsen-Gesellschaft, Munich, May 17, 1980.

[2] Present address: Department of Electrical Engineering and Computer Science, Princeton University, Princeton, New Jersey 08544.

[3] A division of Midwest Research Institute; operated for the U.S. Department of Energy under Contract EG-77-C-01-4042.

programs, carried out in several countries, are attempting to
lower the cost of silicon solar cells by introducing new processes, by reducing the number of manufacturing steps, and by
automation. Simultaneous applied research programs are aimed
at developing new photovoltaic semiconductors and devices that
can be manufactured into photovoltaic panels at lower cost
than for silicon. These programs are justified by economic
analyses that indicate a realistic chance for photovoltaics to
replace conventional electric generating capacity (2).

Three types of technologies are currently being pursued to
develop photovoltaic converters for economic production of
electricity: concentrators, flat-plate silicon, and thin
films. In concentrators, the area of the solar cell itself
may be 10 to 10^4 times smaller than the optical aperture of
the system. Cells may be comparatively expensive but must be
efficient. The key characteristic of a solar cell is its
solar-power to electric-power conversion efficiency, which
depends on both the spectral distribution and the intensity
(see Chapter 11). Since the efficiency of solar cells increases with intensity, the highest values have been measured
under concentrated light. The best results obtained to date
are 20% efficiency in silicon cells (3), 25% in cells based on
gallium arsenide (4), 28% in a converter that employs splitting of the solar spectrum by a dichroic mirror for more
efficient conversion by a pair of silicon and gallium arsenide
cells (5), and 29% for a silicon thermophotovoltaic cell
designed to convert radiation from a black body at 2000°C
(ultimately to be heated by sunlight) (6).

Flat-plate silicon arrays represent the only current
commercial technology, drawing on twenty years of development
for small-scale applications, primarily in spacecraft. Silicon single crystals are used as the starting material, but
less expensive alternatives to single crystals, such as
coarsely cast polycrystalline silicon, are in sight. One
major barrier to the rapid introduction of cast and other non-
single crystal silicon is the lack of understanding of grain
growth, grain-boundary chemistry and impurity diffusion. For
silicon cells, a specific fabrication technology, distinct
from that employed in microcircuit production, has been developed over the years. Due to the small size of the market,
processing steps have evolved incrementally.

The practical aspects of cost (less than $0.70 per peak
watt), performance (10% solar efficiency) and operating life
(20 years) of thin-film solar cell modules have been well-
defined (7), but no such cell has been fully demonstrated in
the laboratory. The lack of an obvious candidate that combines minimal consumption of material with acceptable performance together with module structure that is highly amenable to
automation explains the very broad screening program carried

out at present. These programs are severely hampered by the absence of elementary information about the preparation, reactivity, point-defect equilibria and optical (e.g., absorption edge) and electronic (e.g., conductivity type) properties of the majority of semiconductors of interest. Moreover, the constitution of these semiconductors ranges from amorphous to polycrystalline to single-crystalline, but adequate information on the relation of local and microstructure to electronic properties, reactivity, as well as self- and impurity-diffusion does not exist (8).

II.   OPERATION OF A SOLAR CELL

Absorption of a photon in a solar cell excites an electron from the valence band into the conduction band. Photons are required with energies $h\nu$ equal to or greater than the semiconductor bandgap energy E_g. Excess photon energy, $h\nu - E_g$, excites the electron (or the hole) into the conduction (valence) band. The excess energy is lost by thermalization toward the Fermi-Dirac distribution through rapid emission of phonons. This decay takes place in the sub-nanosecond time range, and the excess energy is converted to heat and therefore is not available for conversion to electricity. Excitation of an electron from the edge of the valence band to the edge of the conduction band, corresponding to a charge transfer, ultimately produces the output voltage of the solar cell.

For momentum-conserving transitions, the absorption coefficient α is 10^4 cm^{-1} or higher because of the high associated density of states in the two bands. Therefore, the absorption length $1/\alpha$ is on the order of 1 μm. Within this distance from the semiconductor surface, a fraction $(1-1/e)$ of the incoming light will be absorbed and converted to excited electron-hole pairs. In other words, in semiconductors with such "direct" transitions, all useful light will be absorbed within a layer a few μm thick. This requirement for a thin active layer constitutes an advantage because of low materials consumption and potentially high production rates per unit area. A number of semiconductors, including GaAs, InP, CuInSe$_2$, Cu$_2$S and amorphous silicon-hydrogen alloys, absorb strongly enough for such thin-film applications.

In the lowest-energy transition in silicon, electron momentum is not conserved. The necessary momentum is acquired from a phonon by what can be described in a particle picture as a three-body (photon, electron, phonon) collision. The absorption coefficient in such an indirect-gap semiconductor is reduced from the direct-gap case by a term that reflects the probability of interacting with a phonon. In silicon, the

absorption coefficient for photon energies slightly above the
band-gap energy (1.11 eV) lies in the range of 10^2 cm^{-1}. At
higher photon energies, direct transitions set in and the
absorption coefficient rises.   At hν = 2.5 eV, α reaches
10^4 cm^{-1}, a value typical of direct-gap materials. The photon
energy range from 1.11 eV (λ = 1.12 μm) to 2.5 eV (λ = 0.50
μm) corresponds to the most important part of the solar spectrum.   In silicon, the absorption lengths for sunlight range
up to 100 μm; thus silicon cells need to be comparatively
thick for complete absorption of the useful spectrum. We will
see shortly that high values of the absorption length impose
the need for very pure materials with a high degree of
crystalline perfection.

Transfer of photon energy to the electron-hole pair is the
first of two key steps in the photovoltaic process.   The
second step is spatial separation of the electron-hole pair
which produces the external current.   This separation occurs
by diffusion to, and drift through, the space-charge region of
the diode junction.   In conventional silicon cells this space
charge arises from the electrochemical potential (Fermi level)
difference at the contact between n-type and p-type silicon
(homodiode).   However, the space charge can also be introduced
at other types of contacts: between two different semiconductors (heterodiode), between a thin, transparent metal layer
and a semiconductor (Schottky barrier diode), or between an
electrolyte and a semiconductor (photoelectrochemical diode).

The lifetime τ of the photoexcited electron-hole pair typically is 10^{-9} to 10^{-8} s for direct-gap semiconductors (9),
and 10^{-6} to 10^{-4} s for indirect-gap silicon (10).   With typical drift velocities of 10^{-5} cm s^{-1} and space-charge widths
of 10^{-5} cm, an electron or hole will drift through the space-
charge junction in about 10^{-10} s.   Thus, an electron-hole pair
generated within the space-charge region will be separated before recombining.   The situation is different when the pair is
generated in the neutral bulk of the semiconductor.   Charge
separation begins by diffusion toward the space-charge junction.   The characteristic diffusion distance before recombination is the diffusion length L which is given by:

$$L \equiv (D\tau)^{1/2} = \left(\frac{kT}{q}\mu\tau\right)^{1/2} \qquad \qquad [1]$$

where D is the diffusion coefficient, kT/q the thermal potential, and μ the charge mobility.   For a typical value of
$\mu = 10^3$ cm^2 V^{-1} s^{-1}, L is on the order of the absorption
length, 1/α. In other words, the probability can be high for
recombination of charges, especially those photoexcited by
light of long wavelengths (i.e., low α) in indirect-gap semiconductors like silicon.

The examples of chemical research applied to photovoltaic conversion listed in this paper address silicon, and deal to a large extent with the need for inexpensive materials that are of adequate chemical purity and crystalline perfection to exhibit a high diffusion length. Many impurities, particularly transition metals, and structural imperfections (point defects, dislocations, grain boundaries and free surfaces) produce electronic levels within the band gap. These levels can act as recombination sites which serve to reduce the diffusion length. Although it is not the only important variable that affects solar-cell performance, the diffusion length provides a convenient basis for the discussion of specific problems within the context of this paper. To provide a background for this discussion, I will describe briefly the manufacturing steps of a silicon solar cell.

III. FABRICATION OF A SILICON SOLAR CELL

The starting materials for the production of silicon is quartzite, usually in the form of sand. The quartzite is reduced to silicon in an arc furnace, where the arc operates in a liquid phase (above the melting point of SiO_2, 1780°C) between a carbon electrode and liquid silicon which is collected at the bottom. The purity of this metallurgical-grade silicon is roughly 99%, whereas the total, electrically active impurity content of silicon used for solar cells is about 10^{-5}%. The required purification is carried out by converting the silicon to chlorosilanes in a fluidized-bed reactor; trichlorosilane is then distilled to high purity and is subsequently reduced by H_2 on a silicon rod that serves as a resistive heater.

The product is the so-called semiconductor-grade polycrystalline silicon. From this material, single crystals are grown by either the Czochralski or float-zone technique (11). The round, single crystals are ground to precise diameters (currently, 4.0 in. = 10.16 cm) and cut into wafers about 0.25 mm thick. The cutting loss ("sawdust" or kerf) also corresponds to about 0.25 mm. After cutting, the wafers are etched to remove the mechanically damaged surface layer and to introduce surface texture; the etchants are either mixtures of HF, HNO_3 and acetic acid or a concentrated aqueous solution of NaOH, depending on crystal orientation and desired texture. Next, the p-n junction is formed by diffusing a donor impurity (usually phosphorus) into a shallow (≤1 μm) surface layer of the p-type wafer. This process is carried out in a tube furnace at ~900°C with a dilute $POCl_3$ mixture as the source of the dopant impurity. Before the diffusion step, the back side

and the edge of the wafers are "masked", for instance with a
native oxide, prepared by thermal oxidation. After diffusion,
this mask is removed, a metal-contact layer is applied to the
back, and a metal-contact grid together with an antireflection
coating are applied to the front surface. The cells then are
assembled, electrically interconnected and packaged into
modules.

This is the manufacturing sequence for conventional
silicon modules. To reduce the cost of such modules, many
development programs are being carried out on alternative
procedures with fewer processing steps (12). In addition,
more fundamental research programs are examining different
concepts, particularly thin-film solar cells. In the following sections, several areas of chemical research are described
that are contributing to the very practical problem of reducing the cost of silicon cells.

A. Preparation and Purification of Silicon

The present (Siemens) process for the purification of
silicon, described above, suffers from a large byproduct
stream of $SiCl_4$, high power consumption, and a low silicon
deposition rate. Research is being carried out on hydrogenation of $SiCl_4$ to the more easily purified $SiHCl_3$:

$$3SiCl_4 + 2H_2 + Si \rightarrow 4SiHCl_3 \qquad\qquad [2]$$

with emphasis on the dependence of the kinetics on temperature, pressure, concentration, catalyst, particle size, surface area and fluidization. Similarly, the kinetics of
hydrochloration of silicon:

$$Si + H_2 + 2Cl_2 \rightarrow SiHCl_3 + HCl \qquad\qquad [3]$$

are under investigation. With the goal of higher rates of
silicon production, reduction of $SiCl_4$ with sodium in a
hydrogen-argon plasma or with metallic zinc in a fluidized bed
reactor are also under study. In the latter process, the $ZnCl_2$
byproduct is reconverted to Zn by electrolysis. Reduction of
SiF_4 with sodium is also under consideration, and a theoretical study of the kinetics of the reactions of silicon halides
with alkali metals is being carried out. A further modification of the Siemens process is concerned with SiH_2Cl_2, or a
mixture of SiH_2Cl_2 and $SiHCl_3$, with emphasis on the kinetics
of equilibration between these two silicon halides. Yet
another approach to the production of silicon is pyrolysis of
silane in a fluidized-bed reactor or possibly in a weightless
environment (i.e., in earth orbit) (13).

One approach to carrying out both production and purification of Si in a single step uses electrolysis of a molten solution of SiO_2 in a Na_3AlF_6 (80%)-LiF (20%) electrolyte (14).
In this case, the SiO_2 starting material is ~99.8% pure
mineral. Electrolysis temperatures lie between 800°C and
900°C using high-purity carbon as the anode and liquid tin as
the cathode where silicon crystallites are produced. The use
of a highly ionic electrolyte in combination with a metallic
cathode is very effective in removing impurities, and the
impurity content of the silicon thus produced is on the order
of 10^{-3}%. The properties of the electrolyte (ionic equilibria, impurity effects) and the reaction kinetics at the
cathode are of particular concern in this process.

B. Growth of Silicon Sheets

Commercial solar cell substrates are cut from silicon
crystals. However, several techniques have been attempted to
grow silicon in sheet form directly from the liquid to avoid
the laborious cutting step and associated kerf losses (15).
Shaped growth of crystals was attempted more than 50 years ago
(16), but experiments on ribbon growth of semiconductors date
from the 1960's. Several techniques for growing self-
supported ribbons (as opposed to films on substrates) have
been investigated (15). In the Stepanov technique, a shaping
die floats on the liquid silicon and a liquid film is pulled
through a slot in the die by a seed crystal. In the dendritic
web technique, two coplanar silicon dendrites containing the
same twin plane [111] grow into a supercooled melt and are
pulled from the melt at their growth rate while the liquid
meniscus between them freezes to a film. In another technique, a ribbon is pulled horizontally from the surface of the
liquid silicon. The edge-defined film-fed growth uses a slot-
like capillary that is introduced into the melt; the liquid
rises to a seed crystal located above the top of the capillary. In a more recent development (17), two parallel filaments of a foreign material are connected with a cross-member
or bridge. The cross-member is dipped into the melt and is
then withdrawn. At the meniscus, silicon freezes to the
cross-member; the solid silicon then acts as a seed for further growth as the two filaments are continuously withdrawn
from the melt.

There are specific materials problems associated with each
technique, in particular with regard to the reaction of
silicon with shaping dies (15). For instance, silicon carbide
inclusions are frequently observed in sheets grown by the
edge-defined film-fed procedure when a carbon die is used.
These inclusions give rise to dislocations, twins, or

polycrystals, all of which reduce the diffusion lengths of the
majority and minority carriers.

Two constraints are common to all sheet-growth techniques.
The first is the need for shaping by control of the capillary
force which establishes the meniscus. The second involves a
limit of the rate of growth due to the requirement that the
heat of fusion be removed through conduction along, and radiation away from, the sheet in order to maintain a stable solid/
liquid interface.

C. The Nature of Deep-Level Impurities

As mentioned earlier, many impurities can introduce "deep"
levels within the semiconductor band gap. A level is considered to be deep when its energy separation from the valence
and conduction band edges is much larger than the thermal
energy. Deep levels positioned near the center of the energy
gap have comparable capture probabilities for electrons and
holes. Therefore, deep impurities can act as efficient
recombination centers which reduce the diffusion lengths of
the charge carriers (10).

Two important aspects of deep levels are of interest here:
experimental verification of the theoretically predicted
energy and recombination kinetics of a deep level; and the
chemical nature of deep levels. Are they similar to the impurities that cause them? If so, can a deep-level impurity be
neutralized by pairing or complexing with a second impurity?

In a molecular orbital picture, such pairing would correspond to the formation of a bonding orbital within the valence
band or an antibonding orbital within the conduction band,
thus removing the original level from the band gap. If this
were the case, the detrimental effects of impurities in silicon (and other semiconductor materials) could be alleviated by
adding other impurities instead of carrying out costly purification. However, current theoretical evidence indicates that,
at least for substitutional impurities without d-level participation, the deep impurity level is the antibonding, host-
like level in the impurity-host "molecule" (18). Clearly, a
host-like level will not be affected by addition of a small
concentration of another impurity. This theoretical result
indicates that impurities must be removed from the semiconductor bulk.

An alternative to complete purification may lie in the
preparation of a polycrystalline material in which impurities
are rejected into the liquid, and ultimately into the grain
boundary, during grain growth.

D. Impurities in Semiconductor Grain Boundaries

Grain boundaries are discontinuities in the crystal lattice and, as a rule, appear to introduce electronic states into the band gap. In silicon, the density of these states can be reduced by diffusing hydrogen into the grain boundary (19). This treatment is most effective when atomic hydrogen is employed, hence it is assumed that H atoms attach themselves to nonbonding Si orbitals at the grain boundary. (A typical experimental condition is to flow H_2, excited by a Tesla coil, over the sample in a tube furnace at 400°C). The effect is thus similar to the impurity-pairing mechanism discussed in the preceding section. This "saturation of dangling bonds" (as it is called in semiconductor terminology) is not practically useful since it must precede the junction-diffusion step carried out at 800°C to 900°C; during that step, the hydrogen is driven out again. For reasons that are not yet understood, H-atoms diffuse only slowly through the highly doped n-type surface layer introduced by junction formation.

Cast polygrain silicon has been found to exhibit a wide variety of electrical activity of its grain boundaries. This activity -- in terms of a wide range of charge-carrier recombination rates -- cannot be correlated uniquely with the grain-boundary orientation. This fact has led to detailed investigations of grain boundaries using surface analysis techniques (20). With a combination of such techniques, it is possible to measure the electron-beam induced photocurrent in a solar cell made from polycrystalline silicon. Low photocurrent intensity near a grain boundary indicates a high charge-recombination rate. Using the electron beam of a scanning Auger apparatus, it is possible to leave the sample in the ultra-high vacuum-analysis chamber, fracture it along a grain boundary of interest, and study both the elemental composition and distribution of very small quantities of impurities in the boundary with Auger electron spectroscopy (AES) and secondary ion mass spectroscopy (SIMS).

Two main results have emerged from such studies (21). First, impurities are not necessarily distributed in the form of a grain-boundary film, but may be segregated into several localized phases (for example, one phase containing metal oxides, the other silicon carbide). Second, highly pure grain boundaries may promote recombination more actively than boundaries that contain impurities, particularly oxide phases. This finding opens the possibility of controlling impurity rejection during grain growth such that the rejected impurities "passivate" the grain boundaries.

IV. SUMMARY

In this brief review, I have described a number of chemical problems from the most applied area of quantum conversion of solar energy, the technology of silicon photovoltaic cells.

ACKNOWLEDGEMENTS

I would like to thank my colleagues, T. F. Ciszek, J. D. Dow, L. L. Kazmerski and J. M. Olson for many stimulating discussions.

REFERENCES

1. Much of the published work regarding the materials aspects of the U.S. Photovoltaic Program can be found in the Proceedings of the annual IEEE Photovoltaic Specialists Conferences; in publications of the Electrochemical Society; in the Proceedings of the Annual Photovoltaic Advanced Materials Review Meetings, published by the Solar Energy Research Institute; and in the Proceedings of the Low-Cost Solar Array Project Integration Meetings, published by the Jet Propulsion Laboratory.
2. For an introduction to photovoltaic conversion see H. J. Hovel, "Solar Cells", Vol. 11 of "Semiconductors and Semimetals" (R. K. Willardson and A. C. Beer, eds.), Academic Press, New York, 1975.
3. H. T. Weaver, R. D. Nasby and C. M. Garner, Technical Digest, Internatl. Electron Dev. Mtg., Washington, D. C., Dec. 8-10, 1980; IEEE, New York, 1980, pp. 190-193.
4. R. Sahai, D. D. Edwall and J. S. Harris, Jr., Appl. Phys. Lett. 34, 147 (1979).
5. A. L. Moon, L. W. James, H. A. Van der Plas, T. O. Yep, G. A. Antypas and Y. Chai, Conf. Rev., 13th IEEE Photovoltaic Specialists Conference, Washington, D. C., June 5-8, 1978; IEEE, New York, 1978, pp. 859-867.
6. R. M. Swanson, in ref. (3), pp. 186-189.
7. L. M. Magid and P. D. Maycock, Conf. Rev., 14th IEEE Photovoltaic Specialists Conference, San Diego, Calif., January 7-10, 1980; IEEE, New York, 1980, pp. 887-892.
8. An excellent overview of photovoltaic materials is given by K. J. Bachman, in "Current Topics in Materials Science" (E. Kaldis, ed.), Elsevier/North-Holland, Amsterdam, 1979, Chapter 6.

9. J. I. Pankove, "Optical Processes in Semiconductors", Dover Publ., Inc., New York, 1975, p. 133.
10. A. S. Grove, "Physics and Technology of Semiconductor Devices", John Wiley & Sons, New York, 1967, pp. 140-143.
11. G. F. Feigl and A. C. Bonora, in ref. (7), pp. 303-308.
12. P. D. Maycock, in ref. (7), pp. 6-12.
13. "Silicon Material Technology Development", in Proc. 15th Project Integration Meeting, Pasadena, Calif., April 2-3, 1980; Jet Propulsion Laboratory Publication 80-27, Report DOE/JPL-1012-44, pp. 81-112.
14. J. M. Olson, Extended Abstracts, Electrochemical Society Spring Meeting, St. Louis, Mo., 80-1, 74 (1980).
15. T. F. Ciszek, "Silicon and High-Melting Metals", Vol. 5 of "Crystals: Growth, Properties, and Applications" (G. Freyhardt and H. Grabmaier, eds.) Springer-Verlag, New York, to be published (1981).
16. E. von Gompertz, Z. Phys. 8, 194 (1922).
17. T. F. Ciszek and J. L. Hurd, Extended Abstracts, Electrochemical Society Spring Meeting, St. Louis, Mo., 80-1, 823 (1980).
18. H. P. Hjalmarson, P. Vogl, D. J. Wolford and J. D. Dow, Phys. Rev. Lett. 44, 810 (1980).
19. C. H. Seager and D. S. Ginley, J. Appl. Phys. 52, 1050 (1981).
20. L. L. Kazmerski, P. J. Ireland and T. F. Ciszek, J. Vac. Sci. Technol. 17, 34 (1980).
21. L. L. Kazmerski, Appl. Surface Sci., to be published (1981).

DISCUSSION

Prof. J. R. Bolton, University of Western Ontario:
You showed efficiency curves which maximized at ~ 28%. Yet Shockley and Quiesser derived a figure of 31% (AMO) for an ideal photovoltaic cell. This would be ~ 34% for AM2. What assumptions are introduced which lower the efficiency to the values you indicated?

Dr. Wagner:
All projected (or "theoretical") maximum efficiency values above 25% for single cells contain optimistic assumptions. Today, when near-ideal photocurrents can be achieved, this assumption usually is manifest in the open-circuit voltage being a larger fraction of the energy gap than demonstrated in practice. Therefore, differences in maximum obtainable efficiencies reflect differences in projected open-circuit voltages.

Dr. M. D. Archer, University of Cambridge:
You mentioned that a silicon cell of 23% efficiency has
been developed. Could you tell us whether that efficiency
was achieved under an irradiance of 1 sun or a higher
irradiance? Also what do you think of the feasibility of
operating silicon cells economically under concentrated
sunlight?

Dr. Wagner:
Silicon cells designed for concentrators have reached
about 20% conversion efficiency under 20- to 50-sun concentration. At 1 sun, the efficiency of such cells is
typically 18%.

Dr. B. N. Baron, IEC, University of Delaware:
The U.S. Department of Energy projects polysilicon at
$14/kg in 1986. Is this possible?

Dr. Wagner:
I consider it technically possible to reach the cost
goal. The timing, however, will depend primarily on non-
technical factors such as a commitment to make the
required investments in pilot and production plants.

Dr. A. Heller, Bell Laboratories:
What is the nature of the electrolyte in the new SiO_2
electrowinning process?

Dr. Wagner:
The electrolyte usually employed is 80 wt% Na_3AlF_6-20 wt%
LiF

Dr. Heller:
Would you please compare the performance of $CuInSe_2$/CdS
and Cu_2S/CdS cells?

Dr. Wagner:
The Cu_2S/CdS and $CuInSe_2$/CdS solar cells are the most
efficient thin-film devices produced so far, each having
reached 10%. While the Cu_2S/CdS cell has been made with a
thickness of 30 μm (which is probably not economical for
large-scale applications), the total thickness of the
$CuInSe_2$/CdS cell is ~5 μm. On the other hand, there is
some concern about the availability of adequate quantities
of In for large-scale production of the $CuInSe_2$-based
cell. Finally, it appears from very preliminary data that
the $CuInSe_2$-cell may be less susceptible to atmospheric
degradation than the Cu_2S-cell.

Dr. H. Tributsch, C.N.R.S., Laboratoires de Bellvue:
You made a clear distinction between indirect bandgap
semiconductors, into which light can penetrate deeply, and
direct bandgap semiconductors, in which the penetration
depth is shallow. There are also semiconductors in which
an indirect transition is combined with an extremely high
absorption coefficient (e.g., transition metal dichalcogenides). What do you think about them?

Dr. Wagner:
Such materials would make ideal photocurrent producers if
the indirect transition were to lead to long minority-
carrier lifetimes and diffusion lengths, so that the
latter is much larger than the absorption length. Unfortunately, it is not yet possible to make a reliable
prediction of the photovoltage from bulk semiconductor
properties, so that the overall power-conversion efficiency of such an ideal absorber still needs to be determined experimentally.

Prof. R. T. Ross, Ohio State University:
Your comment about the quantitative inaccuracy of one of
your early slides prompts me to solicit your help in
resolving a misunderstanding about the role of semiconductors in photoelectrochemical cells. You noted that the
space-charge layer at a sharp p-n junction has a width of
about 0.1 μm, whereas charge is actually collected over a
diffusion length of 10-100 μm. Yet, despite this, the two
depths are usually sketched as being similar. These
sketches, and adaptations of them for semiconductor-liquid
interfaces, have led to the belief on the part of many
electrochemists that it is only those carriers generated
within the space-charge region which can be collected. Do
you agree that this view is rather dramatically incorrect?

Dr. Wagner:
I have made the same observation. However, such minor
lapses all too understandably do occur in the difficult
marriage of electrochemistry and solid-state physics.

CHAPTER 9

ELECTROCHEMICAL PHOTOVOLTAIC CELLS

Rüdiger Memming

Philips GmBH Forschungslaboratorium
Hamburg
Federal Republic of Germany

John J. Kelly

Philips Research Laboratories
Eindhoven
The Netherlands

I. INTRODUCTION

In recent years an ever increasing interest has been directed towards photochemical and photoelectrochemical processes applicable to solar energy conversion. It is well
known that photoelectrochemical cells can be used for solar
energy conversion into electricity as well as for production
of chemical fuels, and the basic processes involved in such
systems have been reviewed recently (1-4). In this paper, we
will restrict ourselves entirely to photoelectrochemical systems that are applicable to conversion of light into electrical energy.

Photovoltaic cells are based on purely solid-state devices
and have been used for a long time, for instance, in various
optical-electronic devices. They consist either of pn-
junctions or of Schottky type junctions, and single-crystal
devices of both types having high conversion efficiencies have
been made. In the preparation of pn-junctions, difficult diffusion techniques are required because the interface must be
close to the surface in order to keep losses sufficiently low
(see Chapter 8). In addition, pn-junctions can be made only
with semiconductors of which both n- and p-type materials are
available. This limits the choice of materials because there
are not many p-type semiconductors.

PHOTOCHEMICAL CONVERSION AND
STORAGE OF SOLAR ENERGY

The application of semiconductor-liquid junctions is very attractive because, in principle, there are many possible ways to produce a junction between a p-type or n-type semiconductor and an electrolyte. All such junctions show photoeffects; in addition, they are easy to make and the incident light can reach the interface without any losses. In these systems, however, other problems arise such as the stability of the electrode, which limits the selection of materials.

In this paper we will review the principle of photoelectrochemical cells and some photovoltaic systems. The main emphasis, however, is on discussion of factors which determine the stability of semiconductor electrodes and limit the conversion efficiency. These problems are discussed in connection with some recent results obtained in our laboratories.

II. ENERGETICS OF THE SEMICONDUCTOR-LIQUID INTERFACE

Such interfaces show certain similarities to a purely solid-state device, i.e., to a pn-junction or a semiconductor-metal interface. The difference is that a Helmholtz double layer exists at the solid-liquid interface which formed by covalently bonded surface groups (e.g., hydroxyl groups) or by adsorption of ions. The potential distribution consists of two parts, a potential difference (U_H) across the Helmholtz layer and another across the space-charge region (U_{SC}) below the surface (Fig. 1a). In most cases U_H remains constant against an externally applied potential and also upon addition of a redox system, i.e., any variation of the electrode potential (U_E) leads only to a change of U_{SC}.

This is the usual electrochemical description of the semiconductor-liquid interface as derived from the results of many experiments. In connection with electron-transfer processes this potential diagram is not sufficient. Therefore, an energy diagram is used which describes the electron energy levels at the interface (Fig. 1b). Equilibrium exists if the electrochemical potential is constant throughout the whole system. The electrochemical potential of an electron is represented in solids by the Fermi level (E_F) and in redox-couple solutions by the Nernst potential ($E_{F,el}$). The former is usually measured relative to vacuum and the latter is on the the normal hydrogen electrode (nhe) scale. The two scales can be related to each other (5) and differ by a constant factor of about 4.5 eV (Fig. 1b). Upon contact between a semiconductor and a redox couple in solution, a potential difference across the space-charge layer is formed which leads to a phenomonon known as band bending, as shown for an n-type semiconductor in Figure 1b. The position of the energy band

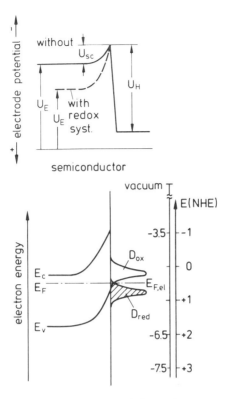

FIGURE 1. Potential diagram (a) and energy diagram (b) of
the semiconductor-electrolyte interface.

at the surface is determined by the Hemholtz double-layer and
remains constant as long as U_H is not changed. The electronic energy levels in the redox system are described by
occupied states (D_{red}) (6), and electron transfer is expected
if energy states exist at the same level on both sides of the
interface. The basic concepts of the semiconductor-
electrolyte interface have been reviewed elsewhere (7).

III. THEORETICAL CONVERSION EFFICIENCY
AND THE 'IDEAL' SOLAR CELL

In semiconductor devices the conversion efficiency depends
on the bandgap of the semiconductor (see Chapter 11). Such a
dependence shows a maximum efficiency for a bandgap around
1.5 eV (~840 nm) at which the theoretical efficiency is about
30%. Accordingly, a semiconductor electrode with a bandgap of

around 1.5 eV should be selected, so that under suitable boundary conditions the power output is maximized:

$$P_{max} = i_{ph} \cdot U_{ph,max} \qquad [1]$$

The photocurrent usually depends linearly on the incident light intensity, whereas the maximum achievable photovoltage $U_{ph,max}$ depends on the bandgap E_g (see below). On the basis of these conditions an "ideal" solar cell can be configured, an energy diagram for which is given in Figure 2, in which an ohmic contact of the semiconductor with a metal counter-electrode is shown. At equilibrium (Fig. 2a) the Fermi level

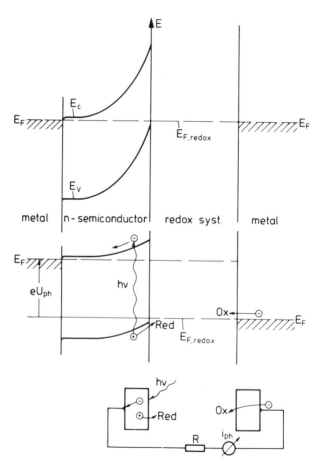

FIGURE 2. Energy scheme of an "ideal" regenerative solar cell. This example is for an n-type electrode at equilibrium (a) and during illumination (b).

is constant over the entire system. Here, a combination of an n-type semiconductor electrode and a redox cycle has been selected so that the band-bending is as large as possible at equilibrium, i.e., the Fermi level is close to the valence band at the semiconductor surface. For simplicity the occupied and empty states of the redox system are not shown.

When such a system is illuminated under open-circuit conditions, electron-hole pairs are created which are then separated by the large electric field across the space-charge layer. This leads to a decrease of band-bending which can be detected by measuring a corresponding photovoltage, U_{ph}, between an ohmic contact at the backside of the semiconductor and a counter electrode as indicated in Figure 2b. The maximum photovoltage is reached when the energy bands are flat. For heavily doped semiconductors the maximum photovoltage is determined by:

$$U_{ph,max} \simeq E_g/e \qquad [2]$$

The intensity dependence of the photovoltage is given by:

$$U_{ph} = \frac{kT}{e} \ln \left(1 + \frac{\Delta p}{p_0}\right) \approx \frac{kT}{e} \ln I \qquad [3]$$

where p_0 is the minority carrier density in the bulk, Δp is its increase under light excitation, and I is the absorbed light intensity (4).

Under short-circuit conditions, the electric field forces the holes of the valence band across the interface to the reduced species of the redox system:

$$Red + p^+ \rightarrow Ox \qquad \text{(anodic process at} \atop \text{n-type semiconductor)} \qquad [4]$$

whereas, the electrons of the conduction band move toward the backside contact and appear again at the counter-electrode where they are used for the reduction of the redox system according to reaction [5]:

$$Ox + e^- \rightarrow Red \qquad \text{(at counter electrode)} \qquad [5]$$

These two reactions are responsible for the cyclic process in a regenerative solid-liquid solar cell. It should be mentioned that, in principle, a p-type electrode can be used as well. In this case a redox system should be selected such that a strong downward band-bending exists at equilibrium (4).

IV. STABILITY OF ELECTRODES AND EXPERIMENTAL EFFICIENCIES

According to the discussion in the previous sections, it would seem to be a relatively simple matter to construct a solid-liquid solar cell with ideal properties. However, investigations of many n-type semiconductors coupled with redox systems have shown that it is difficult to produce a solar cell that works in a regenerative mode. In most cases, anodic dissolution of the semiconductor electrode occurs instead of reaction [4]. For example, a cell consisting of n-CdSe (semiconductor electrode) coupled with Ce^{4+}/Ce^{3+} (redox system) would appear to meet the ideal situation given in Figure 2 with respect to band-bending and the relative positions of the energy levels at the interface. Here, one has excellent overlap between the valence band and the occupied states of the redox system (Fig. 3), and a high rate constant for hole transfer to Ce^{3+} ions is expected. However, experimental investigations have shown that this process cannot compete with anodic dissolution of CdSe, even though the energetic conditions are quite favorable.

Unfortunately, only very few semiconductors, mostly oxides, are available which are initially stable against

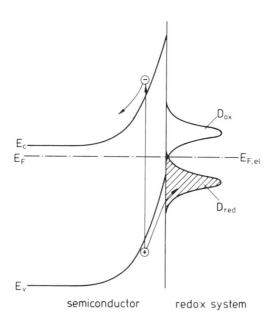

FIGURE 3. Hole transfer from an n-type semiconductor electrode to a redox couple.

anodic decomposition. All of these electrodes studied to date
have very large bandgaps (>3 eV) and, hence, are not interesting as far as solar conversion efficiency is concerned. On
the other hand, it should be possible to stabilize a number of
n-type semiconductor electrodes by using a suitable redox
system. As shown in the last of the examples given in
Table I, some of these cells show rather high conversion
efficiencies. One of the best studied systems is n-GaAs/
(Se_n^{2-}/Se^{2-}) (12,13), for which an efficiency of about 12% has
been reported together with quite good power characteristics
(Fig. 4). For this system the current-voltage curve is nearly

Table I. Stabilized Regenerative Photoelectrochemical Cells

Cell composition	Electrolyte	E_g (eV)	$U_{ph,max}$ (V)	$\eta(\%)$	ref.
n-type					
$(n\text{-CdS})/$ $Fe(CN)_6^{4-/3-}$	H_2O (pH 13)	2.5	1.0	5	8
$(n\text{-CdS})/$ (S_n^{2-}/S^{2-})	H_2O (pH 14)	2.5	0.8	5	9,10
$(n\text{-CdSe})/$ (S_n^{2-}/S^{2-})	H_2O (pH 14)	1.6	0.6	7	10,11
$(n\text{-GaAs})/$ (Se_n^{2-}/Se^{2-})	H_2O (pH 14)	1.4	0.65	12	12,13
$(n\text{-CdS/CdSe})/$ (S_n^{2-}/S^{2-})	H_2O (pH 14)	1.2–1.6	0.96	--	14
$(n\text{-CuInS}_2)/$ (S_n^{2-}/S^{2-})	H_2O (pH 14)	1.3	0.6	?	15
$(n\text{-MoSe}_2)/$ (I_2/I^-)	H_2O (pH 7)	1.7	0.55	3.5	16
$(n\text{-CdS})/$ (I_2/I^-)	CH_3CN	2.5	?	--	17
p-type					
$(p\text{-GaAs})/$ (Eu^{2+}/Eu^{3+})	H_2O (pH 1)	1.4	0.7	--	18
$(p\text{-GaP})/$ (Eu^{2+}/Eu^{3+})	H_2O (pH 1)	2.25	1.1	--	18
$(p\text{-MoS}_2)/$ (Fe^{2+}/Fe^{3+})	H_2O (pH 1)	1.75		--	19

rectangular, at least after surface treatment with RuCl₃ (12) or other metal salts (14-20). Without this treatment, the photovoltage drops even under a weak load. Such behavior is typical for photodiodes if an internal shunt exists within the cell. Parkinson et al. (12) have interpreted this shunt as due to recombination via surface states which are rendered ineffective by treatment with ruthenium.

For all cells in which n-type electrodes are used (Table 1), the observed photovoltages are consistent with the relative positions of energy levels at the interface. However, in each case the photovoltage is much lower than the maximum photovoltage possible for a given band gap (Eq. [2]). In the case of cells made with p-type semiconductors (Table 1) only small photovoltages have been found. The values are even smaller than expected from the band-bending determined at equilibrium.

According to the present results it is quite clear that more information about the properties of the semiconductor-liquid interface is required. In addition, factors that determine electrode stability have to be studied. Some of these problems have recently been investigated and the results are presented below.

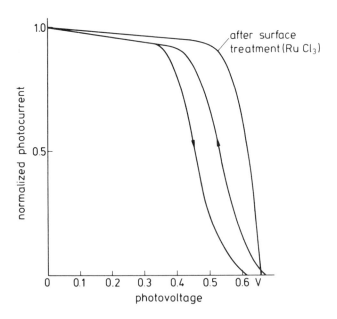

FIGURE 4. Power characteristics of an n-GaAs/ $(Se_n^{2-}/Se^{2-})/C$ solar cell (13).

V. PHOTOELECTROCHEMICAL CHARACTERIZATION
OF SEMICONDUCTOR ELECTRODES

Several authors have discussed the question of whether surface reactions are not only kinetically but also thermodynamically controlled (18,19,21,22). Using simple energy-level diagrams (Fig. 3) one would expect that electron transfer across an interface occurs only if states with equal energy levels exist on both sides of the interface, i.e., the electron-transfer rate should be determined by the density of states of the redox system at the edge of either the valence band or the conduction band. According to this model an anodic charge transfer (oxidation) to a redox system can be favored over a dissolution process even if the potential of the redox couple is more positive than the dissolution potential ($U_{redox} > U_{diss}$). In the case of stable cells (Table 1), however, the energy levels of the redox systems do not overlap with the valence band of the respective n-type semiconductor but occur somewhere in the middle of the forbidden zone; thus the question arises whether such a simple energy diagram is valid. This problem will be discussed below. As far as thermodynamics is concerned, all of the n-type semiconductor electrodes listed in Table 1 are stabilized by redox systems for which $U_{redox} > U_{diss}$. An example is CdS for which at pH 7 $U_{diss} = 0.1$ V, whereas for the $Fe(CN)_6^{3-}/Fe(CN)_6^{4-}$ system, $U_{redox} = +0.44$ V. According to these results it must be concluded that all reactions are kinetically controlled.

More information about important parameters such as stability, quantum efficiency and kinetics can be obtained by typical photoelectrochemical methods, such as capacitance, photocurrent and rotating ring-disc methods. For doped semiconductors one can derive a very simple relation between space-charge capacitance C_{SC} and band-bending U_{SC}:

$$1/C_{SC}^2 = \frac{2}{Ne\varepsilon\varepsilon_o}\left(U_{SC} - \frac{kT}{e}\right)$$

$$= \frac{2}{Ne\varepsilon\varepsilon_o}\left(U_E - U_{FB} - \frac{kT}{e}\right)$$

[6]

which is the well-known Mott-Schottky equation (7) where N is the donor or acceptor density in the bulk of the semiconductor, e is the electronic charge, U_{SC} the potential difference across the space-charge region, U_E the electrode potential, U_{FB} the flatband potential, and ε and ε_o are the dielectric constants of the medium and vacuum, respectively. The capacitance factor $1/C_{SC}^2$ varies linearly with the electrode potential U_E if the applied potential occurs entirely across the space-charge layer as found with most semiconductor electrodes. This behavior is shown schematically for an n-type electrode in Figure 5.

A typical dark current (i_d) vs. electrode potential (U_E) curve is also given in Figure 5. The current increases strongly with potentials that are cathodic with respect to the flatband potential U_{FB} due to electron transfer from the conduction band of the semiconductor to the electrolyte as in reaction [7]:

$$Ox + e^- \rightarrow Red \qquad\qquad\qquad [7]$$

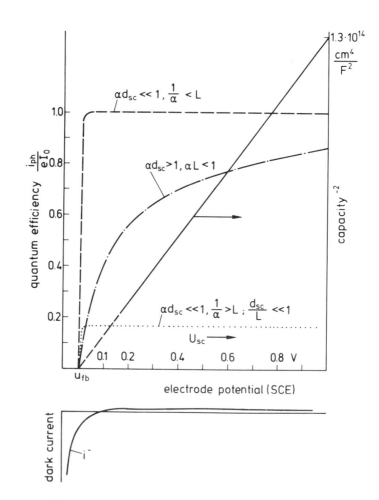

FIGURE 5. Theoretical dependence of capacitance, photo-current and dark current vs. electrode potential.

This current is controlled by the surface density n_S of majority carriers, i.e., the number of electrons in the conduction band, the latter being related to the potential drop across the space-charge layer by:

$$n_S = n_o \exp(-eV_{SC}/kT) \qquad [8]$$

where n_o is the bulk concentration of electrons. The cathodic current (or forward current) of the cell is then given by:

$$i^- = kC_{ox}n_S = kn_o \exp(-eV_{SC}/kT) \qquad [9]$$

This relation holds only within a small range because the Fermi level eventually passes the edge of the conduction band (degeneration). In this cathodic range no photocurrent is expected as long as the change of the majority carrier-density Δn_S induced by light excitation is small compared to n_o.

We now consider an anodic process in which holes (minority carriers) are involved according to reaction [10]:

$$Red + p^+ \to Ox \qquad [10]$$

The corresponding current is then controlled by the hole density at the surface of the n-type electrode. In this case the current will be small due to the low hole density in n-type materials. The hole density can be increased by weak illumination, in which case the corresponding photocurrent consists of two components, one being controlled by diffusion, and the other by the electric field across the space-charge layer:

$$i_{ph} = i_{ph,diff} + i_{ph,field} \qquad [11]$$

The contribution due to diffusion is obtained, according to Gärtner (23), by solving the diffusion equation under the assumption that the change in the population of minority carriers created by light (Δn or Δp) approaches zero near the edge between the bulk semiconductor and the space-charge layer (i.e., Δn or $\Delta p \to 0$ as $x \to d_{SC}$). One thus obtains:

$$i_{ph,diff} = eI_o\left[\frac{\alpha L}{1 + \alpha L}\right]\exp(-\alpha d_{SC}) \qquad [12]$$

where I_o is the incident light flux, α the absorption coefficient, L is the diffusion length of the minority carrier, and d_{SC} the thickness of the space-charge layer which is given by:

$$d_{SC} = 2L_D\left[\frac{eU_{SC}}{kT} - 1\right]^{1/2} \qquad [13a]$$

for a depletion layer (positive U_{SC} values); or:

$$d_{SC} \approx L_D \exp(-eV_{SC}/kT) \qquad\qquad\qquad\qquad [13b]$$

for an accumulation layer (negative U_{SC} values). In the last two equations L_D is the effective Debye length given by:

$$L_D = \left[\frac{\varepsilon\varepsilon_0 kT}{2n_0 e^2}\right]^{1/2} \qquad\qquad\qquad\qquad [14]$$

Accordingly, $I_{ph,diff}$ will decrease with increasing thickness and consequently with increasing band-bending in both directions but more rapidly for an accumulation layer. Hence, the diffusion-controlled photocurrent will drop to zero near the flatband potential. The field-induced photocurrent has been calculated by Gärtner (23) to be:

$$i_{ph,field} = eI_0[1 - \exp(-\alpha d_{SC})] \qquad\qquad\qquad [15]$$

or, in the case of a depletion layer:

$$i_{ph} = i_{ph,diff} + i_{ph,field} = eI_0\left[1 - \frac{\exp(-\alpha d_{SC})}{1 + \alpha L}\right] \qquad [16]$$

For most semiconductors the absorption coefficient does not exceed a value of 10^5 cm^{-1}. Values for the Debye length given by Eq. [14] are $L_D \leqslant 10^{-6}$ cm for a majority-carrier density of $n_0 \geqslant 10^{17}$ cm^{-3} with $\varepsilon = 12$. The corresponding thickness of the space-charge layer, Eq. [13a], is on the order of $d_{SC} \leqslant 10^{-5}$ cm for $U_{SC} \leqslant 1$ V. Accordingly, in most cases, $\alpha d_{SC} < 1$, i.e., Eq. [16] can be simplified to:

$$\phi = \frac{i_{ph}}{eI_0} = \frac{L + d_{SC}}{\frac{1}{\alpha} + L} \qquad\qquad\qquad\qquad [17]$$

where ϕ is the quantum yield of the photocurrent.

Two cases are of interest: (i) $1/\alpha < L$, in which case also $d_{SC} < L$; hence:

$$\phi = \frac{i_{ph}}{eI_0} = 1 \qquad\qquad\qquad\qquad [18]$$

(this is the case shown schematically in Fig. 5), and (ii) $1/\alpha > L$, for which:

$$\phi = \frac{i_{ph}}{eI_0} = \alpha L\left[1 + \frac{d_{SC}}{L}\right] \qquad\qquad\qquad [19]$$

One can discern two limiting cases: if $d_{SC}/L \gg 1$, then Eq. [19] becomes:

$$\phi \simeq \alpha d_{SC} = 2\alpha L_D \left(eV_{SC}/kT - 1 \right)^{1/2} \qquad \text{[20a]}$$

or if $d_{SC}/L \ll 1$, then:

$$\phi \simeq \alpha L \qquad \text{[20b]}$$

Accordingly, there are only two cases in which the quantum yield ϕ actually depends on the thickness of the space-charge layer and consequently on the potential: (i) when $d_{SC} > L$ and (ii) when $\alpha d_{SC} > 1$. The latter case can only occur for low doping levels ($n_0 \leqslant 10^{17}$). The example given in Figure 5 was calculated using Eq. [16]; for comparison, the dark current vs. potential behavior is also shown. The total current during illumination can then be obtained by super-imposing the dark- and photocurrent curves. The final current-potential behavior depends not only on the character-istic data derived for the photocurrent but also on the rate constant k contained in Eq. [9]. All of these parameters eventually determine the power characteristics (i.e., fill factor) of a solar cell.

VI. SURFACE REACTIONS

In order to get more information about the stability prob-lem and the photoelectrochemical properties of semiconductor-liquid interfaces, we have carried out several quantitative experiments using some well-known semiconductors; n- and p-type electrodes will be treated separately.

A. n-Type Semiconductors

We have investigated mainly GaAs, GaP and CdS. All samples were single crystals, their donor densities being in the range of 10^{17} to 10^{18} cm^{-3}. The results are presented in Figures 6, 7 and 8. Photocurrents were obtained using mono-chromatic light for excitation; the wavelength λ_{exc} was selec-ted in a range where the photocurrent becomes independent of any further variation of λ_{exc} (i.e., $1/\alpha < L$). In all cases the quantum yields of the saturation current were close to unity. In the potential range near the flatband potential, the photocurrent-potential behavior of these three semicon-ductor materials differs considerably. For instance, in an electrolyte containing no redox system, the photocurrent de-pends strongly on light intensity (GaAs and GaP in Figs. 6 and 7, respectively). This intensity dependence disappears

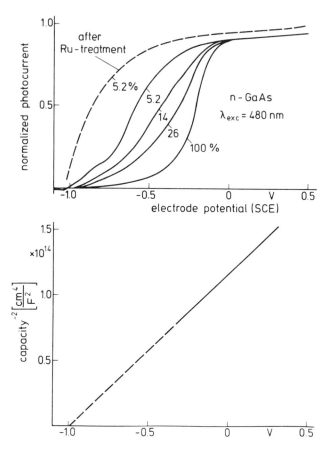

FIGURE 6. Photocurrent and capacitance vs. electrode potential for n–GaAs. Relative light intensities are indicated. For $I = 100\%$, $i_{sat} = 3.8$ mA/cm^2; electrolyte is 0.1 M H_2SO_4.

upon addition of $Fe(CN)_6^{4-}$ as a redox system. The same result was obtained with CdS (Fig. 8). The latter effect does not occur with GaAs although a similar intensity dependence was found in the presence of S_n^{2-}/S^{2-} (24). Such photoelectrochemical behavior can certainly not be interpreted by using the Gärtner model (23) derived in the previous section (see also Fig. 5). As we have proposed earlier, surface states may be responsible by acting as recombination centers or as traps for one kind of charge carrier or the other (18,21).

Additional information can be obtained by measuring the capacitance during illumination, and these results are also shown in Figures 6–8. The measurements were performed at very low light intensities in order to avoid complications. In two

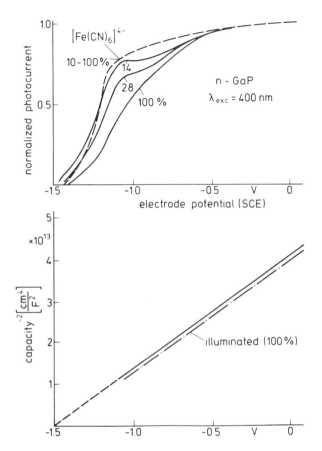

FIGURE 7. Photocurrent and capacitance vs. electrode potential for n-GaP. Relative light intensities are indicated. For I = 100%, i_{sat} = 0.05 mA/cm^2. Concentration of Fe(CN)$_6^{4-}$ = 0.05 M; pH 9.

cases, GaP and CdS, we found a shift of the $1/C^2$ vs. U_E plot towards anodic potentials. With GaAs, however, such a shift did not occur even at higher light intensities. In this case it is not certain whether there are effects at lower $1/C^2$ values which could not be detected, because the signals were neither stable nor reproducible in this range. Furthermore, it should be mentioned that the shifts of the capacitance curves found with GaP and CdS are dependent on the light intensity and saturate at higher intensities. It is interesting to note that this shift is decreased or even disappears when a redox system is added to the electrolyte (Figs. 7 and 8).

The shift of the capacitance curves under light excitation indicate quite clearly that hole traps are present as surface

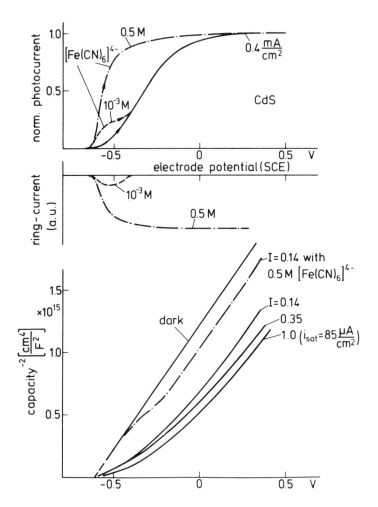

FIGURE 8. Photocurrent, ring current and capacitance vs. electrode potential for n-CdS. Relative light intensities and various concentrations of $Fe(CN)_6^{4-}$ are indicated; solutions at pH 9. Ring currents are displayed in arbitrary units.

states. Since the experiments were performed under potentiostatic conditions (i.e., the potential between the reference electrode and the semiconductor was kept constant), the shift of the capacitance must be interpreted as a change of the potential difference across the Helmholtz layer ΔU_H, as indicated schematically in Figure 9. In the case of CdS, ΔU_H amounts to about 0.3 V. The charge in the surface traps responsible for ΔU_H is given by Eq. [21]:

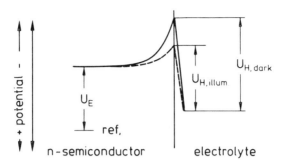

FIGURE 9. Potential distribution for an n-type semi-conductor electrode during illumination.

$$q_{ss} = C_H \cdot \Delta U_H \hspace{4cm} [21]$$

For $C_H \sim 10$ μF cm^{-2} and $\Delta U_H = 0.3$ V, one obtains $q_{ss} \simeq 3 \times 10^{-6}$ coulomb cm^{-2}, i.e., ~1.8 × 10^{13} holes/cm^{-2} are trapped in surface states.

The backward shift of the capacitance curves after addition of a redox system such as $Fe(CN)_6^{4-}$ to GaP (Fig. 7) or CdS (Fig. 8) indicates that the holes trapped in surface states can be transferred readily to the reducing agent (Red). Simultaneously, the anodic dissolution process is suppressed as shown by measurements with a ring-disc electrode (25). In this method the ring current is a measure of the number of reductant molecules being oxidized during the photoelectro-chemical process at the semiconducting disc electrode. An example is shown for CdS in Figure 8b. Quantitative analysis of the ring currents at high $Fe(CN)_6^{4-}$ concentrations indicates that the photocurrent is due entirely to oxidation of the redox system (18,25). Thus, the redox process obviously competes with dissolution of the semiconductor even in a potential range where the latter is very rapid in solutions without an added redox couple. Such a result, however, is obtained only if the saturated photocurrent is kept below the diffusion current of the ions. In the case of much lower $Fe(CN)_6^{4-}$ concentrations (~10^{-3} M), stabilization of the elec-trode likewise occurs (26). Since the ring current decreases above a potential of 0.5 V, stabilization of the electrode occurs only below this potential. Obviously the concentration here is too low to empty the hole traps. Quite similar re-sults were obtained with GaP, although stabilization was ob-served only with samples having a low dislocation density on the surface (i.e., epitaxially grown crystals) (18).

The question now arises regarding the chemical nature of these traps. If they are treated as energy levels within the forbidden zone, as shown in Figure 10, hole transfer via traps (T) may be then described as follows:

$$SC + p^+ \rightarrow T^+ \qquad [22a]$$

$$T^+ + np^+ \rightarrow SC_{diss} \qquad [22b]$$

$$T^+ + Red \rightarrow SC + Ox \qquad [22c]$$

Here it is assumed that the positively charged traps T^+ may either react with the reducing agent (Eq. [22c]) or they capture a further hole which would lead to dissolution (SC_{diss}) of the semiconductor electrode (SC). It has been suggested that such surface states are formed as intermediates in the anodic dissolution process and that the redox process leads to regeneration of surface bonds (4,22). Such a model, however, seems to be too simple for a variety of reasons. First, the number of electrons, and consequently the density of traps, can be very large ($\sim 1.8 \times 10^{13}$ cm^{-2}) compared to the number of surface atoms ($\sim 10^{15}$ cm^{-2}), where the latter value indicates an intrinsic feature of the surface states. Secondly, van Overmeire et al. (28) concluded from rotating ring-disc experiments performed with n-GaP that reaction [22b] is not possible because the competition between anodic dissolution and the redox process is dependent on light intensity. According to these results, the ratio of corrosion to redox current increases with increasing intensity even if the photocurrent is kept below the diffusion current. Instead of Eq. [22b], these authors assumed a second-order reaction of oxidized traps which explained part of their results; however, such a

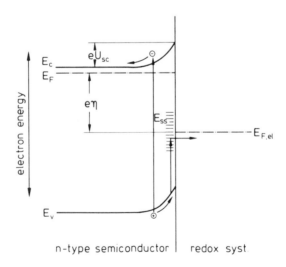

FIGURE 10. Energy scheme for hole trapping. (See text for details).

model implies a rather high mobility of traps within the semiconductor surface.

According to the results discussed above, traps are a favorable route for efficient hole transfer if a redox system of a suitable energy is selected. Effective hole-trapping in surface states inhibits the corrosion current, at least in some cases such as GaP and CdS. The question remains, however, why trapping behavior is not manifested by a shift of the capacitance curve measured with n–GaAs, even though the photoinduced corrosion current is also inhibited (Fig. 6). In the case of GaAs, an inhibition was found even in the presence of a redox system such as S_n^{2-}/S^{2-} (24). In an experiment of the latter type, Heller and co-workers (12,20) improved the kinetics of hole transfer to Se^{2-} by depositing traces of $RuCl_3$ on the surface which also improved the performance of a corresponding solar cell considerably (Fig. 4). We applied the same treatment to our samples, and the electrodes were then examined in acidic solutions without an added redox system. According to the photocurrent–potential dependence obtained with this electrode (Fig. 6), anodic dissolution is also catalyzed by Ru ions adsorbed at the surface. Parkinson et al. (12) interpreted this effect in terms of a lower surface-recombination rate. It is an open question, however, whether catalysis of anodic dissolution can be interpreted in the same way. Difficulties arise because an increasing surface-recombination rate with increasing light intensity has to be assumed and this is in contradiction with generally accepted models of recombination kinetics (29).

There are other known examples, such as $MoSe_2$ (16), where the photoelectrochemical behavior indicates a similar process. Another interesting example is the Fe_2O_3–electrolyte system, which is rather stable in 1 M NaOH even without an added redox couple (30,31). The anodic process (O_2–evolution) is also inhibited, and strong relaxation effects of the photocurrent are observed using modulated light (30,32). Also, the onset of the photocurrent occurs closer to the flatband potential and the relaxation phenomena disappear by fast electron transfer into hole traps if a reducing agent is added (33). According to results obtained with n–ZnO, electron trapping can occur in this system also (34). For example, we observed that the photocurrent starts to rise at potentials more negative than the flatband potential as derived by extrapolation of the linear part of the Mott–Schottky curve. A more accurate measurement of the capacitance near the flatband potential has shown, however, that in this range U_H is changed (i.e., there is no linear region in the $1/C^2$ vs. U_H curve), an effect which has been attributed to surface states (34).

B. p—Type Semiconductors

Effects similar to those reported for n—type materials have also been observed with p-type electrodes. One example (p-GaAs) is shown in Figure 11. The cathodic process, evolution of H_2, sets in about -0.2 V, i.e., ~ 0.6 V below the flatband potential. This photocurrent-potential behavior is independent of light intensity. In contrast to n-GaAs, we found in this case a strong influence of light on the capacitance (Fig. 11) (35). Shifts of the capacitance curve and corresponding changes of $\bar{U}_H$ (Fig. 12) were observed at higher intensities; i.e., electrons are trapped efficiently in surface states, the density of which is also on the order of 10^{13} cm^{-2}. These traps can be emptied by adding a suitable redox system such as $Fe(CN)_6^{3-}$ as proved by the photocurrent measurements and the corresponding reverse shift in the capacitance shown in Figure 11a.

The large overvoltage observed with p-GaAs is caused not only by effective electron trapping in surface states but is also due to the low rate constant k_c for electron transfer from the conduction band either to protons or to H_2O molecules. This was proved by photocurrent measurements performed with ethyl viologen for which the redox potential (EV^{2+}/EV^+) is -0.7 V (sce). Since the position of the conduction band is located at -0.9 V, the redox process is expected to proceed directly via the conduction band. This process was indeed observed, as shown in Figure 11b. The reactions can be summarized as follows:

$$e^- + EV^{2+} \xrightarrow{k_c} EV^+ \qquad\qquad\qquad [23a]$$
$$\text{(or } H^+) \quad \text{(or } 1/2\ H_2)$$

$$e^- + T \xrightarrow{k_1} T^- \qquad\qquad\qquad\qquad [23b]$$

where only the first reaction leads to an observed current. Since these reactions are in competition, direct electron transfer should occur perferentially as soon as the traps T are filled, i.e., at sufficiently high light intensities. This is in agreement with the experimental results obtained with ethyl viologen. As shown in Figure 11b, the onset of the photocurrent depends strongly on the light intensity and reaches nearly the flat-band potential. This result may also be of interest in connection with photochemical production of H_2 since the methyl viologen radical cation MV^+ is capable of reducing H_2O in the presence of a Pt catalyst (36).

The various reactions possible at the p-GaAs electrolyte interface are shown schematically in the energy diagram of Figure 13. Electron transfer via traps as in reaction [24]:

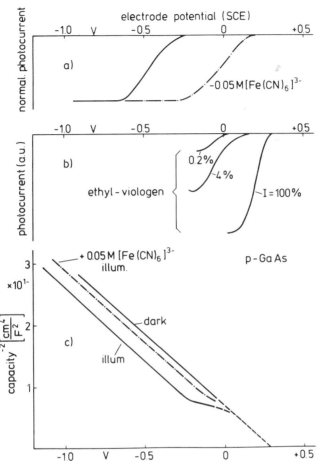

FIGURE 11. Photocurrent and capacitance vs. electrode potential for p-GaAs. Relative light intensities and various concentrations of $Fe(CN)_6^{4-}$ are indicated; solutions at pH 1. For I = 100%, i_{sat} = 5 mA/cm^2 in (b), and 50 µA/cm^2 in (c).

$$T^- + Ox \xrightarrow{k_{ox}} T + Red \qquad\qquad [24]$$

was also observed with a few other redox systems, although the onset of the photocurrent was different for each system. Similar results were obtained with p–GaP. A detailed analysis of the processes at p–type electrodes has been published elsewhere (35). According to the photocurrent data available in the literature, H_2- evolution seems to be inhibited at all known p–type semiconductors (37–39). Also, it was recently shown by Wrighton's group (40) that reduction of methyl viologen at p–Si also occurs near the flatband potential.

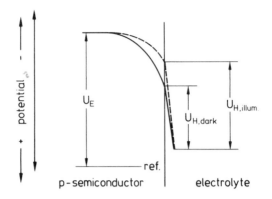

FIGURE 12. Potential distribution for a p-type semiconductor electrode during illumination.

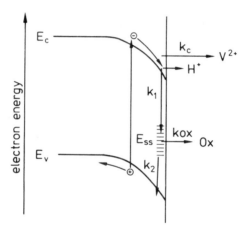

FIGURE 13. Electron transfer and trapping at a p-GaAs
electrode.

It is an interesting question whether p-type semiconductor
electrodes can be applied to practical solar cells, as discussed earlier by Memming (18) and more recently by Bard and
coworkers (41). A p-type electrode is very attractive because
it could be used as a cathode in a solar cell, i.e., reduction
of a redox couple does not have to compete with electrode dissolution. In addition, reactions with several redox systems
compete very well with reduction of H_2O. Various alkyl viologens are of especially great interest because photoelectrochemical reduction occurs even near the flatband potential (at
least above a certain light intensity); i.e., this process can
compete with trapping (Fig. 13). Furthermore, the redox
potentials of simple viologens are rather negative (−0.45 V

vs. nhe) so that at equilibrium a very large band-bending
should be possible. Accordingly, in the case of p-GaAs, a
photovoltage is expected which would approximate E_g/e (the
"ideal" case). There are, however, some problems which have
to be considered. First of all, the reduced species of viologen (V^+) is an organic free-radical which is strongly adsorbed
on electrodes, highly reactive and sensitive to oxygen.
Secondly, the counter-electrode must show a sufficient overpotential for H_2-evolution. Finally, the semiconductor
electrode cannot be used under open-circuit conditions with
strong illumination because it would attain a potential
(near U_{FB}) at which the anodic dissolution current (anodic
dark current) may not be negligibly small.

VII. CONCLUSIONS

The experimental data presented in this paper indicate the
dominant role of surface states which act mainly as traps for
electrons or holes. The densities of these states are large
(~1% of the surface atoms) in aqueous solutions and may be
even greater in organic solutions, as pointed out very recently by Bard and co-workers (42). The chemical origin of
surface states, however, is still unknown.

ACKNOWLEDGEMENT

The authors are indebted to Dr. W. Gissler for sending us
results prior to publication.

REFERENCES

1. A. J. Nozik, Annu. Rev. Phys. Chem. 29, 189 (1978); see
   also, Chapter 10, this Volume.
2. L. A. Harris and R. H. Wilson, Annu. Rev. Mat. Sci. 8, 99
   (1978).
3. H. Gerischer, in "Topics in Applied Physics", (B. O.
   Seraphin, ed.), Vol. 31, Springer-Verlag, Berlin, 1979,
   pp. 115-172.
4. R. Memming, Electrochim. Acta 25, 77 (1980).
5. F. Lohmann, Z. Naturforsch. 22a, 843 (1967).
6. H. Gerischer, in "Physical Chemistry", (M. Eyring, D.
   Henderson and W. Jost, eds.), Vol. 9A, Academic Press,
   New York, 1970, pp. 463-510.

7. R. Memming, in "Electroanalytical Chemistry" (A. J. Bard, ed.), Vol. 11, Marcel Dekker, New York, 1979, pp. 1-84.
8. H. Gerischer and J. Gobrecht, Ber. Bunsenges Phys. Chem. 80, 1635 (1976).
9. A. B. Ellis, S. W. Kaiser and M. S. Wrighton, J. Am. Chem. Soc. 98, 1635 (1976).
10. M. S. Wrighton, A. B. Bocarsly, J. M. Bolts, A. B. Ellis and K. D. Legg, in "Semiconductor Liquid-Junction Solar Cells" (A. Heller, ed.), Proceedings Vol. 77-3, Electrochemical Society, Princeton, N. J., 1977, pp. 138-156.
11. A. Heller, K. C. Chang and B. Miller, J. Electrochem. Soc. 124, 697 (1977).
12. B. A. Parkinson, A. Heller and B. Miller, J. Electrochem. Soc. 126, 954 (1979).
13. A. Heller and B. Miller, Electrochim. Acta 25, 29 (1980).
14. R. N. Noufi, P. A. Kohl and A. J. Bard, J. Electrochem. Soc. 125, 375 (1978).
15. M. Robbins, K. J. Bachmann, V. G. Lambrecht, F. A. Thiel, J. Thomson, R. G. Vadimsky, S. Menerges, A. Heller and B. Miller, J. Electrochem. Soc. 125, 831 (1978).
16. J. Gobrecht, H. Tributsch and H. Gerischer, J. Electrochem. Soc. 125, 2085 (1978).
17. K. Nakatani, S. Matsudaira and H. Tsubomura, J. Electrochem. Soc. 125, 406 (1978).
18. R. Memming, J. Electrochem. Soc. 125, 117 (1978).
19. H. Gerischer, in ref. (10), pp. 1-19.
20. A. Heller, H. J. Lewerenz and B. Miller, Ber. Bunsenges Phys. Chem. 84, 592 (1980).
21. R. Memming, in ref. (10), pp. 38-53.
22. A. J. Bard and M. S. Wrighton, J. Electrochem. Soc. 124, 1706 (1977); see also, ref. (10), pp. 195-209.
23. W. Gärtner, Phys. Rev. 116, 84 (1953).
24. J. Gobrecht and H. Gerischer, Solar Energy Mat. 2, 131 (1979).
25. R. Memming, Ber. Bunsenges Phys. Chem. 81, 732 (1977).
26. B. Miller, S. Menezes and A. Heller, in ref. (10), pp. 186-254.
27. R. Memming, Philips Techn. Rev. 38, 160 (1978/79).
28. F. van Overmeire, F. van den Kerchove, W. P. Gomes and F. Cardon, Bull. Soc. Chim. Belg. 89, 181 (1980).
29. R. A. Smith, "Semiconductors," At the University Press, Cambridge, Mass., 1961, pp. 287-297.
30. K. L. Hardee and A. J. Bard, J. Electrochem. Soc. 124, 1025 (1977).
31. J. H. Kennedy and K. W. Freese, J. Electrochem. Soc. 125, 723 (1978).
32. J. S. Curren and W. Gissler, J. Electrochem. Soc. 126, 56 (1979).

33. P. Iwanski, J. S. Curran, W. Gissler and R. Memming, J. Electrochem. Soc., in press (1981).
34. H. Gerischer, presented at the Snowmass Meeting, August 1979; Surface Sci. 101, 518 (1980).
35. J. J. Kelly and R. Memming, J. Electrochem. Soc., in press (1981).
36. K. Kalyanasundaram, J. Kiwi and M. Grätzel, Helv. Chim. Acta 61, 2720 (1978); see also, Chapters 1, 5 and 6, this Volume.
37. R. Memming and G. Schwandt, Surface Sci. 4, 109 (1966).
38. M. Gleria and R. Memming, Electroanal. Chem. 65, 163 (1975).
39. H. Gerischer, J. Gobrecht and J. Turner, Ber. Bunsenges Phys. Chem. 84, 596 (1980).
40. D. C. Bookbinder, N. W. Lewis, M. G. Bradley, A. B. Bocarsly and M. S. Wrighton, J. Am. Chem. Soc. 101, 7721 (1979).
41. F.-R. F. Fan and A. J. Bard, J. Am. Chem. Soc. 102, 3677 (1980).
42. A. J. Bard, A. B. Bocarsly, F.-R. F. Fan, E. G. Walton and M. S. Wrighton, J. Am. Chem. Soc. 102, 3671 (1980).

DISCUSSION

Dr. R. J. Gale, Colorado State University:
Can you discount the possibility that the shift or dispersion seen in Mott–Schottky slopes is not due to a contribution from a series pseudo–faradaic capacity arising from the photoinduced charge transfer to the redox system, or to trace corrosion?

Dr. Memming:
I don't think so because after addition of a redox system nearly the same Mott–Schottky plot was obtained.

Dr. H. Tributsch, C.N.R.S., Laboratoires de Bellvue:
You have shown experimental data concerning the CdS/ $Fe(CN)_6^{4-/3-}$ system and provided a detailed theoretical explanation. We found experimental evidence for the formation of an insoluble cyano–complex involving Cd and Fe and older literature supporting it. Thus, it is our opinion that the CdS/$Fe(CN)_6^{4-/3-}$ system should be considered a special case and the results should not be generalized.

Dr. Memming:
I did not know this and we will check the results obtained using CdS again.

Prof. R. Reisfeld, Hebrew University, Jerusalem:
   Is it possible to influence the concentration of the surface states by laser treatment of the electrode surface
   which would lead to annealing of the broken bonds?

Dr. Memming:
   I do not think that this is possible. Even if laser
   treatment has a certain influence, after a time some
   layers are probably dissolved and the effect of laser
   annealing is lost.

Dr. A. Heller, Bell Laboratories:
   I would like to reply to that if I may. Laser surface
   annealing has been used to increase grain sizes and thereby decrease the density of grain boundary states in polycrystalline semiconductors. No similar process exists for
   semiconductor-solution interfaces.

Prof. H. Tsubomura, Osaka University:
   We have observed the same phenomena on the effect of illumination of semiconductor electrodes under bias as
   observed from Mott-Schottky plots which Dr. Memming
   described. The interpretation of this effect given by
   Dr. Memming is also quite similar to ours, which is based
   on the assumption of an intermediate of the anodic dissolution reaction which we call "trapped holes".

Dr. Memming:
   It is very interesting to learn of this. I was not at the
   ACS meeting at which your results were presented, so I was
   unaware of this until now.

Dr. W. Kautek, Fritz-Haber-Institut der Max-Planck-Gesell-
schaft, Berlin:
   Photocurrents of illuminated semiconductors add to the
   measured overall potentials, as does the presence of a
   large number of surface states. Mott-Schottky plots would
   not give the proper information on the flatband position
   because of faradaic and surface-charge capacities. Also,
   recent findings of transition metal chalcogenides demonstrate that the bandgap can be exploited for an ideal
   photopotential output in constrast to the systems shown by
   you.

Dr. Memming:
   I wanted to demonstrate that a large density of surface
   states is present which act as electron or hole traps.
   This density may reach values of 10^{13} cm^{-2} which leads to
   a change in the potential distribtion.

Prof. M. Grätzel, Ecole Polytechnique Fédérale, Lausanne:
I was interested in the possibility of reducing viologens on p-type GaAs. Why is this cell run as a redox cell when the possibility exists to achieve photoinduced H_2O-splitting? Here the illuminated p-type semiconductor would have to be coupled with a RuO_2 electrode for O_2 evolution and a Pt catalyst would be employed for H_2 production.

Dr. Memming:
It is certainly more interesting to use a p-type material combined with viologens for producing hydrogen. However a p–GaAs electrode combined with a RuO_2 anode cannot be used for splitting water because the flatband potential is around +0.5 V (nhe) and hence the RuO_2 electrode is not sufficiently shifted towards anodic potentials. In the case of p–GaP one may possibly succeed in splitting water photoelectrochemically.

CHAPTER 10

PHOTOELECTROSYNTHESIS AT SEMICONDUCTOR ELECTRODES[1]

Arthur J. Nozik

Photoconversion Research Branch
Solar Energy Research Institute
Golden, Colorado
U.S.A.

I. INTRODUCTION

The field of photoelectrochemistry is currently enjoying a period of exciting new developments, rapid progress in both theory and applications, and attraction of scientists in a broad range of disciplines. This is because photoelectrochemical systems have shown an excellent capability to perform the three functions required for photochemical conversion and storage of solar energy, namely: (a) efficient absorption of sunlight to produce electrons and positive holes; (b) efficient separation and stabilization of the photogenerated positive and negative charge carriers; and (c) efficient subsequent redox chemistry to yield oxidized and reduced chemical species.

The other approaches to the problem of chemical conversion of solar energy, all of which were discussed at this Conference, have many elements in common with each other and with photoelectrochemistry. Conferences such as this serve the very important function of promoting communication and interaction between the scientists involved in these various approaches. This is most critical, since the three functions listed above will be so difficult and demanding in a practical solar energy conversion system, in which component lifetimes

[1]Work performed under the auspices of the Division of Chemical Sciences, Office of Basic Energy Sciences, U.S. Department of Energy, under Contract EG-77-C-01-4042.

must be on the order of decades; hence, the optimum scheme must incorporate the best features of each approach.

This necessity is apparent, for example, in the case of the solar driven water-splitting reaction in photoelectrochemical and photochemical systems. These two apparently different approaches, one involving semiconductor materials and the other photochemical redox species, are converging in one scheme to a common configuration. This configuration involves small particulate systems; in photoelectrochemistry these are called photochemical diodes (1,2) and in photochemistry micro-heterogeneous redox systems (3). In the latest version of these particulate systems, as described by Grätzel at this and other recent conferences (3), the particles consist of semiconductor powders (TiO_2) that have molecular sensitizers ($Ru(bipy)_3^{2+}$) and catalytic species (Pt and RuO_2) associated with the surface. The molecular sensitizer permits efficient absorption of visible light and the semiconductor permits efficient charge separation and stabilization, while the Pt and RuO_2 catalysts permit efficient reduction and oxidation of water, respectively, to yield H_2 and O_2 (4). Thus, the best qualities of each system (dye, semiconductor, catalyst) are combined to yield a working system for photochemical water-splitting by sunlight. Future interactions between photoelectrochemistry and photochemistry promise to produce equally interesting and significant advances.

In this paper, I will first review the general principles of photoelectrochemistry and photoelectrosynthesis; I will then discuss some new developments in photoelectrosynthesis.

II. BRIEF REVIEW OF PHOTOELECTROCHEMICAL ENERGY CONVERSION

Photoelectrochemical systems are defined as those in which a photoactive semiconductor material is in contact with a liquid (or solid) electrolyte, and a junction is formed between the semiconductor and the electrolyte such that illumination of the semiconductor with light energy equal to or greater than that of the semiconductor band gap produces electrons and holes that spatially separate (because of the junction potential) and subsequently drive chemical reduction and oxidation reactions in the system. Photoelectrochemical systems are distinguished from photogalvanic systems described in detail in this Volume by Archer (Chapter 7); in the latter light is absorbed by molecular pigments in the electrolyte solution.

A general classification scheme for the various types of photoelectrochemical cells is shown in Figure 1. One can first divide the cells into (a) those in which the Gibbs free-energy change in the electrolyte is zero (these are called

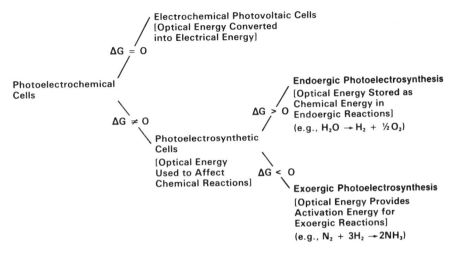

FIGURE 1. Classification of photoelectrochemical cells.

electrochemical photovoltaic cells); and (b) those in which the free-energy change in the electrolyte is non-zero (these are called photoelectrosynthetic cells).

In the photoelectrosynthetic cell, two different redox couples are present and two different redox reactions occur at the electrodes upon illumination thus producing a net chemical change. If the net free-energy change of net electrolyte reaction is positive, optical energy is converted into chemical energy (an endergonic or endoergic process). However, if the net electrolyte reaction has a negative free-energy change, optical energy provides the activation energy for the reaction (an exergonic or exoergic process). In previous discussions the latter situation was termed "photocatalytic" because the light is acting to accelerate a thermodynamically downhill reaction (2,5-7). However, it is also possible to have catalytic effects in the case of thermodynamically uphill reactions that are driven by light. Hence, more appropriate labels for the two types of reactions might be simply endergonic and exergonic photoelectrosynthesis.

The potential advantages of photoelectrochemical energy conversion compared to solid-state photovoltaic systems are listed in Table 1. Electrochemical photovoltaic cells are discussed in this Volume by Memming (Chapter 9), hence the present paper will be concerned only with photoelectrosynthetic cells. Energy-level diagrams for the two types of photoelectrosynthetic cells (5-7) are presented in Figure 2.

For efficient photoelectrosynthesis, three parameters must be optimized simultaneously: (a) the bandgap of the semiconductor must be as small as possible to maximize the absorption

TABLE I.   Potential Advantages of Photoelectrochemical Cells

---

Electrochemical Photovoltaic Cells

    Junction formation is very simple.

    Efficiencies of polycrystalline thin-film electrodes are comparable to those of single-crystal electrodes.

    In situ storage capability can be introduced into the cell.

Photoelectrosynthetic Cells

    For water-splitting, the theoretical conversion efficiency is significantly higher than photovoltaic cells coupled to conventional electrolysis.

    Particulate systems can be used (photochemical diodes).

    Novel reaction products are possible.

    Catalytic effects can be induced on the semiconductor electrode surface.

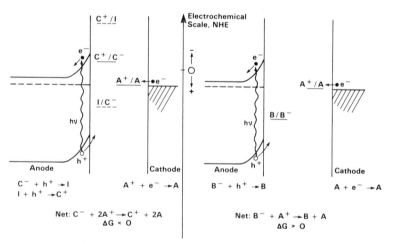

FIGURE 2.   Energy-level diagrams for exoergic (exergonic) and endoergic (endergonic) photoelectrosynthetic cells.

of sunlight but large enough to provide the chemical potential for the desired chemical reaction; (b) the flat-band potential (i.e., the electrode potential at which the semiconductor bands exhibit no band-bending) must be matched to the redox chemistry in the electrolyte so that the desired chemical reaction can proceed at a fast rate with minimum or no external bias voltage; and (c) the semiconductor electrode system must be stable for very long periods of time. It has not yet been possible to find a photoelectrosynthetic system that exhibits all three characteristics together with a high conversion efficiency. Many reviews are available which describe these problems in detail (8-13).

III. NEW DEVELOPMENTS IN PHOTOELECTROSYNTHESIS

A. Energetics of Semiconductor–Electrolyte Interfaces: Band–Edge Unpinning

In the conventional model of semiconductor-electrolyte junctions (8-13) it is assumed that changes in applied potential appear across the semiconductor space-charge layer and that the potential drop across the Helmholtz layer is constant. This means that the positions of the semiconductor band-edges are fixed or pinned with respect to the redox energy levels in the electrolyte, and that they do not change with electrode potential. This leads further to the restriction that only redox couples lying within the semiconductor band gap can undergo photoinduced redox chemistry. ·In the absence of hot-carrier injection (discussed in the next section), photogenerated electrons and holes are injected into the electrolyte at the bottom of the conduction and valence bands at the interface. Hence, redox couples lying outside the band gap are inaccessible to electron reduction or hole oxidation (Fig. 3).

However, recent experiments with Si, GaAs, and MX_2 (M = Mo, W; X = S, Se, Te) indicate that the band-edges can become unpinned, and that applied potentials can shift the semiconductor band positions with respect to the electrolyte redox levels. The effect here is to be able to oxidize or reduce redox couples that lie outside the band gap as determined from values of the flat-band potential obtained in the dark. Several explanations for the unpinning of the bands have been proposed. One explanation is based on the concept of Fermi-level pinning (14) as commonly observed in solid-state Schottky barriers (15). In this model, it is believed that a high density of surface-states pin the Fermi level and produce band-bending which is independent of the redox potential of

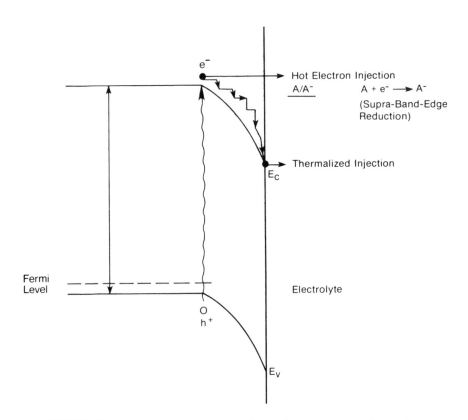

FIGURE 3. Two classes of photogenerated electron injection from a p-type semiconductor into a liquid electrolyte. Hot-electron injection occurs if the photogenerated electron is transferred into the electrolyte to reduce A to A^- before it is thermalized in the space-charge layer via electron-phonon collisions. A thermalized electron, emerging at the conduction band edge (E_c), would have insufficient energy to drive a redox couple A/A^- lying above E_c.

the electrolyte. Photoelectrochemical experiments with $MoSe_2$, GaAs and Si (16,17) show that: (a) a number of A^+/A electrolyte systems with widely different redox potentials produce the same photovoltage for a given semiconductor; (b) certain A^+/A systems produce photovoltages even when the redox potentials are outside the band gap; and (c) photoeffects can be found for redox couples with redox potentials that span a range greater than the semiconductor band gap. If the Fermi level is pinned by surface states, then the system is equivalent to a solid-state Schottky barrier in series-contact with an electrolyte, such that the band edges move with applied potential while maintaining a constant band-bending. This situation is shown in the energy-level diagram of Figure 4. Further research is required to establish the generality and significance of Fermi-level pinning effects in photoelectrosynthesis.

Another explanation for band-edge shifting or unpinning with applied potential applies to small bandgap semiconductors where inversion can occur (18,19). When band-bending in the semiconductor is sufficiently great that the Fermi level at the surface lies closer to the minority carrier band than to the majority-carrier band, then the surface becomes inverted

FIGURE 4. Fermi-level pinning by surface states leading to unpinning of the semiconductor band edges at the semiconductor-electrolyte interface.

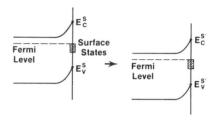

FIGURE 5. Inversion in p-type semiconductors arising from large band-bending. These band diagrams show the relative magnitudes and signs of the charge density under conditions of flat-band, accumulation, depletion and inversion.

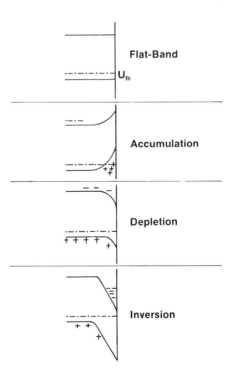

and a high charge density develops in the semiconductor space-charge layer, as depicted in Figure 5. Under conditions of inversion the charge density and capacitance of the semiconductor space-charge layer become comparable to or larger than the charge density and capacitance of the Helmholtz layer (Fig. 6), and additional potential changes applied to the electrode then appear across the Helmholtz layer. This situation results in band-edge unpinning.

The presence of inversion, and subsequent band-edge unpinning, can be followed through capacitance data. In the ideal case, the capacitance decreases in the depletion region (following the Mott-Schottky relationship) and reaches a minimum value at the onset of the inversion region (20). As this region develops, the space-charge density increases rapidly, the effect of which can be followed in the light and in the dark by the dependence of the measured capacitance on the frequency of an applied ac voltage (Fig. 7). At high frequencies (>1 kHz) the electron concentration cannot follow the applied voltage, hence the measured capacitance is flat (and minimized) in the inversion region. However, at low frequency

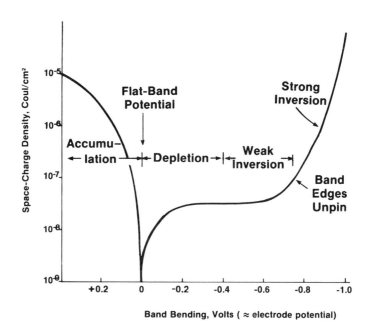

Band Bending, Volts (≈ electrode potential)

FIGURE 6. Calculated space-charge density of p-type Si as a function of band-bending. Band-edge unpinning may occur in the inversion region because of the higher charge density that develops (20).

(<100 Hz) the electrons in the space-charge layer can follow the ac signal and the capacitance increases rapidly with increased inversion. The effect of illumination in these experiments is to increase the measured capacitance in the inversion region at high frequencies such that the low-frequency behavior is produced (20). The characteristic dependence of capacitance on potential, frequency, and light intensity shown in Figure 7 is for p-Si with a thin (~600 Å) oxide layer. The thickness of this layer is another parameter which governs the detailed characteristic behavior of the capacitance in the inversion region (20).

Experiments with p-Si electrodes in acetonitrile and methanol show that the capacitance as a function of potential, frequency, illumination and oxide thickness follow closely the ideal behavior described above for the inversion region of p-Si (18,19,21). Hence, reduction of redox couples (such as anthraquinone, nitronaphthalene, dimethoxynitrobenzene, and dichloronitrobenzene) that lie above (i.e., more negative than) the conduction band edge of p-Si can be explained by

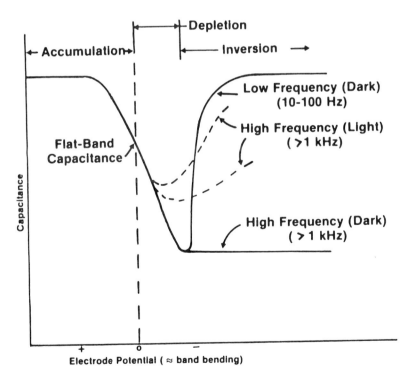

FIGURE 7. Ideal capacitance-voltage behavior as a function of frequency and light intensity for a semiconductor in the accumulation, depletion, and inversion regions (20).

band-edge unpinning that arises from the development of an
inversion layer in the electrode. These supra-band-edge
reductions occur at potentials where the capacitance data
exhibit the characterisics of the inversion region (18,19).

B. Hot-Carrier Injection at Illuminated Semiconductor-Electrolyte Junctions

As discussed in Section III.A, it has generally been
assumed in photoelectrochemistry that the energy of injected
photogenerated carriers is given by the position of the
minority carrier band-edge at the semiconductor-electrolyte
interface. That is, carriers are accelerated to the surface
by the electric field in the semiconductor space-charge layer,
but they lose energy (as heat) in the process via carrier-
phonon collisions. As a result, the carriers are in thermal
equilibrium with the lattice before injection.

In a modification of this model, it has recently been
proposed (22,23) that photogenerated minority carriers that
have not undergone full intraband (thermal) relaxation may
also be injected into the electrolyte, a process called hot-
carrier injection (Fig. 3). This process can occur if the
thermalization time (τ_{Th}) of the photogenerated carriers in
the semiconductor space-charge layer is greater than both the
charge-transfer (or tunneling) time of the carriers into the
electrolyte and the effective relaxation time of the injected
carriers in the electrolyte. These various characteristic
times are shown in Figure 8 for electron injection from a
p-type semiconductor.

FIGURE 8. Hot-electron injection
from a p-type semiconductor showing
the characteristic time constants for
competing de-excitation pathways. τ_T
is the tunneling or charge-transfer
time to the electrolyte of the nonthermalized electron, τ_{Th} is the
electron-thermalization time in the
semiconductor space-charge layer, τ_r
is the effective relaxation time in
the electrolyte to prevent back electron transfer to the semiconductor,
and τ_{CT} is the charge-transfer time
of the thermalized electron.

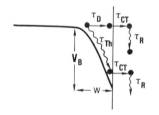

For Hot Electron Injection:

$$\tau_{Th} > \tau_D, \tau_{CT}, \tau_R$$

Values of the various characteristic times have been estimated using classical (6) and quantum mechanical (22,23) approaches. In one model (22), the aqueous electrolyte is treated as a large-bandgap semiconductor, and photoexcited carriers find themselves in the potential well created by the position-dependent potential in the semiconductor-electrolyte barrier. This well has characteristic quantized levels, and carriers can be injected from these levels into the electrolyte. Photogenerated electrons may either cascade down the quantized levels, i.e., thermalize, or they may tunnel through the surface barrier into the electrolyte.

The basic quantum mechanical problem to be solved is first to find the eigenstates for the potential well in the semiconductor. From the solutions of the appropriate equations the properties of the system are deduced in terms of the following characteristic times: τ_{Th}, the time for thermal relaxation of an electron in an excited state in the depletion layer; τ_t, the time required for the carrier to be transferred to an ion in the electrolyte; and τ_r, the time required for the ionic energy level to relax by reorientation of solvent dipolar species surrounding the ion so that reverse tunneling from the electrolyte to the semiconductor is prevented. These calculations have been described in detail elsewhere (22); the results are comparable to those based on both classical and simpler quantum mechanical models (6).

From these analyses and calculations several general criteria for obtaining hot-carrier injection at semiconductor-electrolyte junctions are evident. The overall criterion is that both the tunneling time of the photogenerated minority carriers and the effective relaxation time of the electrolyte must be faster than the thermalization time of these carriers in the semiconductor.

Strong electronic-vibrational interaction in the electrolyte renders the tunneling process from the semiconductor irreversible, obviating oscillations which increase the residence time in the semiconductor and thus enhance the probability of intraband thermalization. This irreversibility has its basic origin in electron tunneling from the semiconductor, in which the electron-phonon interaction is weak, to the electrolyte, where the electron-phonon interaction is strong.

Hot-carrier injection is favored in semiconductor electrodes that have a low effective mass for the minority carrier. This is because a low effective mass will produce more widely spaced quantized levels in the depletion layer which then result in long interlevel thermalization times. Low effective mass of the minority carrier also means that the part of the photon energy that exceeds the band gap will go preferentially to the minority carrier; this will produce hot carriers in the depletion region by virtue of absorption of

photons with energies greater than the band gap. In this
case, hot-carrier injection is more favored in direct bandgap
semiconductors.

Heavy doping will also favor hot-carrier injection because
the depletion-layer thickness will be reduced. This effect
leads to enhanced quantization in the depletion layer, and
hence, longer thermalization times.

The occurrence of hot-carrier injection in photoelectrochemical reactions would be very significant for the following
reasons: (a) the nature of energetically favorable photoinduced reactions at semiconductor electrodes could be more
readily controlled by the electrode potential; (b) the photogenerated carriers would not be in thermal equilibrium in
their respective bands so that quasi-thermodynamic arguments,
such as the use of the quasi-Fermi level to describe the energetics of photoelectrochemical reactions, would not be valid;
(c) the influence of surface states would be restricted to the
class of states originating from chemical interactions of the
electrolyte with the semiconductor surface; and (d) the maximum theoretical conversion efficiency for photoelectrochemical
energy conversion may be much greater compared with the case
of thermalized injection. Initial estimates of the theoretical thermodynamic limit indicate that the efficiency of the
hot-carrier pathway may be as much as twice that of the
thermalized pathway (24).

The existence of hot-carrier injection has not yet been
verified. Experiments designed to probe for hot-carrier
injection by studying supra-band-edge redox reactions (18,19)
were inconclusive because of the band-edge unpinning effects
discussed in Section III.A. Additional research must be done
to establish unequivocally the significance of hot-carrier
injection in photoelectrochemical systems.

C. Derivatized Semiconductor Electrodes

Photogenerated electrons and holes in semiconductor electrodes are generally characterized by strong reducing and oxidizing potentials, respectively. Instead of being injected
into the electrolyte to drive redox reactions, these electrons
and holes may reduce or oxidize the semiconductor itself and
cause decomposition. This is a serious problem for practical
photoelectrochemical devices, since photodecomposition of the
electrode leads to inoperability or to short lifetimes.

A simple model of electrode stability has been presented
wherein the redox potentials of the oxidative and reductive
decomposition reactions are calculated and placed on an
energy-level diagram like that shown in Figure 2 (25,26). The
relative positions of the decomposition reactions are compared

with those of the semiconductor valence and conduction band edges. Absolute thermodynamic stability of the electrode is assured if the redox potential of the oxidative decomposition reaction of the semiconductor lies below (has a more positive value than) the valence band edge, and if the redox potential of the reductive decomposition reaction lies above (has a more negative value) the conduction band edge. This situation does not exist in any of the semiconductors studied to date. More typically, one or both of the redox potentials of the semiconductor oxidative and reductive decomposition reactions lie within the band gap, and hence become thermodynamically possible. Electrode stability then depends upon the competition among thermodynamically possible redox reactions in the electrolyte. This competition in turn is governed by the relative kinetics of the possible types of reactions.

In cases where the redox potentials of the electrode-decomposition reactions are more thermodynamically favored than the electrolyte redox reactions (i.e., oxidative decomposition potentials more negative and reductive decomposition potentials more positive than the corresponding electrolyte redox reactions), the products of the latter have sufficient potential to drive electrode decomposition. Hence this situation usually results in electrode instability, assuming that the electrode decomposition reaction itself is not otherwise kinetically inhibited.

It is frequently the case in photoelectrochemistry that the more thermodynamically favored oxidation reactions also become kinetically favored, so that these reactions predominate. This effect has been used to stabilize n-type semiconductor electrodes by establishing a redox couple in the electrolyte with a redox potential more negative than the oxidative decomposition potential so that the electrolyte redox reaction occurs preferentially over the decomposition reaction and thus scavenges the photogenerated minority carriers. However, this particular stabilization technique can in general be used only for electrochemical photovoltaic cells, since in photoelectrosynthesis the permitted electrode reactions are dictated by the overall chemical reaction of the system.

One method of electrode stabilization involves chemical derivatization of the semiconductor surface (27). It has been shown that covalent attachment of redox species to semiconductor surfaces can effect the kinetics of charge transfer across semiconductor-electrolyte junctions such that less thermodynamically favored reactions predominate over more thermodynamically favored reactions. For example, illumination of n-type Si modified with a ferrocene derivative will not result in the thermodynamically favored oxidation of Si to SiO_2, but rather in oxidation of $Fe(CN)_6^{4-}$ to $Fe(CN)_6^{3-}$ in the electrolyte (27,28). The chemically attached ferrocene species act as hole mediators for n-Si, and serve to channel the photogenerated holes to the $Fe(CN)_6^{4-}/Fe(CN)_6^{3-}$ redox reaction rather than to the Si/SiO_2 reaction (see Fig. 9). This effect can be very important for stabilizing electrodes both in electrochemical photovoltaic cells and in photoelectrosynthetic cells. For example, as shown in Figure 10, derivatization of the semiconductor AB with $R-D/D^+$ could result in the photogenerated charge being channelled through the surface-attached D/D^+ couple to the H_2O-O_2 reaction, rather than to the oxidation of AB to B and A^+; the stepwise and net reactions for this sequence are indicated in Figure 10. Such charge-mediation effects are analogous to the function of the Mn complex in Photosystem II in photosynthesis; in that case, photogenerated holes are also channelled to the H_2O-oxidation reaction, rather than to oxidation of the chlorophyll molecule itself (see also, Chapters 1 and 2).

In addition to enhancing the stability of semiconductor electrodes, chemical derivatization or modification of the semiconductor surface can also be used to increase the catalytic activity of the electrode surface. This has been demonstrated for p-type Si in an aqueous electrolyte (29). Normally it is difficult to evolve H_2 on p-type electrodes, and a large negative deviation from the flat-band potential is required before H_2 evolution is achieved. However, by binding a methyl viologen (MV^{2+}) derivative to the surface, the photogenerated electron is first efficiently captured by the MV^{2+} species attached to the surface, and the reduced MV^+ species in the presence of a catalyst then reduces H^+ to H_2. The overall effect is that H_2 is evolved at a much less negative potential than required for the electrode without surface modification; i.e., the derivatized surface shows electrocatalytic activity.

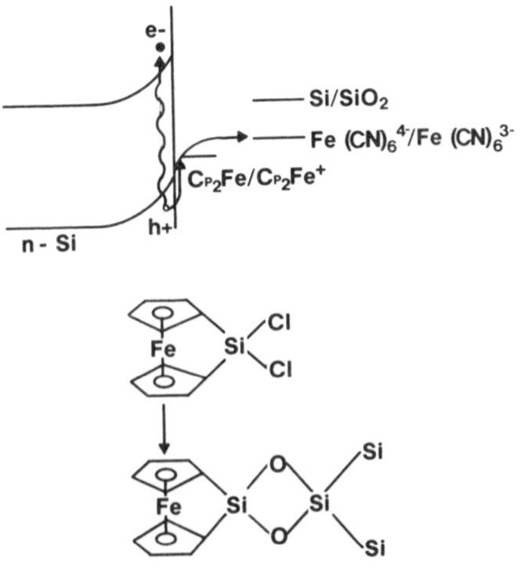

FIGURE 9. n-Type Si derivatized with ferrocene. Such modifications permit oxidation of $Fe(CN)_6^{4-}$ to $Fe(CN)_6^{3-}$ in solution and prevent oxidation of the electrode surface. Photogenerated holes are preferentially captured by the bound ferrocene derivative and are transferred to the redox couple in solution. The structure of the bound ferrocene complex is shown at the bottom of the figure.

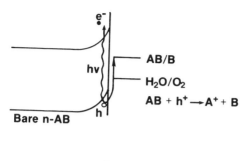

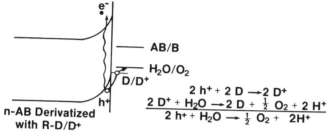

FIGURE 10. General scheme for chemical modification of an n-type semiconductor surface. Photooxidation of H_2O to O_2 rather than self-oxidation of the semiconductor is the favored reaction.

D. Particulate Photoelectrochemical Systems

Because of its inherent simplicity, a photoelectrosynthetic system that consists of microscopic semiconductor particles suspended in an electrolyte is an intriguing and potentially very important type of configuration. Particulate photoelectrosynthetic systems are being studied by several workers (1-4,7) and such particulates have been labeled photochemical diodes (1).

The most efficient photoelectrochemical particles have a heterogeneous structure with an n-type region in contact (through a non-blocking, ohmic-like junction) either to a metal or to a p-type region. The energetics of these heterogeneous structures result in the most efficient systems for spatial separation of photogenerated electrons and holes (1). Electrons are readily removed from the metal or p-type regions, while holes are easily removed from the n-type region to drive reduction and oxidation reactions, respectively, in the electrolyte. Such heterogeneous photochemical diodes have been used to split water (1,10,30), decarboxylate organic acids via a photo-Kolbe reaction (31,32), oxidize organic compounds (33), photoreduce N_2 (34,35), and photoreduce CO_2 (36,37).

In addition to heterogeneous structures, homogeneous semi-conductor particles have also been shown to drive photoelec-trosynthetic reactions (31,32,36,37). In the homogeneous case, band-bending at the liquid interface will be in the same direction everywhere, and this should result in a potential well for the majority charge carrier. This potential well should inhibit majority carrier charge transfer from the semi-conductor to the electrolyte, depending upon the width and height of the barrier. Experimental results comparing homo-geneous and heterogeneous semiconductor particles show that indeed the reaction rates on heterogeneous particles are much greater than those on homogeneous particles (31,32,36,37).

Work on single-crystal $SrTiO_3$ shows that photoinduced water-splitting can occur without platinization of one surface if the water layer in contact with the illuminated surface has a very high OH^- concentration (30). The rate of water-splitting is highest if a photochemical diode is formed by attaching a platinum layer to the back of the crystal, but a finite rate can also be produced without a platinum layer. Even if the $SrTiO_3$ crystal is not reduced (to make it con-ductive), and if the bands are consequently flat, a finite rate of water-splitting can also be observed. Nonplatinized, but reduced, $SrTiO_3$ permits electron injection either by tunneling through the barrier or by photoexcitation over the barrier. Electron- and hole-injection from unreduced and non-platinized $SrTiO_3$ also depend upon the relative rates of bulk diffusion and recombination. The dramatic effect of the OH^- concentration on the rate of H_2O-splitting is not understood.

The principal advantage of homogeneous particles is their simple structures. However, heterogeneous photochemical diodes consisting of metal-semiconductor structures can be made rather easily by in situ photoreduction of metal ions on the semiconductor particles (2,7,31,32). A simple method for producing photochemical diodes containing the n- and p-type semiconductor configuration has not yet been developed.

A particularly interesting example of a heterogeneous photochemical diode is the micro-heterogeneous redox system described by Grätzel and co-workers (3). It has been pointed out (4) that these novel redox systems are essentially dye-sensitized photochemical diodes containing an oxidation cat-alyst (RuO_2) and a reduction catalyst (Pt) on the surface. The dye is $Ru(bipy)_3^{2+}$ which, when excited by visible light, injects electrons into the conduction band of the TiO_2 semi-conductor-support to form an oxidized dye species that sub-sequently oxidizes water to O_2 in the presence of the RuO_2 catalyst. The injected electrons move through the conduction band of TiO_2 band and reduce water to H_2 at the Pt sites.

E. Layered Compounds and Other New Materials

A very active area of research in photoelectrosynthesis involves the study of new semiconductor electode materials. Layered chalcogenide compounds are particularly interesting since the top of the valence band is comprised of metal $d(z^2)$ orbitals rather than the anion 4p orbitals (39). Thus an optical transition near the bandgap involves only metal-to-metal transitions (since the conduction band is also comprised of d-like metal orbitals) and does not disrupt a metal-anion bond. This type of optical transition is believed to be less susceptible to photocorrosion than the usual transitions involving anion-like orbitals in the valence band (39). An alternative explanation for the stability has been proposed which is based on screening of the metal ions by the covalently bonded chalcogenide layers such that solvation of the metal ions by the electrolyte is prevented (40).

Results (41-45) with layered chalcogenides such as MoS_2 (E_g = 1.8 eV), $MoSe_2$ (E_g = 1.4 eV), $MoTe_2$ (E_g = 1.0 eV), WSe_2 (E_g = 1.6 eV), and WS_2 (E_g = 2.0 eV) show that enhanced resistance to photocorrosion is indeed obtained. However, photocorrosion readily occurs at the edges of layered materials (parallel to the c-axis), and this presents problems for the planar face as well, since atomic sized step dislocations in the surface can become photocorrosion sites.

Another class of semiconductor compounds that is receiving attention is the metal oxides. Oxide semiconductors are generally the most stable materials for photo-oxidizing water to O_2, but their bandgaps are too large for practical solar applications. Attempts are being made to reduce oxide bandgaps by creating d-bands above the oxygen 2p-band (46-49). However, one major problem with this approach is that low hole mobilities are expected for these systems.

F. Dye Sensitization

A great deal of work is in progress on improving the visible absorption properties of large bandgap, but relatively stable, semiconductor electrodes by dye-sensitization techniques (50-54). The general mechanism for this approach is shown in Figure 11. A dye (A) with strong absorption in the visible spectrum is bonded to the semiconductor surface. Upon excitation to A*, an electron is transferred to the semiconductor conduction band leaving A^+ at the surface. The electron in the semiconductor subsequently reduces H^+ to H_2 at a metal cathode, while A^+ oxidizes H_2O to O_2 to regenerate A. As seen in Figure 11, the net overall reaction is sensitized photolysis of H_2O into H_2 and O_2.

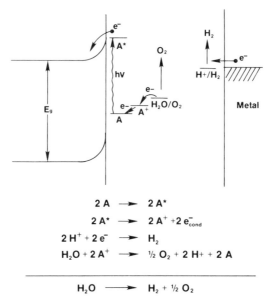

$$2A \longrightarrow 2A^*$$

$$2A^* \longrightarrow 2A^+ + 2e_{cond}^-$$

$$2H^+ + 2e^- \longrightarrow H_2$$

$$H_2O + 2A^+ \longrightarrow \tfrac{1}{2}O_2 + 2H+ + 2A$$

$$H_2O \longrightarrow H_2 + \tfrac{1}{2}O_2$$

FIGURE 11. Energy-level scheme for dye sensitization of n-type semiconductor electrodes.

The problems associated with dye sensitization are: (a) the dyes are generally organic species that have relatively poor long-term stability and thus do not survive repeated oxidation-reduction cycles; (b) the dye layer on an electrode must be very thin to permit charge transfer from the excited dye to the electrode, but this requirement generally precludes the dye layer also exhibiting intense optical absorption; and (c) most dye-semiconductor electrode combinations show very low quantum efficiencies for electron transfer, e.g., 4×10^{-3} for the rose bengal-TiO_2 system (53), although a quantum yield of unity has been reported for monochromatic irradiation of the tris(4,7-dimethyl-1,10-phenanthroline) ruthenium(II) system (54).

To date, however, no system has been found that exhibits long-term stability together with overall solar conversion efficiencies greater than about 1%.

IV. CONCLUSIONS

The photoelectrosynthetic approach to solar energy conversion is very appealing and exciting; rapid progress is being made both in the basic understanding of photoelectrochemical phenomena and in applied systems. The potential advantages of photoelectrosynthesis over photovoltaics coupled to dark electrosynthesis are higher net conversion efficiencies, better engineering designs for solar reactors, and unique catalytic effects that are possible with modified semiconductor electrode surfaces.

The main problem is the lack of a semiconductor material that exhibits both high conversion efficiency and long-term stability. However, one very promising approach involves chemical modification or derivatization of the semiconductor surface to enhance photostability and catalytic activity. This approach is interesting in that it represents a common intersection with some current directions of photochemical approaches to solar energy conversion (3). In the latter, systems are being studied that involve semiconductor particles as catalysts to help drive photochemical redox reactions. These semiconductors act as electron pools to facilitate redox chemistry and catalyze the desired reactions.

It appears that optimum photoelectrosynthetic systems will involve fine particles of chemically modified semiconductors. In my view, this is the most direct way to combine the requirements of high stability, strong optical absorption, and efficient separation of charge carriers; such materials will very likely show enhanced catalytic activity as well. Ultimately, particulate systems should make it easier to optimize the engineering criteria for design and assembly of practical devices that will convert solar energy into fuels and chemicals.

REFERENCES

1. A. J. Nozik, Appl. Phys. Lett. 30, 567 (1977).
2. A. J. Bard, J. Photochem. 10, 59 (1979).
3. M. Grätzel, in "Photoelectrochemistry", Disc. Faraday Soc., General Discussion No. 70, September 1980, Chapter 20, in press (1981); see also Chapters 5 and 6, this Volume.
4. A. J. Nozik, in ref. (3), Introduction.
5. A. J. Nozik, Phil. Trans. R. Soc. (London) A295, 453 (1980).
6. A. J. Nozik, D. S . Boudreaux, R. R. Chance and F. Williams, in "Interfacial Photoprocesses: Energy Conversion and Synthesis", (M. S. Wrighton, ed.), Advances in Chemistry Series, No. 184, American Chemical Society, Washington, D.C., 1980, pp. 155-171.
7. A. J. Bard, Science 207, 139 (1980).
8. A. J. Nozik, Annu. Rev. Phys. Chem. 29, 189 (1978).

9. L. A. Harris and R. H. Wilson, Ann. Rev. Mat. Science $\underline{8}$, 99 (1978).

10. M. S. Wrighton, Accts. Chem. Res. $\underline{12}$, 303 (1979).

11. M. A. Butler and D. S. Ginley, J. Mat. Sci. $\underline{15}$, 1 (1980).

12. H. Gerischer, in "Topics in Applied Physics," Vol. 31 (B. O. Seraphin, ed.) Springer-Verlag, Berlin, 1979, pp. 115-169.

13. M. Tomkiewicz and H. Fay, Appl. Phys. $\underline{18}$, 1 (1979).

14. A. J. Bard, A. B. Bocarsly, F.-R. F. Fan, E. G. Walton and M. S. Wrighton, J. Am. Chem. Soc. $\underline{102}$, 3671 (1980).

15. A. Many, Y. Goldstein and N. B. Grover, "Semiconductor Surfaces," Wiley-Interscience, New York, 1965, pp. 135-136.

16. A. B. Bocarsly, D. C. Bookbinder, R. N. Dominey, N. S. Lewis and M. S. Wrighton, J. Am. Chem. Soc. $\underline{102}$, 3683 (1980).

17. F.-R. F. Fan and A. J. Bard, J. Am. Chem. Soc. $\underline{102}$, 3677 (1980).

18. J. A. Turner, J. Manassen and A. J. Nozik, Appl. Phys. Lett. $\underline{37}$, 488 (1980).

19. J. A. Turner, J. Manassen and A. J. Nozik, in "Photo-effects at Semiconductor-Electrolyte Interfaces" (A. J. Nozik, ed.), ACS Symposium Series, No. 146, American Chemical Society, Washington, D.C., 1981, pp. 253-266.

20. S. M. Sze, "Physics of Semiconductor Devices", Wiley-Interscience, New York, 1969, Chapter 9.

21. M. Klausner, J. A. Turner and A. J. Nozik, to be published.

22. D. S. Boudreaux, F. Williams and A. J. Nozik, J. Appl. Phys. $\underline{51}$, 2158 (1980).

23. F. Williams and A. J. Nozik, Nature $\underline{271}$, 37 (1978).

24. R. T. Ross, personal communication; see also Chapter 11, this Volume.

25. H. Gerischer, J. Electroanal. Chem. $\underline{82}$, 133 (1977)

26. A. J. Bard and M. S. Wrighton, J. Electrochem. Soc. $\underline{124}$, 1706 (1977).

27. M. S. Wrighton, R. G. Austin, A. B. Bocarsly, J. M. Bolts, O. Haas, K. D. Legg, L. Nadjo and M. Palazzotto, J. Am. Chem. Soc. $\underline{100}$, 1602 (1978).

28. J. M. Bolts, A. B. Bocarsly, M. C. Palazzotto, E. G. Walton, N. S. Lewis and M. S. Wrighton, J. Am. Chem. Soc. $\underline{101}$, 1378 (1979).

29. M. S. Wrighton, Fourth DOE Solar Photochemistry Confer-ence, Radiation Laboratory, University of Notre Dame, Notre Dame, Indiana, June 1980.

30. F. T. Wagner, S. Ferrer and G. A. Somorjai, in ref. (19), pp. 159-178.

31. B. Kraeutler and A. J. Bard, J. Am. Chem. Soc. $\underline{100}$, 2239 (1978).

32. B. Kraeutler and A. J. Bard, J. Am. Chem. Soc. 99, 7729 (1977).
33. S. N. Frank and A. J. Bard, J. Am. Chem. Soc. 99, 4667 (1977).
34. C. R. Dickson and A. J. Nozik, J. Am. Chem. Soc. 100, 8007 (1978).
35. M. Koizumi, H. Yoneyama and H. Tamura, J. Am. Chem. Soc., in press (1981).
36. T. Inoue, A. Fujishima, S. Konishi, and K. Honda. Nature 277, 637 (1979).
37. M. Halmann, Nature 275, 155 (1979).
38. G. N. Schrauzer and T. D. Guth, J. Am. Chem. Soc. 99, 7189 (1977).
39. H. Tributsch, Ber. Bunsenges Phys. Chem. 81, 361 (1977); J. Electrochem. Soc. 125, 1086 (1978); Solar Energy Mat. 1, 705 (1979).
40. J. Gobrecht, H. Gerischer, and H. Tributsch, Ber. Bunsenges Phys. Chem. 82, 1331 (1978).
41. J. Gobrecht, H. Gerischer and H. Tributsch, J. Electrochem. Soc. 125, 2085 (1978).
42. B. A. Parkinson, T. E. Furtak, D. Canfield, K. Kam and G. Kline, in ref. (3), Chapter 23.
43. H. J. Lewerenz, A. Heller, H. J. Leamy and S. D. Ferris, in ref. (19), pp. 17–35.
44. T. Kawai, H. Tributsch and T. Sakata, Chem. Phys. Lett. 69, 336 (1980).
45. W. Kautek, H. Gerischer and H. Tributsch, Ber. Bunsenges Phys. Chem. 83, 1000 (1979).
46. R. D. Rauh, J. M. Buzby and S. A. Alkaitis, J. Phys. Chem. 83, 2221 (1979).
47. V. Buruswany and J. O'M. Bockris, Solar Energy Mat. 1, 441 (1979).
48. H. Jarrett, poster presentation in ref. (3).
49. J. Koenitzer, B. Khazai, J. Hormodaly, R. Kershaw, K. Dwight and A. Wold, J. Solid State Chem., in press (1981).
50. A. Hammett, M. P. Dare-Edwards, R. D. Wright, K. R. Seddon and J. B. Goodenough, J. Phys. Chem. 83, 3280 (1979).
51. H. Tsubomura, M. Matsumura, Y. Nomura and T. Amamiya, Nature 261, 5559 (1976).
52. H. Gerischer and F. Willig, Topics Current Chem. 31, 61–84 (1976).
53. M. T. Spitler and M. Calvin, J. Chem. Phys. 66, 4294 (1977).
54. W. D. K. Clark and N. Sutin, J. Am. Chem. Soc. 99, 14 (1977).

DISCUSSION

Prof. H. O. Finklea, Virginia Polytechnic Institute and State
University:
  What experiments are you performing to prove the "hot-
  charge carrier" theory?

Dr. Nozik:
  We are trying to do supra-band-edge reductions with p-type
  semiconductors wherein redox couples lying above (more
  negative than) the conduction band edge are reduced by hot
  electrons under conditions where reduction by thermalized
  electrons would not be possible. To do this, we first
  tried to establish flat-band potentials in the dark and in
  the light to determine the conduction band edges. This
  led to the problem of unpinning of the band edges as discussed in my lecture. Future experiments will be done to
  probe supra-band-edge reactions with pinned band edges.

Prof. Finklea:
  I suggest that you look for changes in rates and mechanisms of photoelectrochemical reactions as a function of
  photon energy.

Dr. Nozik:
  I agree, and we are also planning to carry out such
  experiments.

Prof. B. Holmström, Chalmers University of Technology,
Göteburg, Sweden:
  Regarding the use of dye sensitization for increasing the
  wavelength region that can be utilized by a high bandgap
  semiconductor: There seems to be a dilemma in that a thin
  dye layer can only absorb a small part of incident light,
  and a thick layer would have a high internal resistance.
  Would you please comment on this?

Dr. Nozik:
  You are quite correct; this is a difficulty that is well
  recognized.

Prof. Sir George Porter, Royal Institution:
  The low efficiency of dye sensitization will in many cases
  be due to the competitive process of concentration (self)
  quenching of the dye which, in condensed systems of this
  kind, is usually very fast. Has the rate of electron
  transfer from an excited dye molecule to semiconductor
  surfaces been measured?

Dr. J. A. Richardson, Lawrence Livermore Laboratory:
   I would like to comment on Prof. Porter's question
regarding the rate of electron injection into semiconductor electrodes. We have done preliminary coulostatic
experiments with dye-sensitized TiO_2 electrodes which show
a transient potential excursion in <10 ns, the limit of
our electronics measurement. This is an upper limit by a
<u>direct</u> measurement of the process in question.

Dr. J. J. Katz, Argonne National Laboratory:
   Concentration quenching and excimer formation in concentrated chlorophyll systems may not be a fatal obstacle to
energy transfer. The antenna chlorophyll in photosynthetic organisms is highly concentrated, yet energy transfer demonstrably occurs. If energy transfer is fast relative to other energy-dissipation mechanisms, as it may
well be, the failure of a concentrated system to fluoresce
may be irrelevant to its energy-transfer function.

Prof. H. Tsubomura, Osaka University:
   Concerning the problems of dye sensitization of semiconductor electrodes, I should like to point out that the
generally accepted notion, that electron injection from
excited dye to the semiconductor is low, is not true. The
low efficiency of dye-sensitization is generally due to
the small coverage of dye on the surface of the semiconductor electrodes. We found that the quantum efficiency
of electron injection from the excited dye to the semiconductor can be as high as 40% in the ZnO-rose bengal
system.

   If we use a porous semiconductor electrode, the efficiency
of dye-sensitized photocurrent can be fairly high. (cf.
recent papers by Tsubomura <u>et al</u>. in Nature). So, only
with a proper combination of semiconductor and dye (with
appropriate energy levels for each), can water decomposition be theoretically possible, although such a combination has been difficult to find so far.

Dr. Nozik:
   I think Prof. Mark Spitler (Mount Holyoke College) has
some experiments in progress to determine the quantum
efficiencies of such systems by an elegant technique.

Prof. M. Grätzel, Ecole Polytechnique Fédérale, Lausanne:
   I would also like to comment on dye sensitization of semiconductor electrodes. I believe it is preferable to
employ semiconductors in connection with a sensitizer/
relay couple. If only a semiconductor is employed with

the sensitizer the injection rate has to be very high and
has to compete with back reactions occuring through surface states. The reduced relay can be made long-lived and
there is no danger of back reactions.

Dr. J. Gobrecht, Solar Energy Research Institute:
Because of the similar lifetime of hot carriers in metals
and semiconductors, are the quantum yields for hot-
carrier injection from semiconductors comparable to those
obtained in the well-known process of photoemission from
metals into the electrolyte?

Dr. Nozik:
Theoretically, the lifetime of hot carriers in semiconductor-electrolyte systems can be much longer than in
metal-electrolyte systems. This is because the space-
charge layer that exists near the semiconductor surface
greatly enhances the possibility of carrier injection into
the electrolyte before full thermalization. This enhancement is produced by the high electric field ($>10^5$ V/cm) in
the space-charge region which (a) accelerates carriers to
the surface, and (b) for space-charge layers less than
100 Å, produces quantization of energy levels in the space
charge that greatly inhibit thermalization because of the
necessity of multi-phonon interactions. Hence, hot-
carrier processes can be much more probable in semiconductor electrodes than in metal electrodes, with correspond-
ingly higher quantum yields.

Dr. W. M. Kautek, Fritz-Haber-Institut der Max-Planck-Gesell-
schaft, Berlin:
You proposed that the formation of an inversion layer is
the reason for unpinning of the Fermi-level. Our results
agree with this perfectly. What system did you use?

Dr. Nozik:
We used p-type silicon in acetonitrile/methanol with an
organic redox couple (e.g., anthraquinone).

Dr. Kautek:
I believe the bandgaps of the layered semiconductors which
you gave (~1.5 eV) are too large; they are closer to
1.0 eV.

Dr. W. E. Pinson, Infrared Photo Ltd., Ottawa:
I have two comments: In the 1950's and 1960's, hot-
carrier effects in semiconductors were intensively studied
by semiconductor physicists. I suggest that this literature might be examined to determine whether hot-carrier

effects in semiconducors, e.g., p-Si, had been observed previously. Secondly, I would like to comment that in n-Ge, high electric fields (~kV/cm) can heat the carriers to several thousand Kelvins (with the lattice at 77K). However, in p-Ge, such fields cannot heat holes to more than about 150K, again with the lattice at 77K. The reason is that holes emit optical phonons copiously. To emit optical phonons, only small lattice perturbations, as by chemical impurities in the lattice, are required. The book, "Physics of Semiconductor Devices" by the group at Bell Labs contains a good discussion on these effects.

Dr. H. Tributsch, C.N.R.S., Laboratoires de Bellvue:
V. Gorochov and I have recently studied a semiconductor (PtS_2, $E_g \sim 1$ eV), which shows the phenomenon that you call "pinning" in a nearly ideal way. That is, the onset of photocurrents is shifted in a more or less systemmatic way with the redox potential of the electrolyte. PtS_2 has a $d \to d$ transition; however, ZrS_2 ($E_g \sim 1.68$ eV), which has exactly the same crystal structure and a $p \to d$ transition, does not exhibit this behavior. We attribute the phenomenon to surface states, arising from hole reactions over d-orbitals, which are charged according to the redox potential of the electrolyte. We have proposed the PtS_2 system as a photoelectroanalytical probe.

Prof. H. Kuhn, Max-Planck-Institut für biophysikalische Chemie, Göttingen:
It seems to me that the essential point concerning vectorial charge separation in photosynthesis is using the quantum mechanical tunneling effect in a particular way. There is a high, narrow energy barrier between the photocatalyst and electron donor, and a low, broad energy barrier between the photocatalyst and electron acceptor. The electron, after excitation, moves over the low barrier and is trapped. It cannot tunnel through the broad barrier and in this manner charge recombination is avoided. An electron from the donor finally moves to the oxidized photocatalyst since tunneling through the narrow barrier, even though it is high, is much easier than tunneling through the broad barrier.

CHAPTER 11

PHOTOCHEMICAL ENERGY STORAGE:
AN ANALYSIS OF LIMITS

James R. Bolton

Photochemistry Unit, Department of Chemistry
The University of Western Ontario
London, Ontario
Canada

Alan F. Haught

United Technologies Research Center
East Hartford, Connecticut
U.S.A.

Robert T. Ross

Department of Biochemstry
The Ohio State University
Columbus, Ohio
U.S.A.

I. INTRODUCTION

Photochemical conversion and storage of solar energy has
been a subject of considerable interest in the past few
years. This interest is quite natural as there already exist
at least two successful working systems which are in widespread use: the silicon solar cell for conversion of sunlight
to electricity, and natural photosynthesis for conversion and
storage of solar energy as chemical energy. These two examples illustrate the two major system types: those that convert sunlight to another transient energy form such as electricity (e.g., the solar cell) and those that convert and
store solar energy to a potential energy form such as chemical

PHOTOCHEMICAL CONVERSION AND
STORAGE OF SOLAR ENERGY

energy (e.g., photosynthesis). Our task in this paper is to deal primarily with the latter, namely photochemical energy storage. This includes a number of processes such as molecular energy storage in valence isomerization (1), photosynthesis (2), photoelectrochemical generation of fuels such as hydrogen from water (3), and direct or sensitized generation of fuels in homogeneous solution (4) or on a membrane structure (5). However, to establish the basis for our discussion of thermodynamic limits to photochemical energy storage, we must discuss conversion to a transient energy form as well.

Two kinds of constraints limit the efficiency of photochemical conversion and storage: thermodynamic and kinetic. The former establishes maximum conversion efficiencies for idealized systems while the latter takes into account losses due to nonidealities. The thermodynamic limit arises as an inherent consequence of the quantum nature of the photochemical conversion process and, for a single photochemical system with unfocused sunlight, results in a maximum conversion efficiency[1] of $\eta = 0.31$. At first glance, this result may appear unreasonably low; however, a number of authors have analyzed photochemical conversion processes and have established the validity of this thermodynamic limit with increasing levels of generality.

Trivich and Flinn (6) were the first to study this problem using an elementary model of a quantum converter. Although their model was later shown to be incomplete, they showed that η cannot exceed 0.44. Shockley and Queisser (7) introduced the concept of detailed balance and, using a specific semiconductor model as a quantum converter, showed that for a perfect p-n junction solar cell $\eta = 0.31$. Ross and Hsiao (8), based on earlier work by Ross (9) and Ross and Calvin (10), treated the problem using a chemical thermodynamic approach and showed that the efficiency limit of 0.31 applied to all quantum converters.[2] Haught (11) has recently developed a thermodynamic analysis based on Planck's and Kirchoff's laws, the thermodynamic requirement that the entropy of an interacting system be a maximum at equilibrium, and the principle of detailed balance. He has applied this analysis to both a

---

[1]Throughout this paper, conversion efficiencies, unless otherwise stated, will refer to a single photoconverter at an ambient temperature of 300K exposed to an unconcentrated AMO sun considered to be equivalent to a 6000K blackbody source. AM stands for air mass. AMO corresponds to solar radiation just outside the earth's atmosphere; AM1 corresponds to the solar radiation at the earth's surface with the sun at its zenith on a clear day.

thermal converter and a quantum converter, and for the latter obtained $\eta = 0.31$, a result applicable to all quantum conversion systems. In this paper, we seek to illustrate the nature of the thermodynamic limits by a comparative analysis of the Trivich and Flinn (6), Haught (11), Ross and Hsiao (8), and Shockley and Queisser (7) models and to show the effects of the assumptions and approximations used in each. We shall demonstrate that the last three models are exactly equivalent.

For photochemical systems leading to storage of solar energy, further losses are necessary in order for the products of the reaction to be stable. The storage of part of the absorbed photon energy requires that the reverse reaction be hindered. In this paper, we shall explore the nature of kinetic limits (12) for photochemical energy storage with an examination of the consequences of these limits on a number of systems.

A. The Thermodynamic Limit

In a chemical converter, the net radiant input photon flux (external incident photon flux less radiant photon emission) acts to drive uphill (i.e., $\Delta G > 0$) an atomic or molecular process (e.g., electron-hole excitation, molecular dissociation, etc.) which may be schematically written as (see Fig. 1)

$$A \underset{}{\overset{h\nu}{\rightleftarrows}} B \qquad\qquad\qquad [1]$$

The departure of reaction [1] from its equilibrium at the quantum system temperature[3] T_Q constitutes useful work which can be recovered upon reversion of the photochemical system to equilibrium. A number of assumptions and physical principles are involved in the analysis of photochemical conversion processes:

---

[2]Actually, Ross and Hsiao (8) obtained 0.29, and Shockley and Queisser (7) obtained 0.30. The differences are due to the absorber geometries assumed. Ross and Hsiao used an isotropic (spherical) absorber and Shockley and Queisser used a flat plate with radiation emitted from both sides. In this paper we consider a flat plate with radiation emitted from only one side.

[3]This temperature is usually taken to be the ambient temperature T. Henceforth, in this paper we will use the symbol T to represent both the quantum system temperature and the ambient temperature.

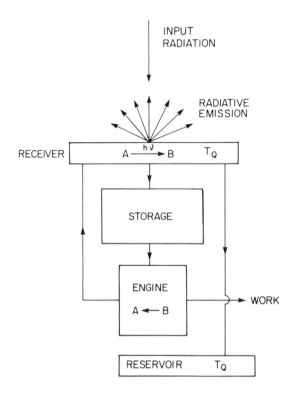

FIGURE 1. Scheme for a general quantum converter of solar energy to useful work.

1. The converter is a threshold device and a perfect absorber: i.e., the photochemical reaction has a threshold photon energy E_g below which the reaction will not occur. All photons with energies above the threshold (i.e., $\lambda \leqslant \lambda_g =$ hc/E_g) are absorbed by the photochemical system and induce the excitation reaction with unit quantum yield. Incomplete absorption can be incorporated into the model but leads to reduced efficiency and, for the purpose of comparison of models, only the maximum efficiency case will be considered here.[4] For $\lambda \geqslant \lambda_g$ no absorption occurs. Multiphoton absorption and the production of multiple excitations by a single high-energy photon are highly improbable events in a chemical system and are not considered.

---

[4]Ross and Collins (13) have presented a more detailed discussion of the perfect threshold absorber.

   2. Both A and B are thermalized at the quantum system temperature T: i.e., all photon absorption with $\lambda \leqslant \lambda_g$ leads to the production of B molecules which are thermalized with the environment. An important consequence of this condition is that for $\lambda \leqslant \lambda_g$, high-energy photons generate no more chemical energy than low-energy photons. Alternatively, high-energy photons could create a second product B' which could have a greater chemical energy content. Within the constraints of the model considered here A $\xrightarrow{h\nu}$ B' would be treated as a separate, independent photochemical conversion reaction operating in parallel with A $\xrightarrow{h\nu}$ B. In the case that B is an excited state of A, the thermalization of B corresponds to establishment of a thermally equilibrated distribution of B molecules among the vibrational and rotational sublevels of B at the temperature T. In a condensed medium this thermalization takes place in a time of ~1 ps or less (14).
   3. The principle of detailed balance is assumed. This requires that for the quantum system at thermodynamic equilibrium with thermal (blackbody) radiation at temperature T, the rate of photon absorption must equal the rate of emission in any wavelength interval.
   4. Kirchoff's law applies. This is closely related to the previous condition, i.e., that the absorption and emission coefficients are equal in any wavelength interval.
   5. Only two routes for the back reaction B → A are considered, namely radiative decay B → A + hν (the inverse of the excitation process, required by detailed balance) and a chemical nonradiative route which leads to the production of work and/or chemical energy storage. More general conditions, which lead to increased losses, are considered in the treatment of non-ideal systems (see Section III).
   6. Equilibrium for the radiation/quantum system is defined by that state for which the total entropy of the radiation field and quantum converter is a maximum. This, of course, is simply one statement of the Second Law of Thermodynamics.

B. A Comparison of Four Models

   1. The Trivich and Flinn Nonradiative Model (6). Ignoring radiative loss by the quantum system (i.e., disregarding the requirements of principles 3 and 6 and assumption 5) gives an oversimplified but intuitive picture of the quantum-conversion process and establishes a "First Law" upper bound to the conversion efficiency. From assumption 1, all photons with $\lambda \leqslant \lambda_g$ are absorbed and each absorbed photon produces an excited-state species of energy $E_g = hc/\lambda_g$. The efficiency η_E for this nonradiative model is then given by:

$$\eta_E = \frac{J_s \cdot E_g}{S} \qquad\qquad [2]$$

where:

$$J_s = \int_0^{\lambda_g} I_s(\lambda)d\lambda \qquad\qquad [3]$$

$$S = \int_0^{\infty} I_s(\lambda)\, \frac{hc}{\lambda}\, d\lambda \qquad\qquad [4]$$

and $I_s(\lambda)$ is the incident photon flux (photons $m^{-2}\ s^{-1}\ nm^{-1}$) in the wavelength band λ to $\lambda + d\lambda$.

The numerator in Eq. [2] is the total flux of excitation energy and the denominator S is the total incident solar power in $W\ m^{-2}$. η_E is plotted vs. λ_g in Figure 2 for AMO and AM1.2 solar radiation. If λ_g is small, then E_g is large but the rate of photon absorption J_s (and consequently the rate of excited-state production) is small and η_E is low. If λ_g is large, J_s, and hence the rate of excited-state production, is large but E_g is very small and again η_E is low as shown in Figure 2. Eq. [2] is a maximum for an intermediate value of λ_g which depends on the form of $I_s(\lambda)$. For a 6000K blackbody sun attenuated by the distance factor $(R_S/R_{EO})^2$, where R_S and R_{EO} are the radii of the sun and the earth's orbit, respectively, η_E has a maximum value of 0.44 at $\lambda_g = 1107$ nm. When the actual AMO radiation is used (15) the result is virtually the same (see the AMO curve of Fig. 2); for AM1.2 radiation (16), $\eta_E = 0.47$ at $\lambda_g = 1110$ nm.

Since the Trivich and Flinn nonradiative model takes account only of the First Law of Thermodynamics and ignores the radiative losses required by detailed balance, η_E is a substantial overestimate of the maximum conversion efficiency; however, this model does illustrate the effect of threshold energy optimization and the severe efficiency penalty associated with the quantum conversion process.

2. **Thermodynamic Model.** Incorporating each of the principles and assumptions 1-6 listed above, Haught (11) has developed a thermodynamic model of quantum conversion which is independent of the specific form of the conversion reaction or quantum system (see Fig. 1).

The net photon flux received by a quantum conversion system is the difference between the input radiant flux absorbed and the radiation emitted. By detailed balance (principle 3) a surface in equilibrium with a radiation field at temperature T_R will emit a photon flux (photons $m^{-2}\ s^{-1}\ nm^{-1}$) of:

$$I_{EMS}(\lambda) = I(\lambda, T_R, 2\pi)\sigma(\lambda, T) \qquad\qquad [5]$$

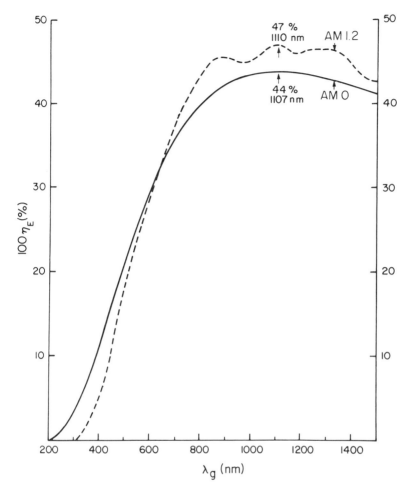

FIGURE 2. Percentage of the input solar energy available as energy of the excited state for the Trivich and Flinn non-radiative model (6). λ_g is the bandgap wavelength (= hc/E_g); AM0 and AM1.2 are solar distributions taken from refs. (15) and (16), respectively.

where $I(\lambda, T_R, 2\pi)$ is the Planck radiation flux and $\sigma(\lambda, T)$ is the surface absorption coefficient (equal to the surface emission coefficient by Kirchoff's law). For a given radiation-matter interaction at equilibrium, from principle 6, the total change in entropy due to the interaction must be zero. Absorption of a photon of wavelength λ represents a thermal energy loss of hc/λ from the radiation field and hence a decrease in the entropy of the radiation field of $hc/\lambda T_R$.

In a quantum conversion system, an amount μ of the absorbed photon energy may be extracted as work from the energy of the excited state. The remaining energy $(hc/\lambda - \mu)$ is dissipated as heat at the quantum system temperature T, resulting in an entropy increase of $(hc/\lambda - \mu)/T$. The quantity μ will be recognized as the partial Gibbs energy or chemical potential of the excited state, but in this treatment it is introduced simply as a parameter which characterizes the quantum system. Equating the entropy loss from the radiation field with the entropy increase of the quantum system, the temperature of the blackbody radiation field, with which a quantum system characterized by μ and T would be in equilibrium, is:

$$T_R = \frac{hc/\lambda}{(hc/\lambda - \mu)} T \qquad [6]$$

Substituting this result for T_R into the explicit form for the Planck radiation flux gives the quantum system emission flux in terms of the quantum system parameters T and μ:

$$I_Q(\lambda,\mu,T,2\pi) = \frac{2\pi n^2 c}{\lambda^4}\left\{\exp\left(\frac{(hc/\lambda - \mu)}{kT}\right) - 1\right\}^{-1} \qquad [7]$$

Using this result, assumption 2, and principle 5, for a quantum system at temperature T, the net absorbed photon flux (in photons $m^{-2} s^{-1} nm^{-1}$) is:

$$J_{NET\ ABS} = \int_0^{\lambda_g} \left\{I_s(\lambda) + I_{BB}(\lambda) - I_Q(\lambda,\mu,T,2\pi)\right\}d\lambda \qquad [8]$$

where $I_{BB}(\lambda)$ is the ambient blackbody photon flux and I_s is the incident solar flux, each for the wavelength interval λ to $\lambda + d\lambda$. From assumption 1, for each net photon absorbed with $\lambda \leqslant \lambda_g$, a unit reaction A $\overset{h\nu}{\rightarrow}$ B takes place and an amount of energy μ is retained as the work available from the excited state. The rate of conversion of photon energy to useful work is then $\mu \cdot J_{NET\ ABS}$; dividing by the incident radiant power, the conversion efficiency of a general quantum converter is:

$$\eta_Q = \frac{\mu \cdot \int_0^{\lambda_g} \left[I_s(\lambda) + I_{BB}(\lambda) - I_Q(\lambda,\mu,T,2\pi)\right]d\lambda}{S} \qquad [9]$$

where S is defined by Eq. [4].

Using Eq. [7] and numerically evaluating the optimum value of η_Q as a function of μ for different values of λ_g for a 6000K blackbody source, the results for the general thermodynamic model (maximum value of η_Q vs. λ_g) are shown in

Figure 3 for AMO and AM1.2. From these calculations the maximum conversion efficiency at AMO for a quantum solar converter is 0.31 and is obtained for λ_g = 949 nm and μ = 0.965 eV.

Consideration of the requirement for emission from the quantum system and the consequences of radiation/converter equilibrium both reduce the quantum system conversion efficiency from that calculated in Eq. [2]. As displayed in Eq. [8], the net flux of photons absorbed is directly reduced by the quantum emission. This effect is relatively small, the emission being only 2.6% of the input flux with λ_g = 949 nm.

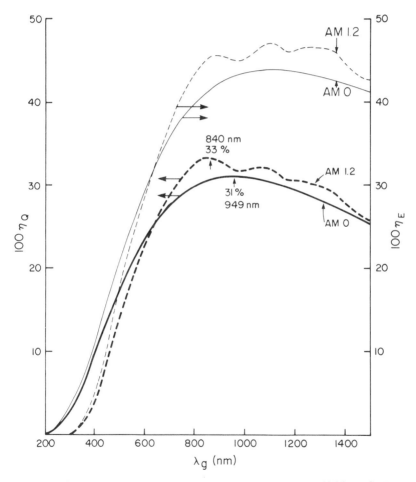

FIGURE 3. Plots of the maximum efficiency (100 η_Q) for an ideal quantum converter as a function of λ_g for AMO and AM1.2 solar distributions (heavy lines). The corresponding efficiencies (100 η_E) for the Trivich and Flinn nonradiative model (Fig. 2) have been plotted in light lines for reference.

The major loss occurs in the fact that the available work μ is considerably less than E_g: for λ_g = 949 nm, E_g = 1.306 eV and μ = 0.965 eV, hence E_{loss} = ($E_g - \mu$) = 0.341 eV. This latter result is an inherent consequence of the quantum radiation conversion process optimized for maximum power. In Eq. [9], μ, the available energy of the excited state, is a free parameter which can be larger, smaller or equal to Eg, the excited state internal energy. Optimizing Eq. [9] for maximum power implicitly determines μ, and for this condition the competing effects of available energy per excitation and quantum system re-radiation always result in μ being less than E_g. Any other values of μ chosen for system operation would give a lower power-conversion efficiency; for example, choosing $\mu > E_g$ would result in a negative conversion efficiency, i.e., net energy loss by the system.

For chemical processes that take place at constant temperature and constant pressure, the change in Gibbs energy μ differs from the internal energy change E_g by $-P\Delta V + T\Delta S$, where ΔV and ΔS are, respectively, the volume and entropy change per reaction. The volume change and the entropy change are each composed of two terms: the standard volume and entropy changes which have fixed values for a given reaction at a specific temperature and pressure, and a volume and entropy of mixing which depend on the constituent concentrations. The volume of mixing term is generally small but the entropy of mixing can be large and accounts for the entropy associated with the dispersal of the newly formed excited state among the unexcited states. Depending on the relative concentrations of reactants and products the mixing terms $-P\Delta V_{mix} + T\Delta S_{mix}$ (physically, the negative of the Gibbs energy of mixing) may be positive, zero or negative, and adjustment of the relative concentrations by changing the degree of excitation is the mechanism by which μ can be made equal to the value obtained from Eq. [9] for maximum power conversion, regardless of the volume and standard entropy change characterizing the specific reaction involved. The energy lost, E_{loss} = ($E_g - \mu$), can never be recovered as useful work since in any subsequent extraction process $\Delta G \leqslant 0$, and μ represents the maximum work that can be obtained in a quantum radiation conversion process optimized for maximum power.

It is important to point out that the energy loss ($E_g - \mu$) is entirely due to a consideration of the Second Law and, hence, does not depend on the specific mechanism of the quantum conversion process. Thus η_Q is the maximum thermodynamic conversion efficiency within the assumptions considered above.

3. **Chemical Potential Model (8).** While in the thermodynamic model the quantity μ is determined implicitly in the evaluation of the conversion efficiency, Ross and Hsiao (8)

carried out their analysis of the quantum system conversion efficiency using explicit calculations of the differences in the chemical potentials between reactant and products in a photochemical reaction. The chemical potential model is just as general as the thermodynamic model; however, for pedagogical purposes we shall consider a special case with the following characteristic (8,10): A photochemical converter is considered to be an ideal chemical system obeying Boltzmann statistics; i.e., a system for which the chemical potential of each consituent has the form:

$$\mu_i(T,P,x_i) = \mu_i^o(T,P) + kT \ln x_i \qquad [10]$$

where $\mu_i^o(T,P)$ is the chemical potential of the standard state and x_i is the mole fraction of component i.

Although the point of view is different, the quantities introduced in the chemical potential model have their counterparts in the thermodynamic treatment, and these parallels will be discussed in this section. Our derivation is parallel to that of Ross and Hsiao (8) but differs in detail because of our assumption of an ideal chemical system.

Referring to Figure 1, B is usually the first excited singlet state of A in chemical systems; thus, we shall develop the argument in terms of a simple two-level system of A and A^*. Since A and A^* are each in thermal equilibrium with the environment at temperature T (assumption 2), they may be treated as separate chemical species:

$$\mu_A = \mu_A^o + kT \ln x_A \qquad [11]$$

$$\mu_{A^*} = \mu_{A^*}^o + kT \ln x_{A^*} \qquad [12]$$

The chemical potential of the excited state A^* relative to the unexcited state A is:

$$\begin{aligned}
\mu &= \mu_{A^*} - \mu_A \\
&= \mu_{A^*}^o - \mu_A^o + kT \ln\left(\frac{x_{A^*}}{x_A}\right) \qquad [13]
\end{aligned}$$

With no external illumination, i.e., with only blackbody radiation at T present, A and A^* must be in thermal equilibrium with each other. Hence, $\mu = 0$ and:

$$\mu_A^o - \mu_{A^*}^o = kT \ln\left(\frac{x_{A^*}^{eq}}{x_A^{eq}}\right) \qquad [14]$$

where x_A^{eq} and $x_{A^*}^{eq}$ are, respectively, the equilibrium mole fractions of A and A^* at T. Substituting this result into

Eq. [13], the chemical potential under steady illumination can be written:

$$\mu = -kT \, \ell n\left(\frac{x_{A*}^{eq}}{x_A^{eq}}\right) + kT \, \ell n\left(\frac{x_{A*}}{x_A}\right) \tag{15}$$

Assuming that the excitation does not appreciably alter the unexcited state population (i.e., $x_A = x_A^{eq} = 1$) Eq. [15] becomes:

$$\mu = kT \, \ell n\left(\frac{x_{A*}}{x_{A*}^{eq}}\right) \tag{16}$$

Neglecting stimulated emission, which is expected to be very small for all solar intensities, the photon emission rate for $A* \to A + h\nu$ is given by the concentration of $A*$ multiplied by the emission probability P_E:

$$J_r = N_{A*}P_E \tag{17}$$

From detailed balance (principle 3) under thermal equilibrium conditions at T, the blackbody-induced rate of emission is:

$$J_r^o = J_{BB} = \int_0^{\lambda_g} I_{BB}(\lambda)d\lambda \tag{18}$$

where $I_{BB}(\lambda)$ is the blackbody photon flux in the wavelength band from λ to $\lambda + d\lambda$. Using Eq. [18] in Eq. [17] gives:

$$N_{A*}^{eq} = \frac{J_{BB}}{P_E} \tag{19}$$

Combining Eqs. [16], [17] and [19] we obtain:

$$\mu = kT \, \ell n\left(\frac{J_r}{J_{BB}}\right) \tag{20}$$

The chemical potential is a maximum when no chemical work is extracted from the system, i.e., when the total photon emission rate equals the rate of excitation, and hence:

$$\mu_{max} = kT \, \ell n\left[\frac{J_s + J_{BB}}{J_{BB}}\right] \tag{21}$$

where J_s (Eq. [3]) is the rate of excitation of the system by an external light source. In the thermodynamic model of Haught (11), this is the value of μ obtained by setting $\eta_Q = 0$ in Eq. [9].

The chemical power produced by a photochemical system is:

$$P = (J_s - J_r) \cdot \mu = J_s(1 - \phi_{loss}) \cdot \mu \qquad [22]$$

where $\phi_{loss} = J_r/J_s$ is the radiative quantum yield. Dividing by S, the incident solar power, the conversion efficiency is:

$$\eta_P = \frac{J_s(1 - \phi_{loss}) \cdot \mu}{S} \qquad [23]$$

From Eq. [20] we get:

$$\mu = kT \ln\left(\frac{J_r}{J_{BB}}\right) = kT \ln\left(\frac{J_r}{J_s} \cdot \frac{J_s}{J_{BB}}\right) \qquad [24]$$

Using Eq. [21] and the assumption that $J_s \gg J_{BB}$ gives:

$$\mu = \mu_{max} + kT \ln \phi_{loss} \qquad [25]$$

Incorporating Eq. [25] into Eq. [23] and setting the derivative of Eq. [19] with respect to ϕ_{loss} equal to zero, the condition for maximum power conversion is:

$$1/\phi_{loss} - \ln \phi_{loss} = 1 + \mu_{max}/kT \qquad [26]$$

from which the approximation:

$$\phi_{loss} \simeq \frac{kT}{\mu_{max}} \qquad [27]$$

can be derived. Using this result in Eq. [25] gives the chemical potential for maximum power conversion:

$$\mu_P = \mu_{max} + kT \ln\left(\frac{kT}{\mu_{max}}\right) \qquad [28]$$

Here, μ_P is the same quantity as μ in the thermodynamic model (see Eq. [9] for maximum η_Q for a given λ_g).
  Using Eqs. [20], [26] and [27] in Eq. [23] gives:

$$\eta_P = \frac{J_s\left[1 - \frac{kT}{\mu_{max}}\right] \cdot \left[\mu_{max} + kT \ln\left(\frac{kT}{\mu_{max}}\right)\right]}{S} \qquad [29]$$

or $\eta_P = 0.31$ for $\lambda_g = 949$ nm at AM0 (6000K blackbody source) and $\eta_P = 0.33$ with $\lambda_g = 840$ nm at AM1.2. In fact, to the extent that the assumption $J_s \gg J_{BB}$ and Eq. [20] are valid, η_P is <u>exactly</u> the same as η_Q, and thus Figure 3 is also a plot of η_P vs. λ_g.

It is instructive to examine the dependence of the values of $E_{loss} = (E_g - \mu)$ and ϕ_{loss} for maximum power obtained with different values of λ_g. These plots are shown in Figure 4. Note that for 400 nm $\lesssim \lambda_g \lesssim$ 1000 nm, i.e., the region of most interest for solar photochemistry, $E_{loss} = 0.40 \pm 0.07$ eV and $\phi_{loss} < 0.03$.

We have presented the case in which ϕ_{loss} is chosen to give an optimal η_Q. Normally ϕ_{loss} is dictated by the characteristics of the absorber and thus it is important to know how η_Q depends on ϕ_{loss}. This was first done by Ross and Calvin (10) and we display a similar plot in Figure 5.

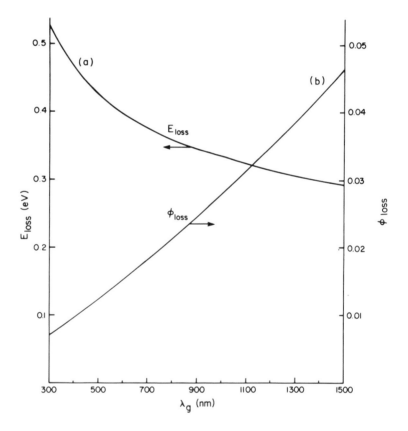

FIGURE 4. Plot of (a) $E_{loss} = (E_g - \mu)$ and (b) ϕ_{loss} vs. λ_g for AM0 solar radiation. The values of E_{loss} and ϕ_{loss} are those which give an optimal efficiency η_Q as a function of λ_g.

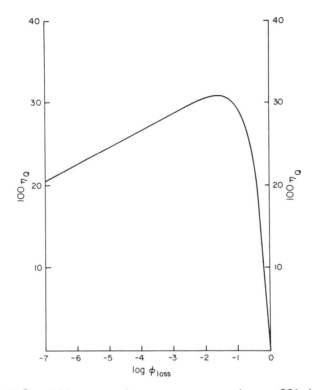

FIGURE 5. Effect on the power-conversion efficiency when ϕ_{loss} is varied [see Ross and Calvin (10)].

Physically, $(E_g - \mu)$, as identified in connection with the thermodynamic model, is equal to $P\Delta V^0 + T\Delta S^0 - \Delta G_{mix}$. For an ideal chemical system $\Delta V_{mix} = 0$; hence, $\Delta G_{mix} = -T\Delta S_{mix}$. The equivalence of $(E_g - \mu)$ with $-P\Delta V^0 + T\Delta S^0 + T\Delta S_{mix}$ is readily demonstrated in the chemical potential model. From Eq. [13] we can write:

$$
\mu = \mu_{A*}^0 - \mu_A^0 + kT \ln\left(\frac{x_{A*}}{x_A}\right)
$$

$$
= \mu^0 + kT \ln\left(\frac{x_{A*}}{x_A}\right)
$$

[30]

The entropy of mixing of n_{A*} moles of A^* at mole fraction x_{A*} with n_A moles of A at mole fraction x_A is:

$$
S_{mix} = -R\left(n_{A*} \ln x_{A*} + n_A \ln x_A\right)
$$

$$
= -k\left(N_{A*} \ln x_{A*} + N_A \ln x_A\right)
$$

[31]

where N_{A^*} and N_A are, respectively, the numbers of A^* and A molecules present. With the creation of one additional molecule of A^* from a molecule of A in a system sufficiently large that x_{A^*} and x_A do not change appreciably:

$$S_{mix} = -k\left[(N_{A^*} + 1) \, \ell n \; x_{A^*} + (N_A - 1) \, \ell n \; x_A\right] \qquad [32]$$

and the change in the entropy of mixing is:

$$\Delta S_{mix} = -k\left(\ell n \; x_{A^*} - \ell n \; x_A\right)$$

$$= -k \; \ell n\left(\frac{x_{A^*}}{x_A}\right) \qquad [33]$$

Using Eqs. [30] and [33] gives:

$$E_g - \mu = E_g - \mu^o + T\Delta S_{mix}$$

$$= -P\Delta V^o + T\Delta S^o + T\Delta S_{mix} \qquad [34]$$

Thus with a general reaction, for which μ^o may be less than, equal to or even greater than E_g, for maximum power $T\Delta S_{mix}$ must assume that value for which $\mu^o - T\Delta S_{mix} = \mu$ has the optimum value given explicitly by Eq. [28] (and implicitly by Eq. [9]), and the mole fractions x_A and x_{A^*} at which maximum conversion efficiency is obtained, are a function of μ^o for the reaction system. If the change in standard entropy and molar volume in the reactions are both zero ($\Delta S^o = 0$ and $\Delta H^o = 0$) as is usually the case for electronic excitation of a moderately large molecule, then $\mu^o = E_g$ and $(E_g - \mu) = T\Delta S_{mix}$, i.e., the term containing the entropy of mixing comprises the Second Law energy loss of the quantum conversion system.

4. **The Ideal Photovoltaic Cell: An Illustrative Example.** Additional insight is obtained by examining the conversion of radiant energy to work by a specific device, viz., the ideal photovoltaic cell as analyzed by Shockley and Queisser (7). We will develop the argument utilizing the treatment of the ideal junction diode presented by Angrist (17).

Consider the current flows shown in Figure 6a, for a p-n junction at thermal equilibrium at T, i.e., with no external illumination. There will be a small number of electrons being excited by thermal energies from the Fermi level at E_f to the conduction band on the p side resulting in a current j_g (electrons $m^{-2} \; s^{-1}$) given by:

$$j_g = A \; \exp\left[\frac{-\left(E_g - E_f\right)}{kT}\right] \qquad [35]$$

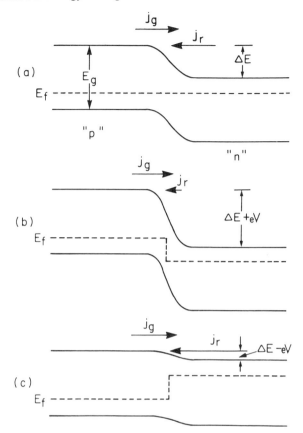

FIGURE 6. Dark currents in a p–n junction diode. E_g is the energy gap between the conduction and valence bands, E_f is the Fermi level, and ΔE is the energy difference between the p and n sides induced by the space charge at the junction and V is the applied voltage: (a) junction with no applied voltage; (b) forward bias voltage; (c) reverse bias voltage applied.

Since the system is at equilibrium, there must be an equal and opposite reverse current j_r which is due to electrons from the n side crossing the junction and recombining with the larger number of holes on the p side.

Applying the bias voltage to the junction as in Figures 6b and 6c, the potential difference between the Fermi level and the conduction band on the p side is unchanged and j_g remains the same. However, the potential barrier for electrons from the n side to reach the p–conduction band is altered and now:

$$j_r = A \, \exp\left[\frac{-(E_g - E_f - eV)}{kT}\right] = j_g \, \exp\left(\frac{eV}{kT}\right) \qquad [36]$$

The net electron current flow will be the difference:

$$j_r - j_g = j_g\left(e^{\frac{eV}{kT}} - 1\right) \qquad\qquad [37]$$

An exactly analogous treatment can be made for the flow of holes. The total current will then be the sum of electron and hole currents, so that the ideal junction current with no external illumination is:

$$j_j = j_0\left(e^{\frac{eV}{kT}} - 1\right) \qquad\qquad [38]$$

where j_0 is the "saturation" or "dark" current.

The equivalent circuit of an ideal photovoltaic cell can be drawn as shown in Figure 7. When the cell is illuminated:

$$j = j_s - j_j = j_s - j_0\left(e^{\frac{eV}{kT}} - 1\right) \qquad\qquad [39]$$

where j_s is the light-induced current and j is the current flowing through the load. From Eq. [39] the short-circuit current (V=0) is j_s and the open-circuit voltage V_{oc} (j=0) is:

$$V_{oc} = \frac{kT}{e} \ln\left(\frac{j_s + j_0}{j_0}\right) \qquad\qquad [40]$$

Comparing Eqs. [40] and [21], we see that μ_{max} corresponds physically to the open-circuit voltage of the ideal photovoltaic cell. The power dissipated in the load is:

$$P = V \cdot j = V\left[j_s - j_0\left(e^{\frac{eV}{kT}} - 1\right)\right] \qquad\qquad [41]$$

Setting the derivative of Eq. [41] with respect to V equal to zero, the voltage under maximum power conditions can be shown

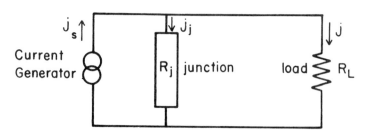

FIGURE 7. Equivalent circuit for an ideal photovoltaic cell.

to be given by:

$$V_m \simeq V_{oc} + \frac{kT}{e} \ln\left(\frac{kT}{eV_{oc}}\right) \qquad\qquad [42]$$

Comparing Eq. [42] with Eq. [28], we see that V_m of the photovoltaic cell corresponds directly with μ. Substituting Eq. [42] into Eq. [41], and noting that eV_{oc} is large compared to kT, j_m, the current through the load under maximum power conditions, is:

$$j_m \simeq j_s\left(1 - \frac{kT}{eV_{oc}}\right) \qquad\qquad [43]$$

Similarly, a comparison of Eq. [43] with Eqs. [23] and [27] shows that kT/eV_{oc}, the fraction of the total current j_s, which at maximum power must occur as electron-hole recombination in the junction, corresponds to ϕ_{loss}, i.e., the radiative quantum loss which must be present for maximum power output.

C. Summary Comparison of the Thermodynamic Models

Three analyses have been used to develop the efficiency of a photochemical system for the conversion of sunlight to useful work. The treatment of the ideal photovoltaic cell displays the physical significance of the quantities involved in terms of measurable currents and voltages. The chemical potential model extends these results to ideal photochemical as well as photovoltaic systems. The thermodynamic analysis demonstrates that the same limits apply for all quantum conversion devices. In Table I we collect the important quantities from the three treatments to show how they relate to one another.

From each analysis the maximum conversion efficiency of a single absorber system with one photoreaction and unconcentrated AMO sunlight is 0.31 as described earlier. **This limitation is thermodynamic and fundamental; it cannot be overcome by the choice of a different photosystem, separation of products, use of catalysts, or any other strategem.**

We have carried out our analysis for a specific set of conditions. Changing the conditions will affect the limiting efficiency. A number of authors have considered the effects of multiple photosystems (8,18), concentration of the solar input (11,18), and the temperature of the photoconverter (18). Each of these leads to a different limiting efficiency (albeit at the practical expense of generating the altered operating conditions), but in each case the conversion efficiency obtained is that produced in accord with the thermodynamic principles discussed in this chapter.

Table I. Comparison of the Important Quantities among the Three Thermodynamic Treatments

Concept	Thermodynamic Model[a]	Chemical Potential Model[b]	Ideal Photovoltaic Cell[c]
Maximum potential (yield = 0)	μ when $\eta_Q = 0$	μ_{max}	eV_{oc}
Maximum power potential	μ for a maximum η_Q at fixed λ_g	$\mu_p = \mu_{max} + kT \ln \phi_{loss}$	eV_m
Requisite yield losses	$\dfrac{I_Q(\lambda)}{I_s(\lambda) + I_{BB}(\lambda)}$	$\phi_{loss} \simeq \dfrac{kT}{\mu_{max}}$	$\dfrac{kT}{eV_{oc}}$

[a]See Eq. [9]
[b]See Eqs. [21],[25]-[27]
[c]See Eqs. [40],[42],[43]

II. PHOTOCHEMICAL ENERGY STORAGE

In Section I, we have seen that photochemical conversion of solar energy results in the production of a photostationary state $A \xrightarrow{h\nu} A^*$, the Gibbs energy of which is available as work which may be used directly (for example, as electricity from a photovoltaic cell) or may be employed to form a stable product, $A^* \to P$, in which chemical energy is stored for subsequent use. The thermodynamic considerations that were developed in Section I limit the efficiency with which sunlight can be converted to the Gibbs energy of the photostationary state.

In this section, we address the storage efficiency associated with subsequent conversion of the Gibbs energy of the photostationary state associated with A^* to chemical energy of a stable product P. Initial consideration of the storage efficiency by Bolton (12) suggested a requirement of an activation barrier for the back reaction ($P \to A^* \to A$) to provide for product stability (see Fig. 8a). For the simple case of a unimolecular reaction, the activation barrier was estimated from unimolecular rate theory to be ~0.6 – 1.0 eV, an energy loss which resulted in a decrease in the photoconversion efficiency from ~31% for the photostationary state A^*

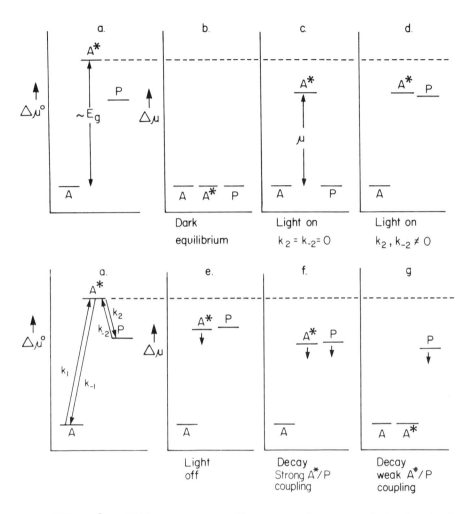

FIGURE 8. Gibbs energy diagrams for a photochemical energy storage system: (a) shows the standard Gibbs energy plot which is appropriate for a molecular picture whereas in (b) through (g) the effects of concentration have been incorporated. See text for further explanation.

to ~20% for the storable fuel, P. In this section we demonstrate that, although an activation barrier for the back reaction is important, it is not relevant to the question of a limiting conversion efficiency. In principle, the energy-loss penalty for storage can be vanishingly small; however, in practice, for real (nonideal) systems, loss processes such as a significant back reaction will produce a substantial decrease in the obtainable efficiency. We shall first treat the ideal case of photochemical storage and then examine the consequences of loss processes represented in nonideal cases.

An energy-level diagram of an elementary photochemical system allowing for storage is shown in Figure 8. Three levels are involved: the initial or ground state A, the photo-excited state A^* radiatively coupled with the ground state, and the product state P kinetically coupled with A^* via a chemical reaction path. Excitation by constant incident radiation raises the population of A^* to a photostationary level which then reacts to form the stable product P. The diagrams in Figure 8 are drawn in terms of the Gibbs energy and thus take account of the entropy associated with each state. Figure 8a displays the standard Gibbs energy of the three states for a defined set of standard temperature and pressure conditions at unit concentration. However, it is the difference in the Gibbs energy ΔG, not the standard free energy ΔG^O, which constitutes the available energy of a chemical system. The Gibbs energy for a given state is given by:

$$G_A = G_A^O + f(x_A) \tag{44}$$

where G_A^O is the standard Gibbs energy per mole of state A and $f(x_A)$ is a function of the mole fraction x_A of the state. For an ideal chemical system:

$$G_A = G_A^O + RT \ \ell n \ x_A \tag{45}$$

and by dividing by Avogadro's number N_O the Gibbs energy per molecule for an ideal chemical system is given by Eq. [11].[5] The difference in Gibbs energy per molecule between two states (e.g., A^* and P) is given by:

$$\Delta\mu(A^*,P) = \mu_{A^*} - \mu_P = \Delta\mu^O(A^*,P) + kT \ \ell n\left(\frac{x_{A^*}}{x_P}\right) \tag{46}$$

where x_{A^*} and x_P are the mole fractions of A^* and P respectively. When $x_{A^*} = x_P$ the Gibbs energy change per molecule is $\Delta\mu^O$; however, depending on the relative concentrations of A^* and P, $\Delta\mu$ can be larger, smaller or even of different sign than $\Delta\mu^O$.

---

[5]The treatment that follows is conventionally couched in terms of G, the Gibbs energy per mole. The discussion is conducted here on a per molecule basis for ease of integration with the results of the chemical potential analysis in Section I.

The $\Delta\mu$ diagrams for an elementary photochemical conversion system with storage are displayed in Figures 8b–8g and represent a series of different operating conditions; Figure 8a displays the $\Delta\mu^o$ diagram for reference. Also, in Figure 8a are displayed the interaction paths accessible to this simplest photochemical storage system: A^* is radiatively coupled to A with rate constants k_1 and k_{-1} and kinetically coupled to P with forward and reverse rate constants k_2 and k_{-2}, respectively; in this system no direct path exists between P and A.

At equilibrium (with no radiation incident), the Gibbs energy of all three states must be the same, as shown in Figure 8b. The relative populations at equilibrium ($\Delta\mu = 0$) depend on the values of $\Delta\mu^o$ and the form of the function $f(x)$. For an ideal chemical system, from Eq. [46], the mole fraction of A^* at equilibrium is given by:

$$x_{A^*}^{eq} = x_A^{eq} \exp\left(\frac{-\Delta\mu^o(A^*,A)}{kT}\right) \qquad [47]$$

where the superscript (eq) denotes conditions of equilibrium. If $\Delta S^o = 0$ and $\Delta V^o = 0$, as is usually the case for electronic excitation of a large molecule, then $E_g = \Delta\mu^o(A^*,A)$; this equivalence has been indicated by the level of E_g in Figure 8a.

With radiation incident, but with no interaction between A^* and P (i.e., $k_2 = k_{-2} = 0$), x_{A^*} is increased, resulting in an increase[6] of the Gibbs energy of A^* relative to A (Fig. 8c). The Gibbs energy difference between A^* and A, $\Delta\mu^o(A^*,A)$, is just μ of the photostationary state as developed earlier, and has a value determined explicitly by Eq. [28] (and implicitly by Eq. [9]) for a given intensity and spectral distribution of the radiation. As discussed in connection with Eqs. [9] and [34], μ is always less than E_g.

With k_2 and $k_{-2} \neq 0$, the concentration of P will increase until A^* and P are in equilibrium and $\Delta\mu(A^*,P) \simeq 0$, as shown in Figure 8d. The Gibbs energy per molecule of the storage state P relative to the ground state A is then μ, the same as the value achieved by the intermediate state A^*. When the illumination is removed, A^* is no longer radiatively driven to the photostationary Gibbs energy μ, and, as shown in Figure 8e, relaxes to the dark equilibrium state shown in Figure 8b. The minimum relaxation rate for $A^* \to A$ is that due to the intrinsic radiative emission from A^* given by Eq. [7].

---

[6]In principle x_A will be decreased; however, we assume that the concentration of A is in great excess relative to that of A^* or P, and hence, $x_A \sim 1$.

For the optically thick flatplate collector geometry considered in this review, the number of nonradiative reactions occurring in the system is proportional to the system volume, while the radiative input and losses are a function only of the system surface area. When the net reaction rate $P \rightarrow A^*$ is large compared to the radiative relaxation of A^*, i.e., when P and A^* are closely coupled:

$$N_0 V\{k_{-2}[P] - k_2[A^*]\} \gg A_1 J_r \qquad [48]$$

Here V is the system volume, [X] is the concentration of state X with the incident radiation present, J_r is given by:

$$J_r = \int_0^{\lambda_g} I_Q(\lambda, \mu, T, 2\pi)d\lambda \qquad [49]$$

N_0 is Avogadro's number and A_1 is the surface area over which radiative losses can occur. For reduced losses and improved storage efficiency, the area A_1 surrounding the storage volume may be less than A_0, the receiver surface area, and is thereby explicitly identified in Eq. [48]. Since P and A^* are closely coupled, P and A^* will remain near equilibrium with $\Delta\mu(A^*,P) \sim 0$ throughout the subsequent relaxation, as shown in Figure 8f. The combined mutual relaxation rate of P and A^* is determined by the radiative emission from A^*:

$$R = A_1 J_r = A_1 J_s \phi_{loss} \qquad [50]$$

where R is in units of molecules s^{-1} and J_s is defined by Eq. [3]. Multiplying Eq. [50] by μ gives the Gibbs energy decay rate with close coupling:

$$R_{decay} = A_1 J_s \phi_{loss} \cdot \mu \qquad [51]$$

The rate of energy storage $R_{storage}$ is given by the numerator of Eq. [9] or [29] multiplied by the receiver area A_0. Hence, using Eqs. [27] and [28]:

$$\frac{R_{decay}}{R_{storage}} = \frac{A_1 J_s \phi_{loss} \cdot \mu}{A_0 J_s (1-\phi_{loss}) \cdot \mu} = \frac{A_1 \phi_{loss}}{A_0 (1-\phi_{loss})} \qquad [52]$$

From the results of Eq. [29] and Figure 4 at maximum conversion efficiency for unconcentrated AMO sunlight, we obtain $\phi_{loss} = 0.026$. Thus at the time when the light is turned off, $R_{decay} = R_{storage} \cdot (A_1/A_0) \cdot 0.027$ and decreases thereafter as μ and I_Q (Eq. [7]) are reduced. From these results, the energy-storage lifetime with close coupling is quite limited and thus this system is not suitable for long-term storage. It may be noted that the close coupling case is, in principle, no different from using the state A^* as the storage state. However, by using two states, E_g for the absorber can be matched to the radiation intensity and spectral distribution for maximum conversion efficiency, whereas the state P can be chosen for maximum storage density, independent of its E_g (or degeneracy).

When the net reaction rate for $P \rightarrow A^*$ is small compared with the radiative decay of A^*, i.e.:

$$N_o V\{k_{-2}[P] - k_2[A^*]\} \ll A_1 J_r \qquad [53]$$

P and A^* are weakly coupled; thus, when illumination is removed, A^* will decay rapidly to equilibrium with A, leaving the accumulated energy stored in state P as shown in Figure 8g. For this case, the maximum rate of Gibbs energy loss from the storage state P is:

$$R_{decay} = N_o V k_{-2}[P] \cdot \mu \qquad [54]$$

which can be chosen to provide the desired product lifetime. However, since the forward and backward rate constants for the storage reaction are related by the equilibrium constant:

$$\frac{k_2}{k_{-2}} = K_2 = \exp\left(\frac{-\Delta\mu^o(A^*,P)}{kT}\right) \qquad [55]$$

k_{-2} cannot be chosen arbitrarily small without adversely affecting the storage rate and thus inducing a concomitant decrease in the stored Gibbs energy. Under the conditions of Figure 8g, the excitation rate $k_2[A^*]$ is small compared with the decay rate $k_{-2}[P]$ and Eq. [53] becomes:

$$N_o V k_{-2}[P] \ll A_1 J_s \phi_{loss} \qquad [56]$$

However, for effective storage with illumination present the storage rate must be equal to the photon absorption rate less the emission rate, i.e.:

$$N_o V(k_2[A^*] - k_{-2}[P]) = A_1 J_s(1 - \phi_{loss}) \qquad [57]$$

Combining Eqs. [56] and [57] we obtain:

$$\frac{k_2[A^*]}{k_{-2}[P]} \gg \frac{1 - \phi_{loss}}{\phi_{loss}} \qquad [58]$$

Substituting Eq. [55] into Eq. [58] gives Eq. [59]:

$$\frac{[A^*]}{[P]} \gg \frac{1 - \phi_{loss}}{\phi_{loss}} \exp\left(\frac{-\Delta\mu^o(A^*,P)}{kT}\right) \quad\quad [59]$$

or, using Eq. [46]:

$$\frac{[A^*]}{[P]} = \exp\left(\frac{\Delta\mu^o(A^*,P) - \Delta\mu(A^*,P)}{kT}\right)$$

$$\gg \frac{1 - \phi_{loss}}{\phi_{loss}} \exp\left(\frac{-\Delta\mu^o(A^*,P)}{kT}\right) \quad\quad [60]$$

and the Gibbs energy of the storage state P will be less than that of A^* by an amount:

$$\Delta\mu(A^*,P) \gg kT \ln\left[\frac{1 - \phi_{loss}}{\phi_{loss}}\right] \quad\quad [61]$$

For the case of maximum conversion efficiency in which ϕ_{loss} = 0.026, Eq. [61] yields $\Delta\mu(A^*,P) \gg 0.094$ eV.

This Gibbs energy loss is obtained for a simple, weak-coupling photochemical system where the values of k_2 and k_{-2} are the same with and without illumination. Using a catalytic system, however, the reaction rates can be large for effective storage when the illumination is present and small for long product lifetimes when the illumination is absent, in which case the Gibbs energy loss of Eq. [61] is eliminated. The catalytic system requires that in the absence of the catalyst a significant activation barrier exist for the storage reaction $A^* \rightarrow P$. With the catalyst present the activation barrier is suppressed and the reaction rates are large, providing storage at the radiative Gibbs energy input rate. With the catalyst absent, the activation barrier is present, and k_{-2} is very small allowing a long storage lifetime. To accomplish this task, Ross (19) has proposed a "chemical rectifier" in which the activation barrier is controlled by the sign of $\Delta\mu(A^*,P)$, i.e., the catalyst is active for $\Delta\mu(A^*,P) > 0$ and is inactive for $\Delta\mu(A^*,P) < 0$.

III. NONIDEALITIES IN REAL PHOTOCHEMICAL STORAGE SYSTEMS

Until now, we have considered the efficiency of photochemical conversion and storage of solar energy only for ideal systems in which there are no losses due to nonradiative transitions. In these systems, the photoproduct P can be returned to the ground state A only via the radiative reversal $(A^* \rightarrow A)$ of the transitions by which solar energy is captured (see Fig. 8a).

In real (nonideal) systems, nonradiative transitions return the photoproduct to the ground state by pathways that are neither useful nor required by thermodynamics. Such a pathway may lead directly from A^* to A, or it may be a "short-circuit" leading from some point in the energy-utilization pathway to the ground state (e.g., $P \rightarrow A$).

Whether or not nonradiative transitions are important, ϕ_{loss} for an otherwise optimal device is still approximately kT/μ_{max} (see Eq. [27]) or ~0.026. However, one should note that when nonradiative transitions occur, ϕ_{loss} includes both radiative and nonradiative pathways. Even if nonradiative transitions do not change ϕ_{loss} for maximum efficiency (i.e., $\phi_{loss} \simeq 0.026$) there is, however, an effect on μ and, through μ, on the maximum efficiency itself. Ross and Hsiao (8) have shown that for this case the optimal μ is given by:

$$\mu \simeq \mu_{max} - kT \, \ln\left(\frac{\mu_{max}}{kT}\right) - kT \, \ln \alpha \qquad [62]$$

where α is the ratio of total $A^* \rightarrow A$ transitions to the radiative $A^* \rightarrow A$ transitions.

Ross and Collins (13) developed this concept further, and analyzed the quantitative effect of various proportions of nonradiative decay. Their measure of the efficiency of nonradiative processes is a quantity κ, defined as the ratio of nonradiative to radiative decay rates characteristic of the material used as the absorber, A. In an optically thin absorber $\alpha = 1 + \kappa$. Because the total amount of nonradiative decay is proportional to the amount of absorbing material, the value of α in a practical device will be somewhat greater than this. In an approximately optimal device having the geometry assumed throughout this paper, it can be shown (20) that:

$$\alpha \simeq 1 + 3\kappa \qquad [63]$$

We will use this relationship later in this section.

Now let us consider the case of chemical energy storage with nonradiative decay from P to A, as well as from A^* to A (see Fig. 9). In the absence of an energy-storage pathway (i.e., $k_2 = 0$), the total rate of decay from A^* is:

$$-\frac{d[A^*]}{dt} = \left(k_{-1} + k_{L1}\right)[A^*] \qquad [64]$$

where k_{-1} is the rate constant for the radiative decay of A^* to A, and k_{L1} is the rate constant for the loss of A^* by nonradiative pathways. In this case κ is given by Eq. [65]:

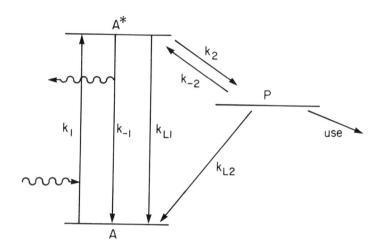

FIGURE 9. Photochemical energy storage with nonradiative losses. (See text for explanation).

$$\kappa = \frac{k_{L1}}{k_{-1}} = \frac{1}{\phi_f} - 1 \qquad\qquad [65]$$

where ϕ_f is the quantum yield of fluorescence in the absence of the storage pathway. We see that a material having a very low intrinsic quantum yield of fluorescence is a poor choice for a solar energy converter.

Now let us admit a chemical storage pathway (i.e., $k_2 \neq 0$) and consider additional nonradiative losses due to a "short-circuit" in which an early intermediate can return directly to the ground state (e.g., P → A) without passing through A^* (see path L2 in Fig. 9). One possibility for such a short-circuit exists when P is a triplet state of the absorber; here inter-system crossing back to the (singlet) ground state can introduce a substantial loss. Another possibility is when P is a charge-transfer state in which an electron (or hole) has moved from the light-absorbing molecule to a nearby site; subsequent tunneling of the electron (or hole) back to the ground state of the absorber will preclude energy storage. In this case, the total rate of return to the ground state via nonradiative loss pathways is:

$$\frac{d[A]}{dt} \text{ (nonradiative)} = k_{L1}[A^*] + k_{L2}[P] \qquad\qquad [66]$$

The impact of path L2 will depend on the relative amounts of P and A^*, which will be determined in part by the magnitude of the other rate constants. However, as we showed earlier, the storage efficiency is maximal when k_2 and k_{-2} are very large, and we will now consider this special case.[7]

When k_2 and k_{-2} are large, A^* and P will be in virtual equilibrium, and:

$$[P] = \frac{k_2}{k_{-2}} [A^*] \qquad [67]$$

Thus the total rate of nonradiative loss via paths L1 and L2 is the same as if all loss occured directly from A^* to A, with a rate constant of:

$$k_{L'} = k_{L1} + \frac{k_{L2}k_2}{k_{-2}} \qquad [68]$$

The ratio of the rates of nonradiative loss in the presence and absence of pathway L2 is:

$$\frac{\kappa(L1 + L2)}{\kappa(L1)} = 1 + \frac{k_{L2}k_2}{k_{L1}k_{-2}} \qquad [69]$$

Combining Eqs. [62], [63], and [69], and assuming that $\kappa \gg 1$, we find that the change in the optimal chemical potential induced by pathway L2 is:

$$-\Delta\mu \simeq kT \ln\left[1 + \frac{k_{L2}k_2}{k_{L1}k_{-2}}\right] \qquad [70]$$

When the loss via L2 exceeds the loss via L1 ($k_{L2}k_2 \gg k_{L1}k_{-2}$), this loss can be rewritten as:

$$-\Delta\mu \simeq kT \ln\left(\frac{k_{L2}}{k_{L1}}\right) + \Delta\mu^o(A^*, P) \qquad [71]$$

where $\Delta\mu^o(A^*, P)$ is the difference in the standard Gibbs energies of A^* and P.

A specific numerical example may be helpful in interpreting this result: Suppose that $k_{L1} \simeq 10^9$ s^{-1}, $k_{L2} \simeq 10^6$ s^{-1}, and $\Delta\mu^o(A^*, P) \simeq 0.40$ eV. Substituting these values into Eq. [71], we find that the loss in chemical potential is:

$$-\Delta\mu \simeq 0.059 \log_{10}\left(\frac{10^6}{10^9}\right) + 0.40 \simeq 0.22 \text{ eV}$$

---

[7]A more general treatment is being developed for publication elsewhere.

If the values of the other rate constants in this scheme (Fig. 9) are optimized, the resulting decrease in chemical potential will be the most important consequence of path L2. Such pathways are probably difficult to avoid, and Eq. [70] probably represents the major loss in photochemical systems for conversion of solar energy.

IV. TWO EXAMPLES OF PHOTOCHEMICAL ENERGY STORAGE

Since the water-dissociation reaction and the reaction of photosynthesis are of considerable interest as routes for photochemical solar energy storage, we will now examine, for these two reactions, the consequences of the concepts we have developed.

The water-dissociation reaction: $H_2O \rightarrow H_2 + 1/2\ O_2$ ($\Delta G^O = 237$ kJ mol^{-1}, $E^O = 1.228$ eV) has been given a great deal of attention as a possible scheme for photochemical storage of solar energy. We have chosen this reaction to illustrate the effects of the thermodynamic limitations developed here on the achievable wavelength thresholds for driving this reaction with sunlight. To derive these limits we invoke the following assumptions:

1. The reaction is sensitized by a dye (or dyes) which, in the primary photochemical step, transfer only one electron per photon absorbed.

2. The threshold wavelength for the reaction is given by:

$$\lambda_{max} = \frac{hc}{\left(E/n + E_{loss}(\lambda)\right)} \qquad [72]$$

where E is the electrochemical potential (in eV) of the reaction under the given conditions, $E_{loss} = (E_g - \mu)$, where μ is given by Eq. [28], and n is the number of photosystems utilized to carry out the reaction (i.e., n is the number of photons absorbed per electron transferred from H_2O to H_2).

3. The conditions for the reaction are absorber temperature T = 300K, unconcentrated AMO sunlight and standard states (whence, the electrochemical potential is E^O).

Since E_{loss} is a function of λ (see Fig. 4), λ_{max} was computed using Eq. [72] and an iterative method. For n = 1, $\lambda_{max} = 778$ nm whereas for n = 2, $\lambda_{max} = 1352$ nm (see Fig. 10). As noted earlier, E_{loss} is likely to be considerably larger than the minimum value set by thermodynamics (see Fig. 4); this will shift λ_{max} to shorter wavelengths. It is clear from Figure 10 that this blue-shift will result in a significant drop in energy-storage efficiency for n = 1, but will have little effect on the efficiency for n = 2. On the other hand,

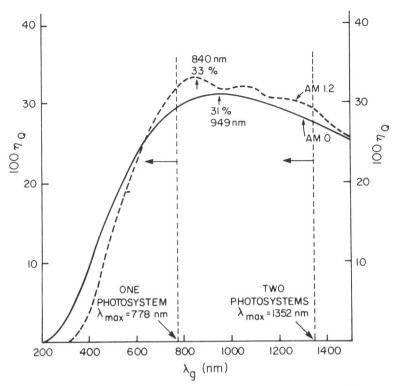

FIGURE 10. A plot of η_Q (%) vs. λ_g (see Fig. 3) showing the calculated threshold wavelengths for one and two photosystems. These wavelengths represent thermodynamic limits and inclusion of any additional loss pathways will blue-shift the λ_{max} values.

the simplicity of the n = 1 system may compensate for the loss in efficiency when all the nonidealities of the system are considered. As noted by Bolton (12) the standard electrochemical potentials for a number of other possible fuel-generation reactions are very similar to that of the water-dissociation reaction. Hence, Figure 10 is generally applicable to those reactions as well.

Our second example is the reaction of photosynthesis: $CO_2(g) + H_2O(\ell) \rightarrow 1/6\ C_6H_{12}O_6(s) + O_2(g)$, for which $\Delta G^o = 496$ kJ mol^{-1} and $E^o = 1.24$ V. Experimentally, λ_g for photosynthesis is ~700 nm which corresponds to a photon energy of 1.77 eV. For photons utilized at this wavelength $E_{loss} \simeq 0.38$ eV and, hence, $\mu = 1.39$ eV if only one photon were used to drive each electron in the reaction. Thus, if photosynthesis had to be operated with just one photosystem, only 0.15 eV would be available to allow for nonidealities. Clearly this

is not enough, since in fact $E_{loss} \sim 0.8$ eV in the primary
step of photosynthesis (12). However, if we employ two photosystems (i.e., two photons per electron transferred), then
only 0.62 eV has to be supplied from each photon which, at
700 nm, leaves 1.15 eV for loss processes giving ample scope
for nonidealities. Since neither we nor nature are likely to
devise the almost perfect photochemical energy storage system,
we can understand why nature developed the more complex two-
photosystem scheme.

V. CONCLUSIONS

In this paper we have presented a comprehensive analysis
and comparison of three equivalent approaches to the thermodynamic limits on conversion of sunlight to useful work. We
have extended this analysis to a consideration of the problem
of photochemical energy storage and have shown that, in
principle, there are no additional penalties arising from
thermodynamics which would impede long-term storage. However,
we have noted the stringent conditions which must be met to
achieve this condition.

We have also examined in a qualitative way the effect of
nonidealities such as nonradiative decay from the excited
state to the ground state and back reaction of the photoproduct directly to the ground state.

Finally, we have looked at the consequences of these considerations on two important photoconversion systems: photochemical water-splitting and photosynthesis. For the water-
dissociation reaction we have derived the threshold wavelengths dictated by thermodynamics for processes involving one
and two photosystems, and have shown the effect that introduction of nonidealities has on the efficiency of energy storage
in each case. For photosynthesis we have shown that although
one photosystem is, in principle, possible with reasonable efficiency, little leeway would be provided for nonideal losses,
and the two-photosystem scheme adopted by nature is apparently
necessary in view of the actual losses of real systems.

ACKNOWLEDGMENTS

JRB thanks the Natural Sciences and Engineering Research
Council of Canada for a Strategic Grant in Energy. RTR ac-
knowledges research support from the U.S. Department of Energy
through the Solar Energy Research Institute.

REFERENCES

1. G. Jones, II, P. T. Xuan and S. H. Chiang, in "Solar Energy, Chemical Conversion and Storage" (R. R. Hautala, R. B. King and C. Kutal, eds.) Humana Press, Clifton, N.J., 1979, pp. 271-298; P. S. Mariano, T. L. Rose, A. A. Leone and L. Fisher, ibid. pp. 299-332; R. R. Hautala, R. B. King and C. Kutal, ibid. pp. 333-369; G. Jones, II, S. H. Chiang and P. T. Xuan, J. Photochem. 10, 1 (1979).
2. M. Calvin in "Solar Energy, Chemical Conversion and Storage" (R. R. Hautala, R. B. King and C. Kutal, eds.), Humana Press, Clifton, N.J., 1979, pp. 1-30; J. R. Bolton, ibid. pp. 31-50; N. K. Boardman and A. W. D. Larkum in "Solar Energy" (H. Messel and S. T. Butler, eds.), Pergamon Press, Oxford, 1975, pp. 123-181.
3. A. J. Nozik, Annu. Rev. Phys. Chem. 29, 189 (1978); Phil. Trans. Roy. Soc. Lond. A295, 453-470 (1980); A. J. Bard, Science 207, 139 (1980); A. J. Bard, J. Photochem. 10, 59 (1979); M. S. Wrighton, Accts. Chem. Res. 12, 303 (1979); see also A. J. Nozik, Chapter 10 in this Volume.
4. M. Grätzel, Chapter 5, and also J.-M. Lehn, Chapter 6, in this Volume; A. Moradpour, L'actualité Chim. pp. 7-18 (1980).
5. P. P. Infelta, M. Grätzel and J. H. Fendler, J. Am. Chem. Soc. 102, 1479 (1980); W. E. Ford, J. W. Otvos, and M. Calvin, Proc. Nat. Acad. Sci. USA 76, 3590 (1979).
6. D. Trivich and P. A. Flinn, in "Solar Energy Research" (F. Daniels and J. A. Duffie, eds.), University of Wisconsin Press, Madison, Wisconsin, 1955, pp. 143-147; see also L. H. Schaffer, Sol. Energy 2, 21 (1958).
7. W. Shockley and H. J. Queisser, J. Appl. Phys. 32, 510 (1961).
8. R. T. Ross and T.-L. Hsiao, J. Appl. Phys. 48, 4783 (1977).
9. R. T. Ross, J. Chem. Phys. 45, 1 (1966); 46, 4590 (1967).
10. R. T. Ross and M. Calvin, Biophys. J. 7, 595 (1967).
11. A. F. Haught, United Technologies Research Center, Internal Report No. UTRC 80-9 (1980); Book of Abstracts, Third Int. Conf. Photochem. Conv. Stor. Solar Energy (J. S. Connolly, ed.), Boulder, Colorado, August 1980, SERI/TP-623-797, pp. 443-445.
12. J. R. Bolton, Science 202, 705 (1978).
13. R. T. Ross and J. M. Collins, J. Appl. Phys. 51, 4504 (1980).
14. N. J. Turro, "Modern Molecular Photochemistry", Benjamin/ Cummings Publishing Co., Menlo Park, Calif., 1978, p. 174.
15. M. P. Thekaekara, Appl. Opt. 13, 518 (1974).
16. K. W. Böer, Sol. Energy 19, 525 (1977).

17. S. W. Angrist, "Direct Energy Conversion",2nd Ed., Allyn and Bacon, Inc., Boston, 1971, pp. 190-200.
18. R. V. Bilchak, J. S. Connolly and J. R. Bolton, Proceedings of the 1980 Annual Meeting, American Section of the International Solar Energy Society (G. E. Franta and B. H. Glenn, eds.), Vol. 3.1, 1980, pp. 84-87; Book of Abstracts, Third Int. Conf. Photochem. Conv. Stor. Solar Energy (J. S. Connolly, ed.), Boulder, Colorado, August 1980, SERI/TP-623-797, pp. 447-449.
19. R. T. Ross, J. Chem. Phys. 72, 4031 (1980).
20. R. T. Ross, unpublished results.

DISCUSSION

Dr. J. J. Katz, Argonne National Laboratory:
You say that it is the entropy of mixing of excited states in a matrix of ground states that is responsible for the energy loss. With modern lasers it is easy to achieve a condition where 100% of the molecules are in the excited state. What happens then to your entropy of mixing loss? I also have a comment: Your analysis does not establish a thermodynamic efficiency limit for photosynthesis because nowhere have you defined a reversible thermodynamic system, which is of course essential for such an analysis.

Prof. Bolton:
From Eq. [13] the difference in chemical potential between the states A and A* of the reaction A → A* is:

$$\mu = \mu_{A*}^{o} - \mu_{A}^{o} + kT \ \ell n(x_{A*}/x_{A})$$

As shown by Eq. [31] the k $\ell n(x_{A*}/x_A)$ term is the entropy of mixing associated with the formation of an A* state at mole fraction x_{A*} and the destruction of an A state at mole fraction x_A. The entropy-of-mixing term is thus determined by the relative mole fractions of the excited and unexcited states, and the case you describe ($x_{A*} = 1$, $x_A = 0$) would be unboundedly large.

Since ΔG must be negative for a reaction step to proceed, the maximum chemical potential, i.e., Gibbs energy per particle, that can be achieved by a reaction system is the Gibbs energy of the driving force. For a photochemical system the driving force is the input radiation field, whose Gibbs energy per photon is the photon energy minus the product of the absorber temperature and the photon entropy. If the Gibbs energy per photon is significantly greater than $\mu_{A*}^0 - \mu_A^0$, then $kT \ln(x_{A*}/x_A)$ can be substantial and x_{A*} may be very much larger than x_A. It is this situation which you must be referring to in your question. The points to note are: first, that the Gibbs energy of the photostationary state will always be less than that of the input radiation field; and second, that the entropy of mixing is taken into account in the concentration-dependent terms of Eq. [13]. For a general case this is the parameter which can be varied by adjustment of x_A and x_{A*} to give the minimum value of $(E_g - \mu)$ for maximum conversion efficiency.

With respect to your comment, just as the ideal Carnot engine operates with heat input from a single temperature source discharging waste heat to a fixed temperature reservoir, the ideal, reversible quantum converter operates with monochromatic radiation input and a discharge reservoir and fixed temperatures.

Both Haught and Ross have analyzed the reversible quantum converter under maximum efficiency and maximum power conditions. The problem considered in this paper, however, involves a radiation source with a distribution of frequencies which introduces inherent irreversibilities for a single quantum converter in the same way that a distributed temperature source introduces irreversibilities in the operation of an ideal heat engine. These irreversibilities are an integral part of the conversion of the distributed spectrum of solar radiation to useful work by a quantum device, and the efficiency calculated here is the maximum power-conversion efficiency for solar radiation by such a quantum converter employing a single threshold reaction. For strict application to photosynthesis, which involves two photosystems, the analysis can be extended to incorporate two or more reactions as has been considered by Ross and Hsiao (8) and by Bilchak, Connolly and Bolton (18).

Dr. E. Berman, ARCO Solar, Inc.:
How does maximum efficiency increase if photons of energy less than the bandgap energy are upgraded to the bandgap, i.e., by being absorbed by a blackbody and re-emitted as photons near the bandgap?

Prof. Bolton:
Frequency upshifting by a thermal absorber (at 300K) to make use of photons with energies below the bandgap is

analogous to the two-collector quantum system for which Ross and Hsiao (8) calculate a maximum conversion efficiency of 41%.

Prof. R. Reisfeld, Hebrew University:
First, I would like to compliment you on your excellent work and very elegant presentation. My question is, would it be possible to circumvent the thermodynamic limit of 31% by using two photons (either simultaneous or consecutive)? Secondly, would you compare the analogy between quantum and thermal conversion as far as thermodynamics is concerned, taking into consideration that the total solar spectrum can be used in the latter?

Prof. Bolton:
We have restricted ourselves to conditions typical of solar radiation where the intensities are many orders of magnitude smaller than those where the probability of the simultaneous absorption of two photons is insignificant.

In answer to your second question: Haught has shown that the single, ideal thermal converter is much more efficient (54%) than the single, ideal quantum converter (31%) (Contributed Paper XI-2). However, it may be that when reasonable nonidealities are considered in Carnot vs. quantum engines, the difference may narrow. There is a need for further research on this point.

Dr. M. D. Archer, Cambridge University:
You have couched your argument in terms of the chemical potential available from diffuse AMO irradiation. As μ increases with the irradiance, 4-photon water photolysis at high irradiance could presumably be run at a threshold wavelength longer than 778 nm. Conversely, photosynthesis is presumably required to run forward at a lower irradiance, and this brings the threshold from a 4-photon process uncomfortably close to 1.8 eV.

Prof. Bolton:
You are correct in that the thermodynamic wavelength threshold of 778 nm for the 4-photon water photolysis reaction is dependent on light intensity. For example, at 10 suns λ_{max} = 810 nm while at 0.1 sun (an intensity at which photosynthesis is probably optimized) λ_{max} = 753 nm. At 0.1 sun and 700 nm, E_{loss} = 0.44 eV; thus μ = 1.36 eV which leaves only 0.12 eV for nonideal losses. Therefore our argument for the necessity of two photosystems in photosynthesis is even stronger at 0.1 sun.

Dr. Archer:
   I would like to make a comment about the thermalization of excited states which is assumed in your analysis. The injection of hot electrons from semiconductors to solution, mentioned by Dr. Nozik, is one example of the utilization of nonthermalized excited states. The creation of these in an absorber produces less of an entropy increase than does the creation of a thermalized population, and the thermodynamic availability of the nonthermalized population is greater than that of the thermalized system. If thermalization could be avoided at all subsequent steps in the conversion process, the maximum conversion efficiency would be greater than for a thermalized system. In the limit, it would approach the efficiency achievable by an infinite array of quantum systems of different bandgaps.

Prof. Bolton:
   This is a very interesting idea; however Ross recently published a paper [Appl. Phys. Lett. $\underline{35}$, 707 (1979)] proving that a device without thermalization cannot be more efficient than an ideal device in which the excited state is thermalized, providing that all of the products of light absorption ultimately reach a common thermalized state. If high- and low-energy photon products are processed separately, then there may be an increase in efficiency. [See, however, last page of this Discussion, Ed.].

Dr. J. S. Connolly, Solar Energy Research Institute:
   With regard to Dr. Archer's last comment, I would like to point out that there exists a large body of literature on nonthermalized photoprocesses in organic crystals, viz., singlet fission from the higher vibronic levels of tetracene to yield two triplets (see Contributed Paper II-14). This process is quite analogous to hot-carrier injection as discussed by Dr. Nozik.

Prof. Bolton:
   Yes, these processes certainly exist. Indeed, there is an experimental test of whether an excited state is thermally equilibrated, and that is to measure the absorption spectrum in the region of emission. This spectrum multiplied by the Planck law relation at the temperature of the experiment (e.g., 300K) should reproduce exactly the observed emission spectrum. If it does, then the emitting state is thermally equilibrated with its surroundings.

Dr. M. Neuman-Spallart, Institut de Chimie Physique, Ecole Polytechnique Fédérale, Lausanne:
   In a system where two photons per electron are needed you

concluded that the maximum wavelength that could be used
is 1352 nm. Was that calculated on the basis of two excitations with the same λ_g? If so, how does the efficiency
change in a system using two photons of different wavelengths? How does the efficiency change with the relative
position of the ground state redox potentials of the two
systems?

Prof. Bolton:

As Ross and Hsiao (8), Bolton (12) and Bilchak, Connolly
and Bolton (18) have shown there is some expected increase
in efficiency for two absorbers operating at different
bandgap wavelengths. However, photosynthesis works at two
relatively close-lying wavelengths (680 nm and 700 nm).
In the case we considered, λ_g was assumed to be the same
for each photosystem.

The relative position of the ground-state redox potentials
must be tuned to the requirements of the reaction that has
to be performed. We have assumed that tuning has been
taken into account, so that only thermodynamic losses are
considered.

Dr. V. N. Parmon, Institute of Catalysis, Novosibirsk:

The thermodynamic restrictions which you discussed imply a
closed cycle for the "working body" of the conversion
device. For this reason the restrictions discussed are
totally valid for photovoltaic cells and for storage of
solar energy in the form of intramolecular energy (e.g.,
isomerization). However when discussing photoproduction
of fuels (like H_2 from H_2O), in particular cases the
thermodynamic cycle, specifically the "working body"
(e.g., H_2O) is open. In such cases the efficiency of conversion under idealized conditions might appear greater
than discussed, just like the efficiency of a thermal
engine can exceed the efficiency of a Carnot cycle in
particular cases of thermodynamically open systems.

Prof. Bolton:

Such cases as you describe involve energy input at a different temperature than the energy extraction, e.g.,
electrolysis of water at high temperature and recombination via a fuel cell at low temperatures. In such cases
the ΔG obtained at the low temperature can be greater than
the electrical energy input in the high temperature electrolysis, the increased useful work in fact being derived
from the heat absorbed at the high temperature less the
heat rejected at the low temperature. Upon closure of the
photochemical fuel cycle any such increase in the useful

work derived in the photoconversion step is "paid for" in
the recombination stage through heat transferred from a
high to a low temperature, change in molar volume, etc.,
and the efficiency of the photoconversion process will be
given by the results presented here.

Dr. T. Cole, Jet Propulsion Laboratory:
Would it not be more accurate to describe this system as
being in the steady state rather than in an equilibrium
state and then treat the problem by nonequilibrium thermodynamics?

Prof. Bolton:
We do treat this system as being in the steady state, and
the formal analysis is indeed based on a kind of nonequilibrium thermodynamics. It is a fairly simple kind of
nonequilibrium situation, because we assume that all of
the vibrational and other sub-states of A, A*, and P are
in thermal equilibrium, so that the only nonequilibrium
transitions are those between these species. The most
commonly used formalism for nonequilibrium thermodynamics
(see S. R. DeGroot and P. Mazur, "Nonequilibrium Thermodynamics", North-Holland, 1962) assumes linear relationships between fluxes and thermodynamic forces and cannot
be applied to chemical systems far from equilibrium.

Dr. A. B. Kuper, Consultant, Diamond Shamrock Corp.:
It seems that the situation in silicon does not fit the
assumptions of radiative loss and detailed balance.
Silicon, being an indirect gap absorber, is excited via a
virtual state and decays mainly nonradiatively. Doesn't
this pose a problem for the model?

Prof. Bolton:
In our model we have assumed that the only routes from the
excited state to the ground state are radiative decay and
nonradiative decay to produce chemical work. It is this
highly idealized system which gives the maximum conversion
efficiency. When nonradiative decay routes are considered
[see ref. (13)] a term, $kT \ln \kappa$, must be added to the
expression for the chemical potential. κ is the ratio of
the nonradiative direct decay rate to the radiative decay
rate. This will reduce the overall efficiency. For example, when $\kappa = 1000$, $\eta = 0.26$ at 950 nm. It is important
to note that ϕ_{loss} is equal to the sum of the radiative
and nonradiative quantum yields when nonradiative decay is
important.

Dr. K. Kalyanasundaram, Ecole Polytechnique Fédérale, Lausanne:
   For water photolysis at a threshold wavelength of 780 nm,
   in your 4-photon scheme, the maximum efficiency is ~30%.
   To get the extra 5%, one can go to an 8-photon scheme; but
   from a practical point of view, won't the nonideal losses
   be so severe as to decrease the overall efficiency in the
   latter case?

Prof. Bolton:
   The thermodynamic threshold wavelength for a 4-photon
   water dissociation reaction is 778 nm vs. 1352 nm for an
   8-photon system. From Figure 10 you can see that the
   efficiency falls off very rapidly as λ_g decreases below
   778 nm, whereas it remains high over a wide range below
   1352 nm. Thus the 8-photon system is much more tolerant
   towards nonidealities than is the 4-photon system.

Dr. W. M. Ayers, Exxon Corporate Research Laboratory:
   Would you clarify the value of the low-temperature reservoir in your analysis?

Prof. Bolton:
   We have assumed the reservoir to be at 300K.

Dr. Ayers:
   At constant temperature and pressure, the work available
   from the system will be due to ΔG of mixing of the baths
   at different chemical potentials. You might construct a
   cyclic process in which the Gibbs energy change for each
   step in the process (i.e., a bath of greater chemical
   potential in contact with an electrode or work-extracting
   device, etc.). Also, the Gibbs energy change for a fuel-
   producing system should be treated as mass loss of a component at constant chemical potential in an open system
   rather than as a pressure change of the system.

Prof. Bolton:
   In calculating the maximum possible efficiency, we have
   assumed that the energy extraction process is a reversible
   one with no loss of available energy. The effect you
   speak of is one of the irreversibilities which would reduce the efficiency from this maximum in any real system.

Prof. J. Rabani, Hebrew University:
   On the basis of the thermodynamics, the efficiency increases realized in a system using many-photon processes
   may be smaller than the losses due to back reactions in
   the more complicated kinetics which must take place. In
   view of this, did you consider the case of using light of

shorter wavelengths for photochemical storage, and leave
the longer wavelengths for photovoltaic cells, all in the
same system?

Prof. Bolton:
As discussed in my reply to Dr. Berman, with a two-
collector quantum system, Ross and Hsiao (8) calculated
that the theoretical conversion efficiency can be as high
as ~41% compared with 31% for a single collector system.
It may well be, as you suggest, that the increased gain
and simplicity of two such direct processes may provide
greater net output than the use of multiple photosystems
to drive an energetic fuel reaction.

Dr. Connolly:
With regard to Prof. Rabani's question of the incremental
gains that can be realized in multiple absorber systems,
we have shown (Contributed Paper XI-3) that by cascading
ideal absorbers with discrete bandgap wavelengths, power-
conversion efficiencies (η_p) approaching 61% can be
calculated for eight photosystems. However, the trade-off
between enhanced η_p and added complexity optimizes at
about three photosystems in an ideal photovoltaic system.

Prof. M. Grätzel, Ecole Polytechnique Fédérale, Lausanne:
Why is a ~3% loss due to fluorescence or nonradiative
relaxation required to obtain the optimum storage yield?

Prof. Bolton:
The reason why a ϕ_{loss} of 2%-3% is desirable can be seen
in Figure 5 where η drops as ϕ_{loss} approaches unity; however, the reason why η drops as $\phi_{loss} \to 0$ is that as the
quantum yield for conversion approaches unity, the population of the excited states becomes smaller and hence μ
decreases.

Dr. Connolly:
I have suggested elsewhere (see General Discussion) that
one of the objectives of our research in solar photochemistry should be to optimize the utilization of excited
singlet states rather than triplets. There are several
reasons for this. One is that the inherent efficiencies
of singlet-state photochemistry ought to be higher since
triplet states in most visible-light absorbing molecules
are ~30% lower in energy than their parent singlet excited
states. Another reason to avoid triplet states is that it
is imperative for the quantum yields of degradative processes (ϕ_d) in solar conversion schemes to be extremely
low (10^{-6} to perhaps as low as 10^{-10}, depending on the

assumptions made) in order for a system to have a 1/e
lifetime of ~10 years. If an excited singlet state lives
long enough to fluoresce or to undergo intersystem crossing, the probability of adverse photochemistry increases
to the point that ϕ_d may become the limiting factor in an
otherwise favorable system. This second aspect, I believe, has not been fully appreciated. Returning to the
first point, however, is it possible that the $S_1^* \rightarrow T_1$
energy loss could be subsumed in the inherent thermodynamic loss of ~0.4 eV? If this were possible, then
chemistry out of the lower energy triplet state would not
necessarily imply a lower solar conversion efficiency.

Prof. Bolton:
The $S_1^* \rightarrow T_1$ loss cannot be counted as part of E_{loss}. The
reason is that E_{loss} cannot be placed on a ΔG^0 plot.
E_{loss} includes the entropy of mixing and therefore is a
property of the ensemble, and can be displayed only on a
ΔG plot.

Prof. R. T. Ross, Ohio State University:
I have comments on several of the questions that have been
raised during the preceding discussion: Dr. Berman asked
about the efficiency of a device in which a blackbody was
placed behind a single-threshold quantum converter. The
blackbody would be heated by low energy photons passing
through the absorber, and some of the blackbody's radiation would be in the form of photons of energy high enough
to be used by the threshold device. This is a cute idea,
and it can produce an efficiency higher than that of a
single-threshold absorber alone. However, I don't believe
that the effect will be significant in any practical device. In unconcentrated sunlight, the temperature of the
blackbody would be less than 1000K, and there would be an
insignificant amount of radiation above the ~1 eV threshold of a practical quantum converter. Now, if you sur-
round the whole thing with a perfect mirror, with a hole
just large enough to let the sunlight in, you might get a
measurable increase in efficiency. A better idea would be
to put a thermal absorber-emitter in front of the quantum-
utilizing device [see P. Wurfel and W. Ruppel, IEEE Trans.
Electron Devices ED-27 745 (1980)].

There was a comment that silicon doesn't have anything
like a 2% fluorescence quantum yield because nonradiative
relaxation is so much more rapid. Quite so. The optimal
ϕ_{loss} of 2% applies to the total of radiative and nonradiative losses. In silicon, the ratio of nonradiative
to radiative losses is about 600, and so you don't see

much fluorescence. These nonradiative losses cost you the kT ℓn(600) that Jim Bolton referred to, and it also costs you a bit more due to the absorber thickness. We discuss this in ref. (13)

With regard to photolysis of water, I like 4 quanta per oxygen molecule. Jim and I are in complete agreement about the thermodynamics, so any differences expressed here represent different degrees of pessimism about our ultimate ability to limit other losses.

In response to John Connolly's concern about the loss of energy in going from excited singlet to triplet levels, I don't see this as any problem. There are some real kinetic advantages in having a state with a lifetime that gives you more time to do chemistry. The singlet-triplet energy drop isn't a loss from the point of view of thermodynamics, because the energy decrease can buy you a corresponding increase in concentration. The only problem is that the triplet has to be quite tight against transitions back to the ground state, or Eq. [71] can kill you.

Finally, some acknowledgements. I personally am very grateful to Art Nozik for supporting my work in this area for the past couple of years. For Alan Haught and myself, I would like to thank Jim Bolton for putting up with us all week. And for all of us, I would like to express our appreciation to John Connolly for providing us with the environment which has stimulated a lot of scientific creativity within this Conference.

[Editor's note: Regarding the question raised by Mary Archer about the efficiency of hot carriers in solar energy convertors, Ross and Nozik (submitted for publication, 1981) have recently shown that a quantum converting device with hot electronic bands can achieve a theoretical efficiency of ~66%. This substantially exceeds the ~32% maximum efficiency of a quantum device operating at thermal equilibrium and the ~52% maximum efficiency of an ideal thermal conversion device].

GENERAL DISCUSSION

Dr. J. S. Connolly, Solar Energy Research Institute:
I would like to exercise a Chairman's perogative and pose a general question which touches on a number of facets discussed at this Conference. It seems to me that to preclude inherent singlet-triplet relaxation losses and thus to enhance the efficiency of synthetic photochemical solar energy conversion, we should be utilizing excited singlet states directly rather than triplets. Despite the kinetic advantges mentioned by Prof. Ross, there also remains the very serious problem of degradative photoprocesses.

This problem has been "solved" in photosynthesis; electron transfer appears to occur via the excited singlet state of the special pair in 3-10 ps. Normal intersystem crossing cannot be involved because the energetics are wrong (Contributed Paper II-14) and the kinetics are too slow (Contributed Paper III-3). In solutions of monomers, however, net electron transfer from neutral porphyrins, chlorins and bacteriochlorins to neutral acceptors appears to take place only out of triplet states, even when oxidation of excited singlets is energetically more favorable. The latter observation has been explained by Gouterman and Holten [Photochem. Photobiol. 25, 85 (1977)] on the basis of the spin of the charge-transfer (CT) complexes which result upon collision of the excited donor with an acceptor. Thus singlet CT states are shorter lived than triplet CT states and only the latter can undergo net electron transfer.

My questions are: (1) What makes the "special pair" in photosynthesis so special in this regard? (2) It is likely that we can utilize excited singlet state energies in model systems such as the linked porphyrin-quinone complexes studied by the Loach and Bolton groups? Who would like to lead this off?

Dr. J. J. Katz, Argonne National Laboratory:
It appears at this time that photochemistry from the photoreaction center special pairs occur from the singlet state rather than from a triplet state as is the usual case in preparative organic photochemistry. Triplet states are more common in preparative photochemistry perhaps because dilute systems are used and long-lived intermediates are required. In the highly concentrated natural systems, singlet photochemistry seems entirely inappropriate. Conventional organic photochemistry may

341

thus not be an appropriate guide to the photochemistry of photosynthesis.

Prof. Sir George Porter, Royal Institution:
It is not generally true that singlet excited states cannot result in charge separation. The work of Weller and others has given a number of examples. The competition in the excited singlet complex is between charge separation and internal conversion to the ground singlet (which is forbidden from the triplet complex). Provided the charge-separation rate can be enhanced, it can occur from the singlet and it can be enhanced by (a) repulsion of identical charges as, for example, in the excited chlorophyll-methyl viologen (MV^{+2}) reaction; (b) favorable conformations which enhance the Franck-Condon electron transfer, and (c) high concentrations of carriers which remove the charge from its original pair. This latter effect is probably the most important in the highly concentrated and organized natural photosynthetic system.

Prof. P. A. Loach, Northwestern University:
It should first be realized that the in vivo system, e.g., photosynthetic bacteria, has four bacteriochlorophylls as well as two bacteriopheophytins and two ubiquinone molecules strung out across the membrane such that charge separation occurs across the entire bilayer as a result of the primary event. These are tightly bound in a protein where distance and orientation are, no doubt, extremely important.

On the other hand, in simple model systems of interacting monomeric bacteriochlorophyll and a monomeric acceptor such as benzoquinone, one cannot prevent a close encounter between the donor and acceptor pair thus allowing a very rapid back reaction. The multiple components of the in vivo system facilitate separation of charge so that each back reaction is somewhat less favorable than the forward reaction.

Some eight years ago, we began a synthetic program of modeling the reaction center by preparing covalently-linked porphyrin dimers and trimers and covalently-linked porphyrin-quinone complexes. We felt these latter compounds could provide a close, but not too close, approach of the quinone to the excited singlet state of the porphyrin so that charge separation could occur, but the back reaction would be retarded, particularly in a solvent of high dielectric. We have demonstrated such charge separation (see Contributed Paper I-3) and have some evidence suggesting that it may arise from the excited singlet state of the porphyrin.

Dr. T. L. Netzel, Brookhaven National Laboratory:
I'd like to re-emphasize the question posed at the start of this discussion and then show that we have obtained some data relevant to this problem. The problem concerns the fact that quinones have been shown by a number of researchers (e.g., Gordon Tollin and Maurice Windsor) to efficiently quench chlorophyll and porphyrin singlet excited states, but no ionic products have been detected. Presumably any ion pairs formed are in the singlet state and recombine at very fast rates ($>10^{11}$ s^{-1}). This raises the question whether any charge transfer (CT) photoproducts produced from a singlet excited state will be detectable. Holten and Windsor were able to obtain some ionic products when the excited singlet state of bacteriopheophytin a was quenched by methyl viologen. This is an interesting case because the ionic photoproducts are both positively charged. However, this is not the case for uncharged quenchers such as quinones.

We present data in Contributed Papers II-2 and II-3 which show that CT photoproducts between MgP and H$_2$P (P is an alkyl substituted porphyrin) are detectable. The MgP$^{+}_{\bullet}$-H$_2$P$^{-}_{\bullet}$ CT product is formed in <6 ps and has a lifetime of 200 ps in CH$_2$Cl$_2$ and 1.3 ns in N,N-dimethylformamide. This diporphyrin (synthesized by Chris Chang at Michigan State University and Bookhaven National Laboratory) is held in a cofacial configuration ~4 Å apart by two covalent bridges. The linking chains are attached at alternate rather than at adjacent pyrroles.

These results demonstrate that a dimeric electron donor (such as P865 in bacterial photosynthesis) is not obligatory for producing a rapid ($>10^{11}$ s^{-1}) S$_1$ → CT reaction and stable ionic products. It is tempting to speculate that this fortunate circumstance, which is completely analogous to the electron transfers in natural photosynthetic systems, is due to both the small nuclear distortions involved and the large exothermicity, ~1.5 eV, of the back reaction. The combination of these two factors would slow the back reaction because it would be nonadiabatic. An alternate possibility is that the electronic coupling between the diporphyrin π → π* singlet state and the CT product state is larger than that between the CT product state and the diporphyrin ground state. However, there is no basis at present for assuming that this latter possibility is true.

A likely explanation for the failure to achieve stable ionic products for quinone quenching of singlet excited states is that (as Gouterman and Holten have noted) the molecular distortions due to reducing a quinone are likely to be larger than those due to reducing a porphyrin-like molecule. In the presence of suitable electronic coupling, large distortions increase radiationless decay processes such as a back electron transfer.

At this time I would like to comment on Paul Loach's and Jim Bolton's recent results which show low yields of ionic photoproducts in a porphyrin-quinone (P-Q) molecule whose donor and acceptor moieties are bound by a single, covalent chain. At present there is insufficient evidence to know whether or not these ionic products are formed from a triplet or singlet excited state. This is important because it is no trick at all to get stable ionic products from porphyrin triplet states. The literature has many such reports. Indeed, the fluorescence-lifetime data on Loach's quinone-porphyrin shows a component with a lifetime of ~1.5 ns which could reflect intersystem crossing followed by ion production from the triplet.

With respect to probing the stability of $P^{\cdot+}-Q^{\cdot-}$ photoproducts, we are planning some picosecond absorption experiments on directly bound porphyrin-quinone molecules synthesized by John Dalton (City of London Polytechnic). He has made a meso-tetra(p-benzoquinone) porphin analogous to meso-tetraphenylporphin; the mono-quinone form also exists. These two compounds are, respectively, nonfluorescent and very weakly fluorescent. The interesting question is, what is the lifetime of the CT product formed as a result of quenching the singlet state?

Prof. N. N. Lichtin, Boston University:
We have been examining the factors (structure and medium) which control the efficiency of net electron transfer in quenching of triplet and singlet states of thiazine dyes in homogeneous solution. Particularly relevant to Dr. Connolly's question is our observation of electron transfer in quenching of methylene blue (S_1) but we have not as yet measured efficiencies accurately nor have we eliminated the possibility that quenching passes into the triplet manifold. We have extensive data on quenching of methylene blue (T_1) by several types of quenchers. Most relevant to the question raised are results obtained with organic quenchers including the ground-state dye. The principal (or only) quenching mechanism in these cases is reversible electron transfer. The systematics of the

efficiency of net electron transfer are beginning to emerge (see Contributed Paper VII-6).

Prof. Loach:
  I would like to reply to one of Dr. Netzel's remarks. Because many people have expressed an interest in the mechanism of charge separation in covalently linked porphyrin-quinone complexes, I would like to underscore that we presently have only weak data (fluorescence quenching and kinetics) on the basis of which we have suggested that charge separation occurs out of the excited singlet state of the porphyrin. We are pursuing the full characterization of the charge-separated species by collaborating with Prof. Kenneth Spears (Department of Chemistry, Northwestern University) to measure spectral intermediates that are formed subsequent to a picosecond light pulse. We view this as most important in establishing the sequence of photochemical events.

  Even if the porphyrin cation-radical and quinone anion-radical do arise from the excited singlet state of the porphyrin, the model will still need to be improved because the quantum yield for photochemistry at room temperature is only about 0.01 and the charge-separated species decay in ⩽ 1 ms (at room temperature) whereas the in vivo system operates with a quantum yield of ⩾0.95 and recombination of the charges requires about 100 ms (also at room temperature). But this kind of model system offers many opportunities for modifying the distance and orientation between donor and acceptor in a well-defined manner.

Prof. D. O. Hall, King's College, London:
  On a somewhat related topic, I would like to suggest that the role of Fe-S complexes in the primary reactions of photosynthesis may be important for effective charge separation. There are now several different synthetic Fe-S complexes which may be useful in photochemical studies.

Prof. H. Kuhn, Max-Planck-Institut für biophysikalische Chemie, Göttingen:
  I would like to comment on a question raised earlier by Prof. Porter: The rate of electron transfer from a dye to a semiconductor can be measured by using the monolayer assembly technique. The dye monolayer at the semiconductor surface is covered by an assembly containing an energy acceptor. The sensitizing action of the dye is then diminished by a competing process of known rate. The photographic process in a dye-sensitized AgBr sample can be stimulated in this way and a value of 10^{10} s^{-1} is

obtained for the rate constant of electron injection into AgBr. The fluorescence of a dye at the AgBr surface is completely quenched, showing that electron transfer to AgBr is fast as compared with fluorescence. On the other hand, a mixed monolayer of dye and electron acceptor at the AgBr surface has no sensitizing effect, showing that electron transfer to the acceptor is fast as compared with electron injection. From these findings, again a value of 10^{10} s^{-1} is found for the rate constant [see Steiger, Hediger, Junod, Kuhn and Möbius, Photogr. Sci. Eng. <u>24</u>, 185-195 (1980)].

I also wish to comment on one remark by Dr. Memming regarding the small quantum yield of dye-sensitized photocurrents: Arden and Fromherz in our laboratory investigated an electrochemical cell with a thin film electrode prepared by evaporation of indium-tin oxide and covered by a bimolecular layer of lipid doped with a cyanine dye in direct contact with the electrode; they obtained a value of 7×10^9 s^{-1} for the rate constant of electron injection [J. Electrochem. Soc. <u>127</u>, 370 (1980)]. Recombination of charge carriers was suppressed by using 3 M thiourea as supersensitizer and an appropriate semiconductor with careful selection of the conditions in the evaporation process (pressure, rate, temperature, humidity). The quantum yield of the photocurrent at saturating potential was 0.7. The low quantum yield obtained in the arrangement discussed by Dr. Memming was probably due to an inappropriate semiconductor.

CONTRIBUTED PAPERS

Session I

SIMULATING PHOTOSYNTHETIC QUANTUM CONVERSION

Plenary Speaker: Melvin Calvin

Session Chairman: Sir George Porter, F.R.S.

Posters I-1 through I-12

CHLOROPHYLL-a PHOTOSENSITIZED REDUCTION OF METHYL VIOLOGEN BY HYDRO-
QUINONES. J.R. Darwent, K. Kalyanasundaram and Sir George Porter, FRS
The Davy Faraday Research Laboratory, The Royal Institution, 21 Albe-
marle Street, London W1X 4BS, England.

Quinone/hydroquinone redox couples are intermediates in the electron
transport chain of photosynthesis and are present in chloroplast
membranes as a "plastoquinone pool" operating between Photosystem I
(PS I) and PS II. In model systems of PS II, plastoquinone and other
quinones have previously been used as electron acceptors for electronically excited chlorophyll-a (*Chl), whilst in model systems for PS I
*Chl has been shown to sensitize the transfer of electrons from donors
such as cysteine or ascorbate to acceptors typically methyl viologen
(MV^{2+}). In order to link PS I with PS II in model systems, in a
manner similar to that which operates in vivo, we have studied the use
of hydroquinones as electron donors in an in vitro analogue of PS I.
*Chl was first reacted with MV^{2+} to give $Chl^{\dagger}$ and $MV^{\dagger}$. The subsequent
reaction of $Chl^{\dagger}$ with hydroquinones was slower than the comparable
reaction with alternative electron donors and an exponential relationship was observed between the redox potential of the hydroquinone and
the rate constant for its reaction with $Chl^{\dagger}$. This sequence of reactions stores 40% of the energy available in the Chl triplet state, but
no permanent storage of energy was achieved since $MV^{\dagger}$ back-reacts with
quinone to give a cyclic process.

INTRAMOLECULAR PHOTOCHEMICAL ELECTRON TRANSFER IN A LINKED PORPHYRIN-
QUINONE MOLECULE. Alan R. McIntosh, Tefu Ho and James R. Bolton,
Photochemistry Unit, Department of Chemistry, University of Western
Ontario, London, Ontario, Canada, N6A 5B7.

We have attempted to mimic the "primary" charge separation reaction
occurring in bacterial and green plant photosynthesis through the
approach of an organic model system. We have chosen to synthesize a
single molecule P-Q containing an electron donor (tetra-aryl porphyrin,
P) and an electron acceptor (quinone, Q) linked by a hydrocarbon and
diester linkage with the number of atoms on the linking chain varying
between seven and nine. Both steady state and transient EPR measurements were carried out for concentrations of P-Q between 10^{-4} M and
10^{-3} M in a frozen methanol solution between 12K and 165K. Steady
state EPR signals were observed in the g = 2.0 region at low temperatures, which became very reversible at 165K. According to the results
of control experiments using mixtures of unlinked porphyrin and quinone
molecules, it was shown that the initial charge separation reaction in
the linked PQ molecules must be intramolecular. However, there was
probably, in addition, a low-yield contribution to the spectrum resulting from some electron transfer into the solvent or intermolecular
transfer to the quinone moiety of another P-Q molecule. In further
flash photolysis studies with EPR detection, we have explored the
kinetic behavior of a triplet species observed from a P-Q molecule at
about 12K. The EPR spectrum of this triplet is very similar to that
of the unlinked porphyrin; however, the PQ triplet decays significantly more slowly than the corresponding unlinked porphyrin's triplet
state.

COVALENTLY-LINKED PORPHYRIN-QUINONE COMPLEXES AS MODELS FOR THE PRIMARY EVENT IN BACTERIAL PHOTOSYNTHESIS   J.L.Y. Kong and P.A. Loach, Department of Biochemistry and Molecular Biology, Northwestern University, Evanston, Illinois

We have previously synthesized a series of covalently-linked porphyrin-quinone complexes where the length of the connecting diol varies from 2 to 6 carbons(Kong, J.L.Y. and Loach, P.A., Frontiers of Biological Energetics, 1, 73-82 (1978); Kong, J.L.Y. and Loach, P.A., J.Het.Chem,(1980) in press). The best characterized derivative has a 3 carbon bridge which we have abbreviated ZnP-3-Q. Its structure and purity have been proven by uv/vis absorbance, nmr, and mass spectra, as well as by cyclic voltametry. The fluorescent yield of ZnP-3-Q compared with that of the ZnP alone was .59,.51,.41,.35 and .12 in hexanol, dioxane, acetonitrile, dichloromethane and petroleum ether, respectively. The fluorescence lifetimes for each species have been measured. Photochemical charge separation has been demonstrated in acetonitrile both at room temperature and at 77^0 K. EPR data indicate that the species formed are the cation radical of the ZnP and the anion radical of Q. Because the fluorescence yield is quenched in ZnP-3-Q and because β-carotene and O_2 have no effect on charge separation at room temperature, we believe the excited singlet state is the immediate precursor of the charge separated state. Further characterization of the photochemical activity of these compounds is in progress.

ELECTRON TRANSFER FROM CHLOROPHYLL TRIPLET STATE TO QUINONES IN POLYMER FILMS AND LIPID BILAYERS. Glen Cheddar, John K. Hurley and Gordon Tollin, Department of Biochemistry, University of Arizona, Tucson, Arizona 85721

Laser photolysis is used to study radical formation and decay due to electron transfer between chlorophyll in cellulose acetate films or phosphatidyl choline vesicles and quinones present within the bilayers or in aqueous solutions in contact with the films or bilayers. Cation radical formation in both films and bilayers is closely correlated with triplet yield. With benzoquinone as acceptor, the maximum percent conversion of triplet to radical is 60% in the bilayer system and 30% in the film, compared to a 75% conversion obtained in ethanol solution. Radical decay in the films is second order ($k=1.5 \times 10^6 M^{-1}s^{-1}$, compared to $2 \times 10^9 M^{-1}s^{-1}$ in ethanol). With the liposome system, radical decay is biphasic, consisting of two first order processes. The t1/2 of the fast decay process is approximately 250µs and is invariant to either chlorophyll or quinone concentration. The t1/2 of the slower decay is in the millisecond range, and is dependent on quinone concentration (50ms at low quinone to 5ms at high quinone) but independent of chlorophyll concentration. For equivalent conditions, the rate of the slower decay is at least 2×10^3 times less than it would be in ethanol. Using lipophilic quinones, such as ubiquinone, as acceptors in the liposomes, only the fast decay process is observed. We attribute the fast decay to radical formation followed by recombination between C^+ and Q^- occurring within the bilayer, and the slow decay to these reactions occurring across the bilayer.

Work supported by the Division of Chemical Sciences, U.S. Department of Energy.

QUENCHING AND RADICAL FORMATION IN PORPHYRIN - QUINONE EXCIPLEXES
S.C. Bera, S. Nehari, J. Levy, N. Periasamy and H. Linschitz,
Department of Chemistry, Brandeis University, Waltham, Mass. 02254.

The reaction of excited porphyrins with quinones to form a pair of
ion-radicals has biological relevance and provides a useful in vitro
system for evaluating the factors governing the efficiency of radical
formation, relative to quenching. For the reaction between the porphyrin triplet and a number of substituted p-benzoquinones, flash
photolysis measurements have been made of primary rate constants,
absolute radical yields and radical decay rates, as function of
solvent polarity, redox potential and isotopic substitution. The
triplet-quinone interaction rate constants vary from 1.0×10^{10} M^{-1}
sec^{-1} for chloranil (E° red = 0.02 V vs SCE) to 1.8×10^9 M^{-1} sec^{-1} for
duroquinone (E° red = 0.76 V vs SCE), and are independent of solvent.
The radical yields are zero for all quinones in benzene and rise upon
addition of acetonitrile to a maximum characteristic of the quinone;
φrad = 1.0 for chloranil, decreasing to 0.15 for duroquinone. The
behavior of duroquinone is not altered by perdeuteration. The radical recombination rate decreases as solvent polarity increases,
corresponding to the solvent dependence of the initial radical yield.

This work was supported by the Division of Chemical Sciences, Office
of Basic Energy Sciences, U.S. Department of Energy.

LIGHT-INDUCED ELECTRON-TRANSFER FROM CHLOROPHYLL TO QUINONE. Kazuhiro
Maruyama, Hiroyuki Furuta, and Tetsuo Otsuki, Department of Chemistry,
Faculty of Science, Kyoto University, Kyoto 606, Japan

Electron-transfer process from photo-excited chlorophyll to quinone has
been widely studied because of its relation to the primary process in
photosynthesis. Here light-induced electron-transfer from Phe a to a
variety of quinones was investigated by means of CIDNP and ESR techniques. Photopotential and photocurrent of the similar system were
measured. The characteristics of the CIDNP and the ESR signals were
much dependent upon the concentration of added acid and the temperature
of the reacting system. The distinct CIDNP signals were observed by
adding an acid or lowering the temperature, indicating the light-induced electron-transfer from Phe a to quinone (K.Maruyama, H.Furuta,
and T.Otsuki, Chem.Lett., (1980) 857). By analysis of these CIDNP observations as well as the corresponding ESR signals, an aggregate formation between Phe a and quinone could be strongly suggested when CIDNP
signals were observed. The important contribution of added acid to
the electron-transfer process was confirmed further by measurement of
photocurrent of the similar system. Such a remarkable effect of acid
may reflect the essential role of metal ion such as Mg^{2+} in the intact
reaction center of photosynthesis. Actually, when Chl a, instead of
Phe a, was subjected to the light-induced electron-transfer reaction
with quinone, the distinct CIDNP signals were observed at ambient temperature even in the absence of acid.

ELECTRON TRANSFER FROM CHLOROPHYLL a TO QUINONE IN MONO AND MULTI-
LAYER ARRAYS. J.-P. Dodelet, M. Ringuet, M.F. Lawrence, R.M. Leblanc,
Groupe de recherche en biophysique, Département de chimie-biologie,
Université du Québec à Trois-Rivières, Trois-Rivières, Canada, G9A 5H7

Mono and multilayers of chlorophyll a (Chla)-lecithin have been
prepared on quartz slides, by means of the Blodgett-Langmuir technique,
for fluorescence studies. Self-quenching of the Chla fluorescence has
been observed in Chla-lecithin single layer excited with a laser light
at 632.8 nm. The fluorescence yield is reduced by 50% at a concentration of 7 x 10^{12} Chla molecules.cm^{-2}. Chla fluorescence quenching,
by adding N,N-distearoyl-1,4-diaminoanthaquinone (SAQ), has been
studied, in a single layer, in pure Chla and also at various dilutions
of Chla in lecithin. The results are explained in terms of a dynamic
quenching rather than in terms of a permanent complex formation, at
the ground state, between Chla and SAQ. The fluorescence quenching
has been interpreted as the result of an electron transfer from
excited Chla to SAQ, and rate constants of 8.3 x 10^{-5} cm^2.molecule^{-1}.s^{-1}
and 2.4 x 10^{-4} cm^2.molecule^{-1}.s^{-1} have been found for pure and diluted
Chla, respectively. Ten percent of the diluted Chla fluorescence
always remains unquenchable independently of the quinone concentration.
In multilayers, where SAQ and Chla are in different layers, there is no
fluorescence quenching for pure or diluted Chla even when the chromophores are in two adjacent layers. This happens only if SAQ is not
able to diffuse from one layer to another. A minimum value of 22.4 nm
has been found for the singlet exciton diffusion length in pure Chla
multilayers.

PHOTOPRODUCTION OF HYDROGEN BY UNIT-MEMBRANE VESICLES OF GREEN PHOTO-
SYNTHETIC BACTERIA. Jerome D. Bernstein and John M. Olson, Biology
Department, Brookhaven National Laboratory, Upton, New York 11973.

Unit-membrane vesicles (Complex I) were prepared from Chlorobium
limicola f. thiosulfatophilum strain Tassajara. The vesicles, about
60 nm in diameter, are free of chlorosomes and bacteriochlorophyll
(Bchl) c, but contain bacteriopheophytin c, Bchl a, carotenoid, cytochrome c-553 and photochemical reaction centers. They contain some
hydrogenase activity but it is only in the order of 3 μmole H_2 generated hr^{-1} (mg Bchl a)$^{-1}$. Light-dependent hydrogen production from
Complex I, 450 μmole H_2 hr^{-1} (mg Bchl a)$^{-1}$, has been obtained using
ascorbate as electron donor and exogenous, excess hydrogenase from
Clostridium pasteurianum. The system also includes gramicidin D,
tetramethyl-p-phenylenediamine, methylviologen, dithioerythritol and
an oxygen scavenging mixture. Similar experiments with chloroplasts
have yielded maxiumum rates of 200 or 30 μmole H_2 hr^{-1} (mg Chlorophyll)$^{-1}$ using ascorbate or water, respectively, as electron donor.
Taking into account that chloroplasts contain about 500 chlorophylls
per photosystem I reaction center and that Complex I vesicles contain
about 100 Bchl a molecules per reaction center, the relative rates of
hydrogen production with ascorbate as electron donor are 90 and 40
mole H_2 hr^{-1} (mmole reaction center)$^{-1}$ for the two systems, respectively. Preliminary experiments with Complex I vesicles suggest quantum efficiencies of the order of one percent.

This work was carried out under subcontract XD-9-8392-1 from the Solar
Energy Research Institute, Golden, Colorado.

STOICHIOMETRY OF SYSTEM I AND II REACTION CENTERS AND OF PLASTOQUINONE
IN DIFFERENT PHOTOSYNTHETIC MEMBRANES. Anastasios Melis and Jeanette S.
Brown, Carnegie Inst. Washington, Stanford, California, 94305, USA

The concentration of photochemical centers and of plastoquinone was
measured in several kinds of photosynthetic membranes by optical difference spectroscopy. Photosystem I reaction centers were measured
from the light-induced absorbance change at 700 nm (oxidation of P700).
Photosystem II reaction centers were estimated from the light-induced
absorbance change at 325 nm (reduction of the primary electron acceptor, Q). Spinach chloroplasts and membrane fractions obtained by
French press treatment, mature and light-stressed pea chloroplasts,
and blue-green algal membranes were investigated. No loss of primary
photochemical activity occurred during fractionation of the chloroplasts. The results indicated a large variability in the ratio of
system II to system I reaction centers (from 0.43 to 3.3) in different
photosynthetic membranes. Oxygen-evolving plants may change the ratio
of their photosystems in response to environmental light conditions.

The amount of photoreducible plastoquinone was also measured at 263 nm.
In spinach chloroplasts, 7-8 plastoquinone molecules were found per
reaction center of system II. Most of the plastoquinone pool was
associated with the grana. However, the ratio of chemically determined
plastoquinone to chlorophyll was similar in the grana and stroma
thylakoids.

INTERMEDIATES OF THE PHOTOSYNTHETIC WATER-OXIDATION ENZYME. G. Charles
Dismukes and Yona Siderer, Department of Chemistry, Princeton University, Princeton, New Jersey 08544.

Spinach chloroplasts which are flash illuminated (532 nm, 25 nsec, 75
mJ) at room temperature and immediately quench-cooled in an isopentane
bath at -140°C exhibit a previously unreported EPR signal having an
intensity which depends on flash number. The signal, which is not present in the dark, appears with greatest intensity after the first flash.
and again on flash 5. The spectrum exhibits over 16 readily resolved
hyperfine transitions and possibly as many as 20 with an average separation of 75G to 90G. This is consistent with a binuclear and possibly
a tetranuclear Mn containing enzyme, as evidenced by the similar spectra of binuclear mixed-valence Mn complexes, for example
$(bipy)_2Mn(III)$<$_O^O$>$Mn(IV)(bipy)_2$. Preliminary analysis suggests that
either a binuclear Mn(II)Mn(III) center or a tetranuclear Mn(IV)3Mn(III)
cluster can account for the spectral width. Incubation of chloroplasts
with 0.8 $\underline{M}$ Tris and 0.01 $\underline{M}$ NaCl, which abolishes water oxidation by ex-
tracting Mn ions, removes the signal. Reduction with 0.05 $\underline{M}$ sodium
dithionite also abolishes this signal. Incubation with the electron
transport inhibitor DCMU has no effect on the first flash. These results indicate a location on the oxidizing side of Photosystem II. Extraction with cholate, which has been shown to remove a 65 kd protein
that is functional in the reconstitution of O_2 evolution and which contains 2 Mn ions, also removes the multiline EPR signal. We believe
this new EPR signal is monitoring a mixed-valence oxidation state of
this binuclear Mn protein, or possibly a larger unit containing two
binuclear subunits.

BIOPHOTOLYSIS OF WATER FOR H$_2$ PRODUCTION USING IMMOBILIZED AND
SYNTHETIC CATALYSTS. David O. Hall, Paul E. Gisby and K. Krishna
Rao, Plant Sciences Department, King's College, 68 Half Moon Lane,
London SE24 9JF U.K.

Water can be split into H$_2$ and O$_2$ at ambient temperatures using visible light in the presence of chloroplasts, hydrogenase and electron-
transfer catalysts. Various natural e.g., ferredoxin, flavodoxin,
cytochrome c$_3$ and NAD, and synthetic electron mediators, e.g., methyl
viologen, [Fe-S] and [Mo-Fe-S] clusters, and 'Jeevanu' particles, can
couple the photochemically generated electrons and H$^+$ ions from water
to hydrogenases or Pt to produce hydrogen. At present the in vitro
system, under optimum conditions, can produce 5 litres of H$_2$ per g of
chlorophyll -- the reaction usually lasts 5 to 6 hours (50 μmoles
H$_2$/mg chlorophyll/hour). In the last few years we have isolated
more stable enzymes and membranes than those previously used in the
H$_2$-producing system. Since the biological components lose their catalytic efficiency on exposure to light and O$_2$, we have immobilized the
catalysts to improve their stability, e.g., on charged particles,
alginates, serum albumin, etc. Thylakoid membranes encapsulated in
calcium alginate films are found to be more stable during storage than
non-immobilized chloroplasts. Immobilization does not affect the
photosynthetic oxygen evolution capacity of the chloroplasts though it
does reduce the rate of H$_2$ evolution with hydrogenase. A system where
the catalysts were embedded in alginate evolves H$_2$ on illumination.
Platinum coated on polyvinylalcohol is a better catalyst than PtO$_2$
(Adam's catalyst). The O$_2$ and thermal stability of hydrogenases are
enhanced by binding the enzymes to Sepharose of Spherosil.

P740: A NOVEL, PHOTOCHEMICALLY ACTIVE FORM OF CHL a IN PHOTOSYSTEM I?
Stephen Lien, Solar Energy Research Institute, Golden, Colorado 80401,
USA, G. Brookjans, and A. San Pietro, Department of Biology, Indiana
University, Bloomington, Indiana 47401, USA

A satellite absorption band located at approximately 738 to 742 nm was
observed in the crude preparations of Photosystem I (PSI) particles
isolated from spinach chloroplasts according to a previously described
procedure (S. Lien and A. San Pietro, Arch. Biochem. Biophys. 194, 128-
137, 1979). Further studies showed that the long wavelength absorption
band, designated as pigment 740 (P740), can be greatly enriched (by
differential centrifugation of the crude PSI preparations) relative to
the major red absorption band located between 670 and 678 nm found in
the normal SPI particles. Since the majority of chlorophyll-protein
complexes present in most chloroplasts and subchloroplast preparations
from higher plants exhibit red absorption bands in the region of 670
to 715 nm, the long wavelength absorption band, P740, found in the
crude PSI preparation is of particular interest. The physical properties, the spectral characteristics and photochemical activities of
P740 will be compared to those described previously for: (1) the photo-
transformed 743 nm absorbing, water-soluble chlorophyll-protein complex
from Chenopodium (A. Takamiya, in "Method in Enzymology", A. San Pietro
ed., Academic Press, New York, Vol. 23A, pp. 603-613, 1971); and (2)
the microcrystalline forms of chlorophyll-H$_2$0 adducts absorbing at 740
nm region.

Supported by the Division of Biological Energy Research, Office of Basic
Energy Sciences, U.S. Department of Energy (S. Lien) and a National
Science Foundation Grant (PCM75-03414 A03 to A.S.P.).

CONTRIBUTED PAPERS

Session II

BIOMIMETIC AND MODEL SYSTEMS FOR SOLAR ENERGY CONVERSION

Plenary Speaker: Joseph J. Katz

Session Chairman: Henry Linschitz

Posters II-1 through II-14

ELECTRON TRANSFER IN SYSTEMS OF WELL DEFINED GEOMETRY. R. E. Overfield
and K. J. Kaufmann, Department of Chemistry, University of Illinois,
Urbana, Illinois 61801, USA, M. R. Wasielewski, Chemistry Division,
Argonne National Laboratory, Argonne, Illinois 60439, USA.

Two mesopyropheophorbide macrocycles can be joined via two covalent
linkages to produce a cyclophane. It is possible to insert one or two
Mg atoms into the cyclophane. The Q_y transitions of the macrocycles
are nearly orthogonal. The visible absorption spectrum of the monometal cyclophane is nearly a superposition of the spectra of the
monomers. Emission from the monometal cyclophane arises primarily from
the red most absorbing chromophore. The excited state difference
spectrum shows that both macrocycles are excited. Fluorescence lifetimes of the monometal cyclophane decrease with increasing dielectric
strength. Changes in the fluorescence and the triplet yield parallel
the shortening of the singlet lifetime. Thus the radiative rate is
solvent independent. This is in contrast to what one would expect if
the emitting state had charge transfer character. Since the fluorescence lifetime is dependent on dielectric, the nonradiative relaxation
from the singlet state is due to formation of a radical pair. The decay rate of the postulated radical pair was monitored by observing the
kinetics of ground state repopulation. For the geometry of this cyclophane, electron transfer proceeds relatively slowly ($k = 3 \times 10^9 \text{sec}^{-1}$)
in the forward direction. Modeling calculations indicate that the rate
of annihilation of the radical pair may decrease as the solvent dielectric decreases.

This work was supported by the Department of Energy.

A PICOSECOND SPECTROSCOPIC SEARCH FOR ELECTRON TRANSFER REACTIONS IN
DIMERIC AND TRIMERIC REACTION CENTER MODELS CONTAINING PYROCHLORO-
PHYLL a. T. L. Netzel, Department of Chemistry, I. Fujita, Department
of Energy and Environment, Brookhaven National Laboratory, Upton, NY
11973, R. R. Bucks and S. G. Boxer, Department of Chemistry, Stanford
University, Stanford, CA 94305.

Natural photosystems (photosynthetic bacteria as well as PSI and PSII
in green plants and algae) harness solar energy by rapidly trans-
forming the energy of an absorbed photon into electron transfer products. In this paper we wish to report the results of our studies on
covalently linked dimeric and trimeric RC models comprised of one or
two pyrochlorophyll a (PChl a) subunits as the electron donor and
either pheophorbide a (Pheo a) or pyropheophorbide a PPheo a) as
the acceptor. We have varied (a) the chain length which binds the
electron donor and acceptor, (b) the relative orientation of the donor
and acceptor, (c) the differences in free-energy between the donor excited state and the stored energy of the electron-transfer products,
and (d) the solvent dielectric constant. In no case did an $S_1 \rightarrow CT$
reaction occur with a rate greater than 10^{10} s^{-1}, even when such a
process was favored by over 300 meV. However one trimer,
$(PChl\ a)_2$ ∿∿ (Pheo a), yielded data consistent with an $S_1 \rightarrow CT$
reaction whose forward rate was $9 \times 10^9 \text{ s}^{-1}$ and whose reverse rate
was $3 \times 10^8 \text{ s}^{-1}$. These results contrast strikingly with those
obtained for a cofacial diporphyrin PSII model [T. L. Netzel,
P. Kroger, C. K. Chang, I. Fujita and J. Fajer, Chem. Phys. Lett. 67,
223 (1979)] in which an $S_1 \rightarrow CT$ rate $> 10^{11} \text{ s}^{-1}$ has been observed.
Research performed at BNL and supported by U.S. Department of Energy,
Contract No. DE-AC02-76CH00016.

THE PRIMARY CHARGE SEPARATION AND A BIOMIMETIC MODEL OF PHOTOSYNTHETIC
OXYGEN EVOLUTION. M. S. Davis,[a] A. Forman,[b] I. Fujita,[a] T. L. Netzel[c]
and J. Fajer,[a] Brookhaven National Laboratory, Upton, NY 11973[d]

Recent spectroscopic data indicate that short-lived transients mediate
charge transfer in Photosystems I and II of green plants. The ESR
linewidths, g-values, saturation behavior, as well as the ENDOR transition characteristic of the anion of pheophytin (Pheo) mirror those
observed for I$^-$, the primary acceptor of PSII. The combination of
emf, optical, as well as ESR and ENDOR properties found for I$^-$ and
for Pheo$^-$ in vitro, suggests that P680 is a ligated Chl a monomer
whose function in the phototrap is determined by its environment.
[Davis et al., Proc. Natl. Acad. Sci. USA 76, 4170 (1979)]. This then
leads to the following mechanism for the primary charge separation in
PSII: P680 (ligated Chl a monomer) + Pheo $\xrightarrow{h\nu}$ P680$^+_\cdot$ + Pheo$^-_\cdot$ with an
estimated minimum potential span of ~1.4 eV. Supporting evidence for
this mechanism is found in the picosecond data for a synthetic diporphyrin, where the chloride complex forms a charge-transfer species
with a lifetime of 620 (±20) ps, and spans a redox potential of
~1.6 eV. [Netzel et al., Chem. Phys. Lett. 67, 223 (1979); Netzel et
al., this Conference, preceding paper].

a) Dept. of Energy and Environment; b) Medical Research Center;
c) Chemistry Dept.; d) Work supported by the Division of Chemical
   Sciences, U.S. Dept. of Energy (Contract No. DE-AC02-76CH00016).

PICOSECOND ELECTRON TRANSFER IN DIPORPHYRIN MODELS OF PHOTOSYSTEM II OF
GREEN PLANTS. T. L. Netzel, Department of Chemistry; I. Fujita, C.-B.
Wang and J. Fajer, Department of Energy and Environment, Brookhaven
National Laboratory, Upton, N.Y. 11973.

Current spectroscopic evidence supports the view that the primary
electron donor in photosystem II of green plants is monomeric chlorophyll (M. S. Davis et al., Proc. Natl. Acad. Sci. US 76, 4170 (1979)
and that the primary acceptor is pheophytin (I. Fujita et al., J. Amer.
Chem. Soc. 100, 6280 (1978)). Previous picosecond measurements (T. L.
Netzel et al., Chem. Phys. Lett. 67, 223 (1979)) on a model of the
photosystem II reaction center, a cofacial diporphyrin comprised of a
Mg porphyrin covalently bonded to a free base with two five-atom chains,
provided evidence of a singlet → charge transfer (CT) reaction within
6 ps of excitation. This CT product decayed with a lifetime of 400 ps.
To explore spacial and environmental effects on the rates of electron
transfer, we varied 1) the dielectric constant of the solvent, 2) the
length of the covalent bridges, and 3) the relative orientation of the
porphyrin subunits. While none of the above variations altered the ≤ 6
ps formation time for the CT product, the decay time increased to 2 ns.
These results provide a reasonable biomimetic model for PS II in terms
of kinetic behavior and energy efficiency, and suggest that the molecular architecture of synthetic donors and acceptors may play a significant role in controlling wasteful back reactions.

This work was supported by the Division of Chemical Sciences, U.S.
Department of Energy, Washington, D.C., under Contract No. DE-AC02-
76CH00016.

ELECTROCHEMICAL PRODUCTION OF CHLOROPHYLL a AND PHEOPHYTIN a EXCITED STATES. PHOTOSYNTHESIS IN REVERSE. Michael R. Wasielewski, Rebecca L. Smith and Arthur G. Kostka, Chemistry Division, Argonne National Laboratory, Argonne, IL 60439

The reaction of chlorophyll a^+ (Chl a^+) with either Chl a^- or pheophytin a^- (Pheo a^-) in addition to the reaction of Pheo a^+ with Pheo a^- were studied in butyronitrile (BCN), BCN-1% THF, THF, and DMF. The electrochemically produced radical ion pairs Chl a^+-Chl a^- and Pheo a^+-Pheo a^- react in each solvent to produce a 10^{-7}-10^{-6} yield of luminescent states based on the initial number of radical pairs. The Chl a^+-Pheo a^- reaction produces no observable luminescence in any of the solvents examined. The luminescence maximum for the Pheo a^+-Pheo a^- reaction occurs at 730 nm in each solvent and is strongly red shifted relative to the fluorescence maxima for optically excited Pheo a in these solvents. A similar result is obtained for the Chl a^+-Chl a^- reaction in BCN. However, emission from the Chl a^+-Chl a^- reaction in the other three solvents occurs at 680 nm and corresponds more closely to normal fluorescence from optically excited Chl a. The red shifted spectra are consistent with the formation of excimers. AC voltammetry of Chl a in BCN provides evidence that Chl a is aggregated in the ground state in this solvent. Thus, the Chl a^+-Chl a^- reaction in BCN does not form a true excimer, whereas the Pheo a^+-Pheo a^- reactions in each solvent do.

This research was supported by the Office of Basic Energy Sciences, Division of Chemical Sciences, U.S. Department of Energy.

A PHOTOSYNTHETIC PHOTOELECTROCHEMICAL CELL. R. Bhardwaj, R.-L. Pan and E. L. Gross, Department of Biochemistry, The Ohio State University, Columbus, Ohio 43210.

We have previously developed a solar battery using Photosystem I (PSI) particles (Gross, E.L., Youngman, D.R. and Winemiller, S.L., Photochem. Photobiol. 28, 249 [1978]). We have now shown that isolated chloroplasts are also effective. They are easier to prepare and they have greater light harvesting capacity than PSI particles. Chloroplasts (equivalent to 80 μg Chl.) were aspirated onto a 1 cm^2 diameter Metricel filter which was placed between two compartments containing the electron acceptor and the electron donor respectively. Anaerobicity was maintained by bubbling argon through the cell. When dichlorophenol-indophenol and flavinmononucleotide were used as electron donor and acceptor (with 80 mM Tricine buffer pH 7.0), we observed an open circuit potential of 515 mV, a short circuit current of 3.5 mA, and a maximum power of 380 μW (3.8 W/m^2) across a 100 Ω resistance. The front face power conversion efficiency using white light of 30 W/m^2 intensity was approximately 1%. When anthroquinone sulfonate was used as the electron acceptor at pH 8.2 the open circuit potential was 452 mV and the maximum power was 242 μW. Chloroplasts on the filter remained active at 4^{0}C at least for a week. This research was supported by DOE Grant No. DE-FG02-79ER-10538.

REACTION-CENTER ELECTRODES -- A MODEL SYSTEM FOR SOLAR ENERGY CONVER-
SION. A. Frederick Janzen, Photochemical Research Associates, London,
Ontario, N6E 2V2, Canada and Michael Seibert, Solar Energy Research
Institute, Golden, Colorado 80401, U.S.A.

Reaction-center complexes isolated from the photosynthetic bacterium
Rhodopseudomonas sphaeroides R-26 can be dried as a film on platinum
and semiconductor (SnO_2) electrodes. The light-induced primary charge
separation which occurs across the biological complex couples electrically with the SnO_2 but not with the metal electrode. As the working electrode in a two-electrode photoelectrochemical cell, reaction-
center-coated SnO_2 generates photovoltages as high as 70 mV and photocurrents as high as 0.5 $\mu A \cdot cm^{-2}$ when exposed to light >600 nm. In a
three-electrode configuration with the working electrode poised at
+200 mV versus the saturated calomel electrode, much higher photocurrents (3.7 $\mu A \cdot cm^{-2}$) are observed. The power conversion efficiency is
comparable to that observed in other types of organic photoelectrochemical cells, but the internal resistivity is many orders of magnitude lower. The major limitation so far is the lack of specific orientation of the reaction center with respect to the electrode material. However, an energy-level model to explain the electron transfer
reactions at the SnO_2/reaction-center film/ electrolyte interface is
presented. Reaction-center electrodes may lead to new methods of
probing the primary photochemistry of photosynthesis and, in addition,
may also serve as model systems for future photoelectrochemical devices.

This work was supported in part by the Division of Biological Energy
Research, Office of Basic Energy Sciences, U.S. Department of Energy.

ZINC PORPHYRIN SENSITIZED REDOX PROCESSES IN MICROEMULSIONS.
M. P. Pileni,* Ecole Polytechnique Fédérale, Lausanne, Switzerland

Photoinduced electron transfer from zinc tetraphenylporphyrin to acceptors solubilized either in the lipid interiors or in aqueous bulk
of anionic oil-in-water microemulsions has been investigated by nanosecond laser photolysis. While intimate cosolubilization appears to
decrease considerably the efficiency of electron transfer in the former, greatly enhanced charge separation of redox products has been
observed in the latter.

*Present address: Université P. et M. Curie, Laboratoire de Chimie-
Physique, 11 rue P. et M. Curie, 75005 Paris, France.

PORPHYRIN SENSITIZED PHOTOREACTIONS IN MICROEMULSIONS. R. A. Mackay and
L. E. Weaver, Department of Chemistry, Drexel University, Philadelphia,
Pennsylvania 19104, U.S.A.

We have previously demonstrated that microemulsions may be used as
solvent media for reactions involving oil and water soluble reactants
and sensitizers. In particular, we have examined the effect of pH and
surface charge on the quantum yield of chlorophyll and pheophytin
mediated photoreductions. These studies have now been extended to
synthetic prophyrins, including a number of surfactant porphyrins. For
the latter, a large variation in quantum yield is observed which cannot
be accounted for in terms of the variation in charge of the porphyrin
head group. The results are interpreted in terms of the location and
orientation of the sensitizer in the microdroplet interphase region,
presenting some interesting possibilities for both the development
of artificial photosynthetic systems and application to some aspects of
membrane-mimetic studies.

PHOTOINDUCED ELECTRON TRANSFER REACTIONS BY SEVERAL PORPHYRINS BIS-
PORPHYRINS, PHTHALOCYANINES, PYRIDINOPORPHYRAZINS AND THEIR METALLIC
DERIVATIVES. P. Maillard, P. Krausz, C. Giannotti, Institut de Chimie
des Substances Naturelles, Gif-sur-Yvette 91190 France - S. Gaspard
Laboratoire des Pigments Végétaux et Substances Modèles, Ecole
Normale Supérieure de St Cloud, St Cloud, 92210 France.

Aerobic irradiation, of tetraphenylporphyrins, bis tetraphenylporphyrins, phthalocyanines, tetra t-butyl phthalocyanines, tetracarboxylph-
thalocyanines, tetrapyridinoporphyrazins and some of their metallic
derivatives, with visible light $\lambda > 420$ nm give singlet oxygen by energy transfer and oxygen superoxide by electron transfer the study as
been made by selective spin traps. We use 2-methylpenten and ergosteryl
acetate as singlet oxygen quenchers and also 2,2',6,6' tetramethylpipe-
ridin for the singlet oxygen and 5,5' dimethyl -1-pyrroline-N-oxide as
superoxide spin traps. The intensity of the signal coming from the spin
trapped O_2^- superoxide which is sharply reduced by addition of 10 to
30 % of methanol or water is nearly unaffected by addition of a good
$^1\Delta g$ oxygen quencher such as 1.4 diazabicyclooctane (DABCO), showing
that the formations of oxygen superoxide is not coming from the decomposition of the $^1\Delta g$ oxygen from the direct quenching. The Zn and Pt
phthalocyanines and Cu^{II} porphyrins or bis-porphyrins can induce singlet oxygen formation but they cannot give any oxygen superoxide. We
have shown that these compounds are quenching the Zn TPP oxygen superoxide formation. Two mechanisms can explain these results : either the
very fast decomposition of the superoxide anion O_2^- or the quenching
of the electron transfer from the donor to the oxygen.

COUPLING OF PHOTOCHEMICAL REDOX REACTIONS ACROSS BILAYER MEMBRANE IN LIPOSOME SYSTEM. Mitsunori Sukigara and Kazue Kurihara, Institute of Industrial Science, The University of Tokyo, 7-22-1 Roppongi, Minato-ku, Tokyo 106, Japan

Coupling of photochemical redox reactions through phosphatidylcholine liposome bilayer membrane incorporating purified chlorophyll a was examined in relation to designing light energy conversion and storage system. The concentration change of oxidizing agent such as Cu^{2+}, $Fe(CN)_6{}^{3-}$ and O_2 in the outer solution of liposome was measured during illumination with the light absorbable by chlorophyll a. The enhancement of photoreduction of Cu^{2+} by a reductant such as potassium ascorbate localized in a solution of the opposite side of the membrane suggested that photoredox reactions at both membrane-solution interfaces of a liposome were coupled through the bilayer. Kinetic analysis of the reactions based on a phenomenological reaction scheme was carried out. The decrease of $Fe(CN)_6^{3-}$ concentration was also observed upon illumination without adding any particular reductant in the system. The reaction rate was strongly affected by the addition of carbonylcyanide m-chlorophenylhydrazone (CCCP) to the dispersion. The apparent O_2 consumption was suppressed by the existence of $Fe(CN)_6^{3-}$ in a solution of the inner compartment. These facts suggest the possibility of the coupled redox reaction with positive free energy change at the both membrane-solution interfaces during illumination, through across membrane electron transport facilitated by the addition of CCCP.

COMPARATIVE THEORETICAL CALCULATIONS ON THE ELECTRONIC ABSORPTION SPECTRA OF BACTERIOCHLOROPHYLLIDE a AND BACTERIOPHEOPHORBIDE a ANION RADICALS. J.D. Petke, G.M. Maggiora, R.E. Christoffersen, Departments of Chemistry and Biochemistry, University of Kansas, Lawrence, Kansas, 66045, U.S.A., and L.L. Shipman, Chemistry Division, Argonne National Laboratory, Argonne, Illinois, 60439, U.S.A.

The present study describes the results of ab initio configuration interaction calculations of the electronic absorption spectra of the π-anion radicals of ethyl bacteriochlorophyllide a (Et-BChl $a^{\bar{\ }}$) and ethyl bacteriopheophorbide a (Et-BPheo $a^{\bar{\ }}$). A comparison of the calculated excited states and transition energies points out several features which may be useful in distinguishing one anion radical from the other. In particular, the lowest two excited states, D_1 and D_2, of Et-BChl $a^{\bar{\ }}$ are found to be degenerate and are predicted to contribute to the experimentally observed absorption at 10,000 cm^{-1}. In Et-BPheo $a^{\bar{\ }}$, however, D_1 appears with low oscillator strength (f = 0.002) at 8600 cm^{-1} above the ground state, D_0, while D_2 is predicted to contribute to absorption at 11,000 cm^{-1}. Another difference between the two predicted spectra appears in the low-energy shoulder of the Soret band, where for Et-BPheo $a^{\bar{\ }}$ transitions to D_7, D_8 and D_9 contribute to the spectrum, as opposed to transitions to a pair of states, D_{10}, and D_{11}, for Et-BChl $a^{\bar{\ }}$.

This work was supported in part by the Division of Basic Energy Sciences of the United States Department of Energy, and the Upjohn Company, Kalamazoo, MI.

INVESTIGATIONS OF FUNCTION AND STRUCTURE IN THE PHOTOSYNTHETIC UNIT BY
ELECTRON SPIN ECHO SPECTROSCOPY. M. C. Thurnauer, M. K. Bowman and
J. R. Norris, Chemistry Division, Argonne National Laboratory,
Argonne, Illinois 60439 and T. M. Cotton, Illinois Institute of Tech-
nology, Chicago, Illinois 60616.

We are applying electron spin echo (ESE) spectroscopy to investigate
function and structure in the photosynthetic unit. This work includes
time resolved ESE experiments on photosynthetic algae and studies of
the electron spin echo envelope modulation observed in chlorophyll
anion radicals in solid solution. The two studies are closely related. In the time-resolved ESE experiments we have observed spin
polarized EPR signals which appear to be linked to the primary reactions in photosystem I. We take advantage of several unique properties of ESE spectroscopy to understand the mechanism of the early
photosynthetic reactions. For instance, we have observed an unex-
pected echo phase shift in the signal due to P700$^+$. We propose that
the phase shift is produced by time-dependent radical-pair interactions between P700$^+$ and one of the early electron acceptors. Thus, we
are particularly interested in identifying the early acceptors. Both
chlorophyll and pheophytin (chlorophyll without the central Mg) anions
have been proposed as possible primary electron acceptors in both bacterial and green plant photosynthesis. Therefore, we are extending
the magnetic resonance investigations of chlorophyll-like anions to
include studies by ESE spectroscopy. A comparison of the electron
spin echo envelope modulation observed in chlorophyll anion radicals,
pheophytin anion radicals and anion radicals enriched with ^{25}Mg should
allow us to distinguish between chlorophyll and pheophytin.

CALCULATED EFFECTS OF MAGNETIC FIELD ON FLUORESCENCE OF DIMERIC BAC-
TERIOCHLOROPHYLLS. Charles E. Swenberg, Department of Chemistry, New
York University, New York, NY 10033 and John S. Connolly, Solar
Energy Research Institute, Golden, CO 80401.

It is generally accepted that bacterial photosynthetic reaction centers (RC's) contain a "special pair" of bacteriochlorophyll (BChl)
molecules, probably a dimer, which distinguishes these photoactive
sites from the antenna chlorophylls by providing both a sink for electronic excitation energy and a pathway for primary electron transfer. One of the most important questions that remains to be answered,
however, is how electron transfer in vivo is effected out of an excited singlet state when the preponderance of in vitro data shows that
net light-induced electron transfer involving neutral molecules occurs
only out of triplets, even when oxidation of excited singlets is thermodynamically more favorable. The kinetics of in vivo electron transfer (3-10 ps) would appear to rule out normal intersystem crossing
(~10 ns) to a singly excited triplet of BChl. However, double triplets (^{3}BChl$\cdots$^{3}BChl) might be formed by singlet fission if the triplet
energy of monomeric BChl (E_T) is within a few kT of half the singlet
excitation energy (E_S). This paper discusses the theoretical dependence on magnetic field of the fluorescence quantum yield for two
microscopic models of fully reduced bacterial reaction centers and
other dimeric BChl structures in which net electron transfer is precluded. The results suggest that experimental verification can distinguish between the two models. [This work was supported by the
Division of Chemical Sciences, Office of Basic Energy Sciences, U.S.
Department of Energy].

CONTRIBUTED PAPERS

Session III

QUANTUM HARVESTING AND ENERGY TRANSFER

Plenary Speaker: Christian K. Jørgensen

Session Chairman: Hans Kuhn

Posters III-1 through III-9

COLLECTION AND CONVERSION OF SOLAR ENERGY FOR PHOTOCONVERTING SYSTEMS.

R. Reisfeld, Department of Inorganic Chemistry, University of Geneva,
CH 1211, Geneva 4,* and E. Greenberg, A. Kisilev and Y. Kalisky, *Department of Inorganic Chemistry, Hebrew University, Jerusalem, Israel.

Solar energy will be utilized as a major energy source when its cost
is lowered significantly. One of the possibilities is to concentrate
a large amount of useful solar photons on small areas thus permitting
the use of expensive solar devices such as photovoltaic cells, photoelectrochemical cells or chemical reactions. We have suggested [Natur-
wiss. 66, 1 (1979)] that fluorescent concentrators may have the following advantages: light concentration up to 100 times without tracking, high collection efficiency of diffuse light (especially important
when diffuse light is on the order of 60%; good heat dissipation from
large areas of the collector; and optimization of the concentrated
emitted light to maximum sensitivity of the light converting device.
The requirements of predicted collectors are: high collection efficiency, high absorption in the maximum of the solar spectrum; high
fluorescence efficiency; separation of absorption from emission bands
in order to reduce self-absorption; and good stability. Although
organic dyes possess much higher absorption and larger quantum
efficiencies than inorganic materials, their disadvantages are
photodecomposition and overlap of the emission spectra with absorption. This leads to considerable self-absorption of the luminescent
light along the path of the collector. Inorganic ions in glasses have
much narrower absorption bands and usually lower quantum efficiencies;
however, they are expected to be stable over longer time periods and
can be incorporated as an integral part of photoconverting devices.

LIGHT ABSORPTION AND ENERGY TRANSFER IN CAROTENOPORPHYRINS. Edward J.

Land, Paterson Laboratories, Christie Hospital and Holt Radium
Institute, Manchester M20 9BX, England, René V. Bensasson, Laboratoire
de Biophysique, Muséum National d'Histoire Naturelle, 61 Rue Buffon,
75005 Paris, France, Ana L. Moore, Gary Dirks, Devens Gust and Thomas
A. Moore, Department of Chemistry, Arizona State University, Tempe,
Arizona 85281, U.S.A.

Intramolecular energy transfer in a new class of compounds where a
meso-tetraarylporphyrin is covalently linked to a carotenoid through
an ester group has been investigated. Such studies help elucidate
the geometrical requirements for light harvesting by carotenoids in
photosynthetic membranes and will aid in the design of synthetic
porphyrin-based solar energy conversion schemes.

It has been found that singlet-singlet energy transfer from a 26π
electron nonfluorescent polyene moiety linked to the porphyrin
through the para position of a meso aromatic ring is essentially zero,
whereas porphyrin triplet to carotene triplet intramolecular energy
transfer is very efficient. On the other hand, linkage through an
ortho position (which allows the carotenoid to fold over the face of
the porphyrin π system) results in singlet-singlet energy transfer
from the carotenoid to the porphyrin of ∿25% and a greatly enhanced
rate of population of the carotenoid triplet state. In addition,
substantial quenching of the porphyrin fluorescence by the carotenoid
moiety has been observed.

This work was performed with the support of the U.S. Department of
Energy, Grant No. DE-FG02-79ER10545.

INFLUENCE OF SOLVENT ON FLUORESCENCE PROPERTIES OF CHLOROPHYLLS.

A. Frederick Janzen, Photochemical Research Associates, London, Ontario N6E 2V2, Canada, James R. Bolton, Department of Chemistry, The University of Western Ontario, London, Ontario, N6A 5B7, Canada, and John S. Connolly, Solar Energy Research Institute, Golden, CO 80401.

It is well known that the electronic spectra, both absorption and emission, of isolated chlorophylls are markedly solvent dependent, but systematic studies of these effects in the case of bacteriochlorophyll a (BChl a) are lacking. We have measured absorption and fluorescence spectra and fluorescence lifetimes of BChl a in a variety of solvents. By comparing the fluorescence lifetimes (obtained by time-resolved, single-photon counting) with the calculated radiative lifetimes (from integration of the absorption spectra) we have been able to calculate fluorescence quantum yields. Measurements of short fluorescence lifetimes by the single-photon counting technique can be subject to ambiguities related to the mathematical deconvolution procedure, especially at wavelengths (~800 nm) near the red edge of the photomultiplier spectral response. Analysis of our data obtained to date indicates that the shortest lifetime of ^{1}BChl a* (~2.5 ns) is observed in methanol and the longest (~3.6 ns) in pyridine and THF. The calculated radiative lifetimes in these two solvents are ~17 and ~14 ns, respectively; thus the calculated fluorescence quantum yields range from ~14% (MeOH) to ~26% (pyridine). Intermediate values were obtained in acetone (~18%) and ether (~19%) solvents. [This work was supported in part by the Division of Chemical Sciences, Office of Basic Energy Sciences, U.S. Department of Energy].

ANTENNA ORGANIZATION AND ENERGY TRANSFER IN THE GREEN BACTERIUM CHLORO-FLEXUS.

J. A. Betti*, R. E. Blankenship**, S. G. Sprague*, L. C. Dickinson† and R. C. Fuller*. * Dept. of Biochemistry; † Depts. of Polymer Science and Chemistry, University of Massachusetts, Amherst, MA; ** Dept. of Chemistry, Amherst College, Amherst, MA.

Green bacteria contain an antenna system which is quite different from the membrane bound Chl-protein complexes of purple bacteria and green plants. Antenna functions are carried out by chlorosomes, flattened vesicles attached to the underside of the cytoplasmic membrane. Chlorosomes contain a large amount of BChl c and a relatively small amount of protein. Physical properties of purified chlorosomes were investigated. The absorption spectrum has peaks at 460 nm, 740 nm (BChl c) and 790 nm (BChl a) with intensity ratios of 20:25:1. The molar ratio BChl c: BChl a is approximately 15:1. The fluorescence spectrum shows emission peaks at 755 nm and 807 nm with approximate corrected intensities of 1:1.5 when excitation is into either the 460 nm or 740 nm bands. The CD spectrum shows an asymmetric derivative-shaped CD centered at 739 nm and a negative CD in the 400-500 nm region. Treatment with ferricyanide results in the appearance of a 2.3 G wide ESR spectrum at g = 2.002. Whole cells show fluorescence peaks identical to those in chlorosomes plus a shoulder at longer wavelength; the intensities of the 755 nm and 807 nm emissions are nearly equal. These experiments indicate that the chlorosome pigments are functionally organized such that energy transfer from BChl c → BChl a occurs within the chlorosome prior to energy transfer to the cytoplasmic membrane.

Supported by NSF Grant PCM 7910970.

INTERACTION OF CHLOROPHYLL a AND N,N-DIMETHYLMYRISTAMIDE ON POLY-
ETHYLENE-UNDECANE PARTICLES. G.R. Seely and A.M. Rutkoski, Charles F.
Kettering Research Laboratory, Yellow Springs, Ohio 45387, USA

Chlorophyll (Chl) and surfactants such as N,N-dimethylmyristamide
(DMMA) adsorbed to particles of polyethylene swelled with plasticizers
such as undecane form a model system with interesting and perhaps unique
photophysical properties. These materials were deposited on the particles out of methanol-water under near equilibrium conditions. Most of
the chlorophyll is in a monomeric form (662.5 nm) in the presence of
excess DMMA, but asymmetry of the absorption band suggests the presence
of aggregated forms absorbing at longer wavelengths. Fluorescence
spectra reveal several emitting species. In addition to the fluorescence of monomeric chlorophyll (670 nm at 295 K, 675 nm at 77 K) fluo-
rescences of dimeric or more highly associated forms have been distinguished near 681 and 725 nm at 295 K (683 and 735 nm at 77 K), near
695 nm at 77 K in preparations with high chlorophyll concentration, and
near 705 nm at 77 K in preparations with adsorbed DMMA/Chl less than 4.
Extensive transfer of energy from monomeric chlorophyll to the associated forms is implied by the dissimilarity of absorption and fluorescence spectra, at 295 K as well as 77 K. Adsorption isotherms in
several methanol-water mixtures suggest that a distinct monolayer does
not form, and that partition of chlorophyll onto the particles is somewhat enhanced by increased amounts of chlorophyll and DMMA already
adsorbed. It appears that the particles may have a viscous surface
region in which chlorophyll, DMMA, methanol and water combine in
thermodynamically stable associations.

KINETIC ASPECTS OF TRIPLET ENERGY MIGRATION by J.C. Scaiano, Division
of Chemistry, National Research Council of Canada, Ottawa, Canada,
K1A OR6.

The conversion of radiant energy into chemical products or electricity
requires the occurence of excited state processes at chosen reaction
sites. These processes can only occur with high efficiency if the
energy is absorbed at that site or if migration processes allow the
electronic energy to move towards the desired site, with some of the
chromophores effectively behaving as an antenna. Similar energy
transfer properties are also desirable in macromolecules which are
required to withstand prolonged exposure to sunlight, since energy
migration plays a critical role in polymer photostabilization.

A series of triplet energy migration processes has been examined, in
most cases using laser flash photolysis techniques, and the importance
of energy hopping, molecular mobility, entropy factors and phase
separation will be discussed. The systems examined include: (a) intermolecular quasi-isoenergetic transfer in fluid solution; (b) energy
hopping in molecules with two similar chromophores; (c) migration in
polymers and copolymers; (d) inter-micellar energy migration and
(e) reversible energy transfer between different chromophores of
similar energy.

THE EFFECT OF MOLECULAR WEIGHT ON TRIPLET EXCITON PROCESSES: DELAYED EMISSION OF SOLID POLY(2-VINYLNAPHTHALENE). Nakjoong Kim and <u>Stephen E. Webber</u>, Department of Chemistry, University of Texas, Austin, Texas 78712.

Triplet-triplet annihilation in solid poly(2-vinylnaphthalene) at 77K has been found to depend on molecular weight. Just as in isolated polymers in low temperature matrices, increasing the chain length enhances the intensity of delayed fluorescence. In addition, the delayed fluorescence (which is excimer-like) is shifted by ~1800 cm^{-1} relative to the prompt fluorescence (also excimer-like), implying that triplet annihilation is occurring at trapped triplet sites, yielding primarily trap fluorescence, i.e.

$$T^*_{exciton} + T^*_{trap} \rightarrow S^*_{trap} + S_o \rightarrow 2S_o + (h\nu)^{trap}_{fl}$$

The phosphorescence is also excimer-like, which may imply that essentially all phosphorescence originates from T^*_{trap}.

The reason for the molecular weight dependence is not known at the present time. For isolated polymers the statistics of excitation favor triplet annihilation on longer chain polymers. If the same argument holds here, then the implication is that triplet excitons are largely confined to a single polymer chain in these solids.

This work has been supported by the National Science Foundation (DMR-78-02600) and the Robert A. Welch Foundation (F-356).

ENERGY TRANSFER AND IONIC PHOTODISSOCIATION PROCESSES OF POLYMERS HAVING AROMATIC GROUPS IN SOLUTION.
<u>Hiroshi Masuhara,</u> Satoshi Ohwada, and Noboru Mataga, Department of Chemistry, Faculty of Engineering Science, Osaka University, Toyonaka, Osaka 560, Japan

Primary photoprocesses of poly(N-vinylcarbazole) and its related compounds in solution have been studied by laser photolysis method. In the quencher-free systems a S_1-S_1 annihilation process characteristic of polymers and laser excitation was confirmed, which suggests energy transfer along a polymer chain. When the electron acceptor dimethyl terephthalate is added in N,N-dimethylformamide, ionic photodissociation to polymer cation and acceptor anion occurs efficiently. The geometrical structure of polymer cation is considered, comparing the spectra of monomer, dimer and polymer systems. All the produced ions obeys second-order decay kinetics. In the case of polymers with high degree of polymerization, a fast recombination process is observed in addition to normal homogeneous one. The relative dissociation yield decreases as the degree of polymerization increases (4-1100) and its change is smooth and moderate.

Comparing these results with those obtained in our studies on monomer donor-monomer acceptor systems, charge separation in polymer solution is considered.

SYNTHESIS AND PROPERTIES OF SOME POLYMER-BOUND PHOTOSENSITIZERS
Carl C. Wamser and William R. Wagner, Department of Chemistry, California State University, Fullerton, Fullerton, California 92634.

We have covalently attached a variety of photosensitizers to cross-linked polystyrene beads and investigated the spectroscopic and photochemical properties of the polymer-bound photosensitizers in comparison with monomeric model photosensitizers. Preliminary work has been done with a polymer-bound analog of benzophenone, which effectively photosensitizes the trans-cis isomerization of stilbene, and a polymer-bound analog of naphthalene, which shows fluorescence quenching by amines.

A more detailed study has been completed with p-methyl-α-cyclohexylacetophenone (I) and its polystyrene-bound analog (II). Both forms undergo Norrish Type II photoelimination of cyclohexene. Exhaustive irradiation indicates that II was 6% functionalized, based upon the total yield of cyclohexene evolved. Quantum yields depend upon the solvent (Φ=0.59 in pentane, 0.25 in ethanol for I) and are lower for the polymer-bound analog (Φ=0.42 in pentane, 0 in ethanol). Energy transfer to trans-stilbene was monitored by the quenching of the photoelimination and by the sensitization of the trans-cis isomerization of stilbene. The polymer-bound photosensitizer was somewhat less effective as an energy transfer donor than its monomeric analog, based upon their respective Stern-Volmer quenching constants ($k_Q T$ in pentane: 11,580 M^{-1} for I and 3,960 M^{-1} for II).

CONTRIBUTED PAPERS

Session IV

PHOTOCHEMICAL ELECTRON TRANSFER REACTIONS IN HOMOGENEOUS SOLUTIONS

Plenary Speaker: Vincenzo Balzani

Session Chairman: David Whitten

Posters IV-1 through IV-13

VISIBLE LIGHT-INDUCED ELECTRON EJECTION FROM ANIONS   Marye Anne Fox, James Hohman, and Robert C. Owen, Department of Chemistry, The University of Texas at Austin, Austin, Texas   78712   USA

Upon excitation with visible light, a variety of highly absorptive organic anions eject electrons as a primary photoprocess. If such photolyses are conducted in homogeneous solution in the presence of electron acceptors, the chemical consequences of this electron trans- fer include coupling, rearrangement of the acceptor, or sensitized geometrical isomerization. Under heterogeneous conditions, e.g., at the surface of a doped n-type semiconductor electrode, electron exchange leads to the production of photocurrent ($0.1-5\mu A/cm^2$) and to oxidative coupling.

Recent results from our laboratory show that anions 1-4 are useful in

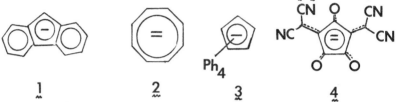

carbanionic photogalvanic cells employing doped or chemically-modified tin oxide or titanium dioxide semiconductors as anodes.

This work was supported by the Division of Chemical Sciences, Funda- mental Interactions Branch Office of Basic Energy Sciences, U.S. Department of Energy.

PHOTOINDUCED ELECTRON TRANSFER REACTIONS WITH SOME CARBONYL AND "Fe$_4$S$_4$" CLUSTERS. G. Mousset, P. Krausz, H. Chanaud and C. Giannotti, Institut de Chimie des Substances Naturelles CNRS, Gif-sur-Yvette 91190 France.

Irradiation with visible light of several polynuclear carbonyl cyclo- pentadienyl complexes of Fe, Mo, Ni and carbonyl complexes of Co, Fe and Mn may leed to electron transfer on methyl viologen and several quinones. In this part of the work we use essentially the $[\pi^5 \, C_5H_5 \, (Co)_3 \, MO]_2$ as donor and chloranil, dichlorodicyanobenzoquinone and the ortho methyl or phenyl quinone as acceptor. We consider two possibility, 1- the photoinduced electron transfer reaction is conducted in pure aceto- nitrile, 2- the reaction is conducted in acetonitrile containing 3 % of water. In the firth case we obtain the formation of the ·CH$_2$ CN radi- cal ; in the second case we obtain the addition of the hydroxyl radi- cal ·OH coming from the water added ; these radicals gives a spin adduct with the spin trap we have used : the 5.5' dimethylpyrroline N-oxide.

With the "clusters" $[N(n-C_3H_7)_4]_2 \, [Fe_4S_4(SR)_4]^{2-}$ R = n-C$_3$H$_7$, C$_6$H$_4$NO$_2$.

We describe the photoinduced electron transfer reaction and its depen- dance on the basicity and the wavelength of the light used, and also with addition of cysteine molybdenum complex as cocatalyst the trans- fer was performed on methyl viologen, acenaphtene quinone and tetra- cyano-ethylene and also on TPP Mn^{III} cl and its dimer. We report also the possible reversible intramolecular electron transfer of the cluster core to the ligand SC$_6$H$_4$NO$_2$.

THE NATURE OF INTERMEDIATES FROM EXCITED-STATE ELECTRON-TRANSFER
REACTIONS: Cr(II), Rh(II), and Rh(I) POLYPYRIDYL COMPLEXES.
M. Z. Hoffman, Department of Chemistry, Boston University, Boston,
MA 02215, USA, N. Serpone and M. A. Jamieson, Department of Chemistry
Concordia University, Montreal, Quebec, H3G 1M8, Canada, Q. G.
Mulazzani, S. Emmi, P. G. Fuochi, and M. Venturi, Istituto di
Fotochimica e Radiazioni d'Alta Energia del C.N.R., 40126 Bologna,
Italy.

The reductive quenching of the excited states of polypyridyl complexes
of Cr(III) and Rh(III) have been shown to lead to the reduction of
water to H_2 through the intermediacy of Cr(II), Rh(II), and Rh(I)
species. These reduced intermediates can be generated conveniently
using radiation chemical techniques thereby obviating the limitations
of photochemical excitation, quenching, and back electron-transfer
reactions. The spectra of $Cr(NN)_3^{2+}$, formed by the one-electron
reduction of $Cr(NN)_3^{3+}$, are reported and analyzed. The reactions of
$Cr(NN)_3^{2+}$ with Fe_{aq}^{3+} occur with rate constants spanning the range 10^8-
10^{10} $M^{-1}s^{-1}$ that can be treated by the Marcus relationships for
electron-transfer reactions. The one-electron reduction of $Rh(bpy)_3^{3+}$
yields $Rh(bpy)_3^{2+}$ which undergoes slow loss of bpy. At pH 10,
$Rh(bpy)_3^{2+}$ undergoes disproportionation with ligand-labilized Rh(II) to
form $Rh(bpy)_2^+$; at pH>10, ligand-labilized Rh(II) reduces $Rh(bpy)_3^{3+}$
resulting in a redox-catalyzed ligand-labilization chain reaction.
The behavior of O_2-sensitive $Rh(bpy)_2^+$, its spectrum, and state of
aggregation are highly dependent upon the pH of the solution, [Rh(I)],
and the nature of the counter-anion. H_2 is produced in neutral solution in the absence of any catalyst with an efficiency of ∿15%.

ELECTRON TRANSFER FROM EXCITED STATES OF PYRENE TO METAL IONS IN
CAFFEINE-SOLUBILIZED AQUEOUS SOLUTIONS. <u>Akira Kira</u>, Yoshio Nosaka, and
Masashi Imamura, The Institute of Physical and Chemical Research, Wako,
Saitama 351, Japan.

Caffeine solubilizes pyrene in aqueous solution by forming a van der
Waals complex. Pyrene thus solubilized was excited with a 347-nm laser
pulse, and the fluorescence quenching rates, triplet-state quantum
yields, and cation yields were measured as a function of the concentrations of added metal salts. Seventeen metal ions were examined.
The formation of the pyrene cation was observed on the oxidation of
the singlet state by Cu^{2+}, Eu^{3+}, Hg^{2+}, and Fe^{3+}, for which the
conversion efficiencies were 0.68, 0.39, 0.15, and 0.2, respectively.
The pyrene cation formation from the triplet state was observed for
Fe^{3+} and Hg^{2+} (efficiencies ≥ 0.9). The enhancement of intersystem
crossing by metal ions was predominant on reactions of the singlet
state with Ag^+, Tl^+, Co^{2+}, Ni^{2+}, Tb^{3+}, and Sm^{3+}. Plots of the
fluorescence quenching rate constants vs. the free energy differences
showed considerable deviation from Marcus theory, and the reasons for
the deviation were argued for some of the metal ions.

THE PHOTOOXIDATION OF CE(III) BY THE LOWEST EXCITED STATE OF URANYL ION.

Tchiya Rosenfeld-Grünwald, Marek Brandeis and Joseph Rabani, Energy Research Center, Hebrew University of Jerusalem, Jerusalem 91000, Israel

Suspensions of metal oxides such as PtO_2 or RuO_2 in aqueous solutions were reported by J. Kiwi and M. Grätzel, (Chimia 33, 289-291 (1979)), to catalyze the oxydation of water by oxidants such as Ce(IV) producing oxygen. We studied the photoelectron transfer from Ce(III) to the lowest excited state of uranyl ion, $*UO_2^{2+}$. Uranyl ion was chosen as it absorbs light in the shorter wavelength range of the solar spectrum, producing a relatively long-lived excited state. Although indirect evidence suggests that photoelectron transfer may take place between Ce(III) and UO_2^{2+}, no direct evidence for such a reaction was previously given. We have now demonstrated the photooxydation of Ce(III) by $*UO_2^{2+}$ in 0.1-1 N H_2SO_4 aqueous solution using a high intensity Nd laser following the formation of the product Ce(IV). The exciting light, $\lambda=353$ nm, is not absorbed by Ce(III). The quantum yield of Ce(IV) formation was found to be unity in both H_2SO_4 and HNO_3 solutions. The back reaction between UO_2^+ and Ce(IV) is second order with $k/\varepsilon(\lambda=380$ nm$) = 2.8\times10^3$ cm sec^{-1} ($\varepsilon(\lambda=380$ nm$) = 1.44\times10^5$ M^{-1} cm^{-1}). Future work is in progress in order to compose an inorganic aqueous system which will store photo energy and evolve oxygen using visible light. Preliminary experiments show that some O_2 can be produced photochemically in these systems.

This work was supported by the Israeli Ministry of Energy.

THE DEGRADATION OF METHYL VIOLOGEN IN THE SYSTEM WATER-Ru(bpy)$_3^{2+}$ - EDTA - Pt.

O. Johansen, J.E. Lane, A. Launikonis, A.W.-H. Mau, W.H.F. Sasse and J.D. Swift, Division of Applied Organic Chemistry, CSIRO, Box 4331, G.P.O., Melbourne, 3001 Australia.

The production of hydrogen during irradiation of aqueous solutions containing (Ru(bpy)$_3^{2+}$, methyl viologen (MV^{2+}), EDTA, and platinum supported on polyvinyl alcohol has been studied (A. Moradpour, E. Amouyal, P. Keller, and H. Kagan, Nouv. J. Chim, 2, 547-549 (1978)).

The initial steady rates of formation and the total yields of hydrogen depend markedly on the method of deposition of platinum on the carrier, on the amount of catalyst present, and on the degree of hydrolysis of the PVA which supports the platinum; other, less important, factors include the molecular weight of the PVA, as reflected in the viscosities of aqueous solutions, and the origin of the support.

As the amount of a given catalyst is increased the rate of gas evolution increases until a maximum is reached; further increases in the amount of catalyst result in lower rates. This effect has been related to the hydrogenation of MV^{2+}. The major product formed during irradiation has been identified as the 1-methyl-4-[4-(1-methyl-piperidyl)] pyridinium cation. Salts of this cation have been isolated from large scale experiments and its structure has been established by independent synthesis. The formation of this product is favoured by catalysts that are prepared from completely hydrolysed PVA; such catalysts produce less hydrogen at lower rates than catalysts prepared from incompletely hydrolysed PVA.

LIGHT-INDUCED ELECTRON TRANSFER REACTIONS OF THE BIS(2,2'-BIPYRIDINE)
(DIETHYL 2,2'-BIPYRIDINE-4,4'-DICARBOXYLATE)RUTHENIUM DICATION
Oddvar Johansen, Anton Launikonis, Albert W.-H. Mau and Wolfgang H.F.
Sasse, Division of Applied Organic Chemistry, CSIRO, Box 4331, G.P.O.,
Melbourne, 3001 Australia.

The photoreactivity of the tris(2,2'-bipyridine)ruthenium dication (1)
has been compared with that of the bis(2,2'-bipyridine)(diethyl 2,2'-
bipyridine-4,4'-dicarboxylate)ruthenium dication (2) in a series of
electron transfer reactions.

In water (pH 5.0) in the presence of EDTA, methyl viologen (MV^{2+}), and
a platinum catalyst (A. Moradpour, E. Amouyal, P. Keller and H. Kagan,
Nouv. J. Chim., 547-549 (1978)) (2) induces hydrogen formation at ca.
1/10th the rate observed with (1). Kinetic studies of the quenching
of (1)* and (2)* by MV^{2+}, benzylviologen and dimethylaniline have been
carried out. In absolute ethanol (2)*, but not (1)*, is subject to
bimolecular quenching by water (bimolecular quenching rate constant
$9.3 \times 10^4 M^{-1}s^{-1}$; Stern-Volmer constant $9.5 \times 10^{-2}M^{-1}$).

Quenching of (2)* by water and, to a lesser extent, electron
withdrawal by the ester groups account for most of the decrease in the
rate of electron transfer from (2)* to MV^{2+}. Quenching of (2)* by
water is at variance with an earlier report that the luminescence of
the bis(2,2'-bipyridine)(dioctadecyl 2,2'-bipyridine-4,4'-dicarboxy-
late)ruthenium dication in dioxane cannot be quenched by the addition
of water (G. Sprintschnick, H.W. Sprintschnick, P.P. Kirsch and
D.G. Whitten, J. Amer. Chem. Soc., 1976, 98, 2337.

NEW PHOTOCHEMICAL REDOX REACTANTS. 8-QUINOLINOLATO
COMPLEXES OF PLATINUM(II) AND IRIDIUM(III). a. Franco
SCANDOLA, Roberto Ballardini, M. Teresa Indelli, and
Graziano Varani, Centro di Fotochimica CNR, Istituto
Chimico, Università di Ferrara, 44100 Ferrara, Italy.

Bis(8-quinolinolato)platinum(II), Pt(QO)$_2$, and tris(8-
quinolinolato)iridium(III), Ir(QO)$_3$, have been synthesized
and their spectroscopic, photophysical and excited-state
redox properties have been investigated.

Pt(QO)$_2$ and Ir(QO)$_3$ exhibit strong absorption in the
visible, intense red luminescence, and long (μs) excited
state lifetime in deaerated acetonitrile or DMF. For both
complexes, the emitting state is assigned as a ligand
π-π* triplet. The metal has the role of (i) shifting the
absorption into the visible, (ii) promoting intersystem
crossing, and (iii) decreasing the triplet radiative lifetime. The excited-state redox properties have been investigated by quenching and laser photolysis, using a free-energy relationship approach. Consistent with the state assigne
ment, both complexes have almost identical excited-state
redox properties. They behave as good photochemical reductants ($E_\frac{1}{2}$ -0.9V vs SCE) in organic solvents and aqueous
micellar solutions.

CHARGE SEPARATION OF PHOTOPRODUCED IONS BY MICELLE-SENSITIZER OR MI-
CELLE-QUENCHER ION-PAIR FORMATION. Russell H. Schmehl and David
G. Whitten, Department of Chemistry, University of North Carolina,
Chapel Hill, North Carolina 27514 USA

An important step in the application of excited state electron transfer reactions to solar energy conversion is efficient charge separation of the redox pair formed upon photolysis. Recently the use of
charged surfactant micelles has been examined as a means of improving
the efficiency of such processes. We have explored the use of ionic
micelles to achieve charge separation via ion-pairing interactions between one of the ions produced and the micelle surface. In homogeneous solution the donor, a tetraanionic porphyrin, and the acceptor, an
alkyl viologen dication, form a ground state ion-pair; further, quenching is extremely rapid and no separation of photoproduced ions is
observed. Upon addition of either sodium dodecylsulfate or cetyltri-
methylammonium chloride at concentrations above the cmc the quenching
rate is reduced by several ordersof magnitude; however in this case
separation of the photoproduced porphyrin radical trianion and the
viologen radical cation is observed. The quenching and back electron
transfer rates depend upon several factors including the hydrophobicity of the viologen, the type and concentration of the electrolyte
used and the surfactant concentration. The results indicate that the
rate differences are due to ion-pairing of either the donor or acceptor with the micelle (depending on the charge of the surfactant) coupled with repulsion of the other species involved in the photoredox
reaction from the micelle.

CHARGE REDISTRIBUTION IN INORGANIC PHOTOREDOX SYSTEMS. Jürg Baumann,
Gion Calzaferri, Marcel Gori, Hans-Rudolf Grüniger, Ernst Schumacher,
Barbara Sulzberger, Institute for Inorganic and Physical Chemistry,
University of Bern, Freiestrasse 3, CH-3000 Bern 9, Switzerland.

We compare different types of charge redistribution in inorganic photoredox systems. It is important to achieve an organized charge redistribution in order to collect several charges within a certain geometrical
arrangement. This is not possible in a homogeneous system. Heterogeneous
inorganic photoredox systems offer several possibilities to build cyclic
photoredox systems in which a more or less organized charge redistribution is obtained by irradiation with visible light. Systems with a pronounced directional charge redistribution can be built by starting from
complexes which show d-π^* transitions along a well defined intramolecular axis. In another experiment we have observed astonishingly high photopotentials on a doped and sintered α-Fe_2O_3 electrode inserted into a
I_2/I^-,KCl solution. Macroscopic directionality in semiconductor/electro-
lyte systems is given because of the electrolyte/semiconductor interface.
Silver ion exchanged zeolites release oxygen from adsorbed water under
irradiation and they form molecularly dispersed silver. Directionality
of the charge redistribution is in this case due to the electrolyte/so-
lid interface.
G. Calzaferri & H.R. Grüniger, Helv.chim.acta 62(1979)1112 and 2547;
G. Calzaferri & H.R. Grüniger, in preparation; J. Baumann, H.R. Grüniger
& G. Calzaferri, Z.Phys.Chemie NF 118(1979)11; M. Gori, H.R. Grüniger &
G. Calzaferri, J.Appl.Electrochem. 10(1980)345; S. Leutwyler & E. Schumacher, Chimia 31(1977)475. - This work was supported by the Swiss National Science Foundation, Grants 4.340.79.04 and 2.157.78.

RAPID PHOTOREDUCTION OF METHYLVIOLOGEN SENSITIZED BY METAL-PHTHALO-
CYANINES <u>Takashi Tanno</u>, RIKEN Co., Ltd., Chiyoda, Tokyo 102, Japan,
Dieter Wöhrle, Universität Bremen, Bibliothekstrasse, NW2, 2800 Bremen
33, BRD(West Germany), Masao Kaneko and Akira Yamada, The Institute of
Physical and Chemical Research, 2-1, Hirosawa, Wako, Saitama 351,
Japan

We have found that metal-phthalocyanines(MPc) show remakedly high activity in photoreductive formation of methylviologen cation radical
($MV^{+\cdot}$) which is capable of reducing H^+ with H_2 evolution in the presence
of some appropriate catalyst. A typical photoreductive system is composed of MPc/DMSO, MV^{2+}/H_2O as an electron acceptor and EDTA/H_2O as an
electron donor. Visible light irradiation of the above solution very
readily gives the formation of $MV^{+\cdot}$. It strongly depends on the central
metal ion. Interestingly enough, MgPc showed the remarkably high rate
of MV^+ formation. The rate of MV^+ formation also showed very strong
dependence on the ratio of organic solvent to H_2O. When DMSO was used,
the maximum rate (3.2×10^{-4} mole/dm^3·min) was observed at DMSO/H_2O =
95/5 (V./V.). The quantum yield for the above case was found to be
47% which is a tremendously high value. Furthermore, there existed a
solvent effect on the MV^+ formation. It was found that, among some polar aprotic solvents, DMF, DMSO and DMAc gave the higher photoactivity.
Finally, the above photoreductive system led to H_2 gas evolution which
was confirmed by gas chromatography.

ENHANCEMENT OF ELECTRON TRANSPORT SENSITIZER EFFICIENCIES BY IONIC
ENVIRONMENTS IN HOMOGENEOUS SOLUTIONS. <u>Shigeo Tazuke</u>, Yoshiaki Iwaya,
Noboru Kitamura, and Takashi Inoue. Research Laboratory of Resources
Utilization, Tokyo Institute of Technology, 4259 Nagatsuta, Midori-ku,
Yokohama, Japan 227

Aiming at high primary charge separation yield($\emptyset_s$) of photoredox processes, we investigated ionic atmosphere effects on electron transferring
sensitizers in homogeneous systems. These are ⊙CO-⊙~N$^+$⊱ (I),
⊙CO⊙~SO$_3^-$(II), copolymers of $CH_2=C(CH_3)COOCH_2CH(CH_2$⊙$)N(CH_3)_2)(CH_2$(1-
pyrenyl)(M_1) with an anionic monomer(III), a cationic monomer(IV), and
acrylamide(V) and also the homopolymer of M_1(VI). Photoreactions investigated are: i) photooxidation of Leuco Crystal Violet(DH) to Crystal
Violet cation(D^+) sensitized by I, II and and benzophenone(BP)in acetonitrile, and ii) photoreduction of methylviologen dication(MV^{2+}) to $MV^+_{\cdot}$
sensitized by III, IV, V, and VI and its monomer model with and without
EDTA in aqueous DMF. In i), the sequence of $\emptyset_s$ is in the order, BP$\approx$II
<<I=1.0 indicating that the charge separation is strongly assisted by
cationic atmosphere. In ii), with EDTA, IV is more efficient than
others by a factor of over 10 whereas IV is comparable to others without EDTA. Enhanced local [EDTA] for I will be responsible for the sequence. Pyrenyl group is scarcely decomposed during irradiation. An
important conclusion is that high $\emptyset_s$ is achieved without using micelles,
vesicles, membranes or heterogeneous catalysts by the use of Coulombic
effect in homogeneous solutions.

This work was supported by Grant-in-Aid for Special Project Research,
Grant No. 412212.

IMPURITIES IN RUTHENIUM COMPLEXES AND ATTEMPTS TO PHOTOCHEMICALLY
CLEAVE WATER. John A. Broomhead, Charles G. Young, Chemistry
Department, Faculty of Science, Australian National University,
Canberra A.C.T. 2600 and Leon A.P. Kane-Maguire, Chemistry Department,
University College, P.O. Box 78, Cardiff CFIXL United Kingdom.

The photochemical cleavage of water (G. Sprintschnik, et al., J.Am.Chem.
Soc., 98, 2337 (1977); 99, 4947 (1977)), uses surfactant ruthenium complexes prepared from cis[RuCl$_2$(bipy)$_2$] (bipy = 2,2'-bipyridine) but has
not been reproduced, perhaps on account of impurities in the original
preparation. The conditions used to prepare cis[RuCl$_2$(bipy)$_2$] have been
investigated and shown to yield [Ru(bipy)$_3$]$^{2+}$, [RuCl$_4$(bipy)]$^-$, and ruthenium carbonyl complexes of the type [RuCl$_2$bipy(CO)$_2$]. Surfactant
complexes prepared from the Ru(II)-Ru(III) complexes have been studied
in monolayer assemblies. Elemental analysis was used to characterise
cisRuCl$_2$(bipy)$_2$ but this would not reveal the presence of [Ru(bipy)$_3$]
[RuCl$_4$(bipy)], or of up to 9% [Ru(bipy)$_3$][RuCl$_4$(bipy)]$_2$. However, the
lability of the chloro ligands in [RuCl$_4$(bipy)]$^{2-}$ and its ease of aerial
oxidation combine to make it unlikely that a Ru(II)-Ru(II) cation-
anion complex impurity is present.

The Ru(II)-Ru(III) complex did not fluoresce in the solid state or in
concentrated solution but its luminescence spectrum in ethanol gave the
characteristic charge transfer phosphorescence of [Ru(bipy)$_3$]$^{2+}$. Mixtures of this complex with (and without) cis[RuCl$_2$(bipy)$_2$] were reacted
with 4,4'-dihexadecyl-2,2'-bipyridine under N$_2$ and gave the orange (or
brown) perchlorate salts. Only the orange product was strongly fluorescent and neither product gave gas on irradiation in aqueous monolayers.

CONTRIBUTED PAPERS

Session V

**PHOTOINDUCED WATER SPLITTING
IN HETEROGENEOUS SOLUTIONS**

Plenary Speaker: Michael Grätzel

Session Chairman: Joseph Rabani

Posters V-1 through V-12

PHOTO REDOX REACTIONS IN FUNCTIONAL MICELLAR ASSEMBLIES; USE OF
AMPHIPHILIC REDOX RELAYS TO ACHIEVE LIGHT ENERGY CONVERSION AND CHARGE
SEPARATION   Pierre-Alain Brugger and Michael Grätzel, Institut de
chimie physique, Ecole Polytechnique Fédérale, Ecublens, 1015 Lausanne,
SWITZERLAND.

The photoreduction of a homologous series of amphiphilic viologens
(C_nMV^{2+}) using Ru(bipy)$_3^{2+}$ and zinc-tetra(-N-methyl-pyridyl) porphyrin
(ZnTMPyP^{4+}) as sensitizers (S) was studied in water and aqueous solution
containing cationic micelles.

$$S \ + \ C_nMV^{2+} \ \xrightleftharpoons[\Delta]{h\nu} \ S^+ \ + \ C_nMV^+$$

Here, C_nMV^{2+} stands for N-alkyl-N'-methyl-4,4'-dipyridinium dichloride
(alkyl:dodecyl($C_{12}MV^{2+}$), tetradecyl ($C_{14}MV^{2+}$), hexadecyl ($C_{16}MV^{2+}$) and
octadecyl ($C_{18}MV^{2+}$)). The forward electron transfer occurs with the viologen present in the aqueous phase. Upon reduction the relay acquires
hydrophobic properties leading to rapid solubilisation in the micelles.
The subsequent thermal back reaction is impaired by the positive surface
potential of the aggregates. This decreases the rate constant for the
back electron transfer at least 1000-fold. The stabilization and yield
of redox products is optimal in a system containing ZnTMPyP^{4+} as sensitizer, $C_{14}MV^{2+}$ as a relay and cetyltrimethylammoniumchloride micelles.
A kinetic model is presented to explain the effects observed and implications for energy conversion systems are discussed. Coupling of the
photo events with colloidal redox catalysts to afford hydrogen production
from water is demonstrated.

ENHANCEMENT OF PHOTOCHEMICAL HYDROGEN GENERATION BY THE USE OF ELECTRON-
TRANSPORTING POLYMER AND PLATINUM COLLOID IN COMBINATION WITH TRIS(2,2'-
BIPYRIDINE)Ru(II) COMPLEXES. Taku Matsuo, Toyoki Nishijima, Keisuke Takuma,
Tetsuo Sakamoto, and Toshihiko Nagamura, Department of Organic Synthesis,
Faculty of Engineering, Kyushu University, Hakozaki, Fukuoka 812, Japan

In order to construct molecular assemblies for photocatalytic generation
of molecular hydrogen, water soluble polymers carrying pendant viologen
groups and colloidal platinum have been prepared, and the photochemical
properties were investigated. Polystyrene-type polymer, which contains
a pendant viologen group for each monomer unit, was defined as PV(100),
while the one with low viologen content (1%) was PV(1). Polysoap-type
viologen polymer (PVC16) with hexadecyl viologen units (10%) was also
prepared. Two ruthenium complexes were used as the sensitizer: tris-
(2,2'-bipyridine)Ru(II)$^{2+}$ (abbreviated to Ru(bpy)$_3^+$) and the amphipathic
derivative (N,N'-didodecyl-2,2'-bipyridine-4,4'-dicarboxamide)-bis(2,2'-
bipyridine)Ru(II)$^{2+}$ or RuC$_{12}$B^{2+}). In the presence of EDTA, the absorption spectra indicated the formation of viologen radical dimers from the
beginning of the photoirradiation of Ru(bpy)$_3^{2+}$/PV(100)- and RuC$_{12}$B^{2+}/
PVC16 systems. The radical dimers were also observed on the irradiation
of the solid films of PV(100). In addition, the ESR spectra of the
radical in PV(100) showed extreme line broadening (2 gauss) due to electron exchange. Thus, electron migration among the pendant viologen
groups was strongly indicated. Hydrogen generation from water was improved by the use of colloidal platinum in combination with the viologen
polymers.

This work was supported by a Grant-in-Aid for Scientific Research from
the Ministry of Education of Japan.

LIGHT-INDUCED HYDROGEN EVOLUTIONS FROM WATER MEDIATED BY
COLLOIDAL PLATINUM HYDROSOLS. Patrick Keller and <u>Alec Moradpour</u>,
Laboratoire de Physique des Solides and Laboratoire de Synthèse
Asymétrique, University of Paris-Sud, 91405-Orsay, France.

Visible-light induced hydrogen evolutions from water using the
$\left(\text{Ru(bipy)}_3^{2+}/\text{methylviologen/EDTA/Pt}\right)$ model system were investigated. A
factor which limited the overall yields of the hydrogen produced by
this photosystem was found to be a methylviologen-consuming process.
This irreversible side-reaction was a platinum-dependent electron-
relay hydrogenation. As a consequence, by simply adjusting the added
amount of the Pt-catalyst, it has been possible to improve, by about
one order of magnitude, the hydrogen-formation efficiencies. For further improvements of these redox catalysts particle-size effects were
studied using samples with narrower size distributions obtained from
the previous polydispersed colloidal hydrosols by centrifugation. However, the resulting optimum values for the hydrogen formation rates
and yields were found to be very similar for all catalysts studied :
this was true for the polydispersed as well as selected small (mean
radius 16 Å as determined by electron microscopy) and large particles
(> 1000 Å). In fact, no <u>net</u> catalyst size-effect on the hydrogen pro-
ductions, involving methyl<u>viologen</u> as an electron mediator, were detected over this wide size-distribution range. Thus, the possibility of
using this particular catalytic system in water splitting processes
which obviously require exceedingly efficient redox catalysts may
presently be questionned.

Ru(bipy)_3^{2+} ⋆ ELECTRON TRANSFER QUENCHING BY VIOLOGEN
SALTS : KINETICS AND THEORETICAL MODELS. <u>Edmond Amouyal</u>
and Bertrand Zidler, Processus Photophysiques et Photochimiques,
Université Paris-Sud, Bâtiment 350, 91405 Orsay, France

We have previously shown that hydrogen is evolving upon visible light
irradiation of aqueous solutions containing Ru(bipy)_3^{2+}, methyl-violo-
gen (MV^{2+}), EDTA and colloidal-platinum catalyst. In such a system,
the first step of H_2 generation consists in an oxidative quenching of
the Ru(bipy)_3^{2+} excited state by MV^{2+}.
In the present study, rate constants k_q for the quenching of Ru(bipy)_3^{2+}⋆
by a series of viologen salts having different redox potential have been
determined in deaerated aqueous solutions at pH5 by laser flash photolysis. Our results show that (i) the k_q value depends on the reduction
potential $E_{1/2}$ of the quencher, namely k_q increases with $E_{1/2}$ up to a
value which is close to the diffusion-controlled limit (ii) for the same redox potential, the molecular structure of the quencher does not
seem to affect k_q significantly (iii) there is no clear relationship
between H_2 production rates and the viologen redox potential.
The quenching data will be discussed in terms of free-energy change,
i.e. in the framework of the theoretical and semi-empirical models
describing electron-transfer reactions. It appears by best fitting procedure that our results are consistent with the Rehm-Weller treatment.

HYDROGEN EVOLUTION FROM WATER BY HYDROGENASE IN THE
PRESENCE OF REDUCED METHYL VIOLOGEN PRODUCED BY IRRADI-
ATION WITH VISIBLE LIGHT. Ichiro Okura, Nguyen Kim-Thuan
and Tominaga Keii, Department of Chemical Engineering,
Tokyo Institute of Technology, Meguro-ku, Tokyo 152, Japan

The enzyme hydrogenase catalyzes the decomposition of water
in the presence of an electron donating agent. The reduced
form of methyl viologen, which has a sufficient redox potential for the decomposition of water, is known to be a specific cofactor for hydrogenase. In this study, the reduction of methyl viologen (MV) with the aid of light energy
was carried out in the presence of suitable reducing agent
such as mercaptoethanol and photosensitizers. And also an
attempt was made to reduce water to hydrogen by the use of
a modified electron transfer system containing a photosensitizer, MV and hydrogenase. The activities of photosensitizers for the reduction of MV were compared and the
order of the activities is as follows: $Ru(dime-phen)_3^{2+} >$
$Ru(Cl-phen)_3^{2+} > Ru(phen)_3^{2+} > Ru(bpy)_3^{2+}$. This is the same
order of the lifetimes of the excited state of the Ru-
complexes. When hydrogenase was added in the system, hydrogen evolved. The higher activity for hydrogen evolution
was observed by the system with the higher activity for
the MV reduction. Hydrogen evolution was also observed by
the irradiation of sunlight.

PHOTOPRODUCTION OF HYDROGEN FROM MINERAL-ORGANIC MIXTURES
Adolph E. Smith,Clair E. Folsome*,and Sherwood Chang, Ames Research
Center, NASA, Moffett Field, Ca. 94035 and * Dept. of Microbiology,
Univ. of Hawaii, Honolulu, HI 96822

In research related to the origin of photosynthesis, we searched for
evolution of hydrogen from various minerals, and mixtures of glucose
and minerals in electrolyte solution. Hydrogen was found to be evolved
from some minerals in aqueous glucose solutions. Titanium dioxide,
ilmenite $FeTiO_3$, and siderite $FeCO_3$ were found to be effective. Titanium dioxide and ilmenite without glucose did not evolve hydrogen. The
irradiation was done with Xenon lamps through Pyrex tubes. The amount
of hydrogen was small being about 2.5 uM per gram mineral for 50 hours
of irradiation, and was done under anaerobic conditions. Siderite
alone was found to evolve hydrogen under these conditions and the
amount was proportional to the surface area but the evolution ceased
after about 40 hours. This led to suspicion of surface contamination.
ESCA showed a substantial part of the surface carbon to be organic.
The source of the hydrogen is under investigation. Silicic acid,
SiO_2 xH_2O suspensions under 254 nm irradiation produced hydrogen. As
this material has OH bonds it is possible that the hydrogen is coming
from these since the band gap is greater than the photon energy at
254 nm. These results suggest that simple organic-mineral mixtures may
be found which evolve hydrogen during exposure to sunlight.

SQUARILIUM DYES AND BIPYRIDINE DERIVATIVES AS PHOTO-
CATALYSTS FOR THE SPLITTING OF WATER IN HOMOGENEOUS
SOLUTION. Martin Forster, Ronald E. Hester, Department of
Chemistry, University of York, York YO1 5DD, Great Britain

Squarilium dyes of the type R_1—⬡—R_2 (I) and 4,4'-
bipyridine derivatives (II) have been investigated by UV-,
IR-, resonance Raman- and emission spectroscopy. Dye I
with $R_1 = R_2 = N\emptyset_2$ has been shown to reduce Ce^{4+} solutions
rapidly with negligible decomposition of the dye. Furthermore, dye I has been shown to mediate the photoreduction of
methylviologen (MV^{2+}) with a quantum yield of $6 \pm 5 \times 10^{-4}$ at
458 nm in solutions containing EDTA as electron donor. Dye
I is proposed as the photocatalyst in the hypothetical
reactions (1)-(3).

$$I + MV^{2+} \xrightarrow{h\nu} I^+ + MV^{+\cdot} \quad (1); \qquad I^+ + \tfrac{1}{2}H_2O \longrightarrow I + \tfrac{1}{4}O_2 + H^+ \quad (2);$$

$$MV^{+\cdot} + H_2O \xrightarrow{PtO_2} MV^{2+} + \tfrac{1}{2}H_2 + OH^- \quad (3)$$

In similar reactions using coloured bipyridine derivatives
II in place of I, $MV^{+\cdot}$ could be generated with a quantum
yield of $4.7 \pm 0.6 \times 10^{-2}$ at 458 nm. Solvent dependence,
long-term behaviour and evidence for the monophotonic course
of this reaction will be presented.

This work was supported by the British Science Research
Council and the Swiss National Foundation.

EXCITED STATE REACTIONS OF RUTHENIUM-TRIS-BYPYRIDINE IN AN ION EXCHANGE
RESIN. Hanna Slama, Jehuda Feitelson and Joseph Rabani, Department of
Physical Chemistry, The Hebrew University, Jerusalem, Israel.

The influence of the electrostatic field in a negatively charged ion
exchange resin (Sp-C50 Sephadex) on the excited Ruthenium-tris-bipyri-
dine ion $Ru(bpy)_3^{2+}$, was studied by laser photolysis. The time dependence of the excited $Ru(bpy)_3^{2+}$ emission and its quenching by various
electron acceptors, Q, were determined. The back reaction between the
reduced acceptors Q^- and the oxidized $Ru(bpy)_3^{3+}$ was measured by following the reappearance of the $Ru(bpy)_3^{2+}$ absorbance at 450-480nm after its
instant bleaching by the laser flash. It was found that: (a) at low
$Ru(bpy)_3^{2+}$ concentrations in the resin gel and at low laser flash intensities the decay is of strictly first order. For a deoxygenated
resin the rate constant is $k_e = 1.54 \times 10^6 sec^{-1}$. Both at higher concentrations of $Ru(bpy)_3^{2+}$ in the resin and at higher laser intensities,
when excited neighbor interactions become probable, a second order component is observed in the decay kinetics. (b) Quenching by Fe^{3+} and
Cu^{2+} is a dynamic process with no preformed D-A complexes present in
the resin. The rate constant for excited $Ru(bpy)_3^{2+}$ quenching by Fe^{3+}
was found to be $k_{Fe} = 3.7 \times 10^9 M^{-1}sec^{-1}$ and by Cu^{2+} $k_{Cu} = 4 \times 10^8$
$M^{-1}sec^{-1}$. (c) The back reaction between the trivalent $Ru(bpy)_3^{3+}$ and
Fe^{2+} of Cu^+ ions is also accelerated by the "built in" proximity of
the reactants in the resin. $k'_{Fe} = 3.4 \times 10^8 M^{-1}sec^{-1}$ and $k'_{Cu} = 4.1 \times 10^9 M^{-1}sec^{-1}$.

SOLID PHASE PHOTOENERGY CONVERSION ON CELLULOSE. Masao Kaneko and
Akira Yamada, The Institute of Physical and Chemical Research, Wako-shi,
Saitama, 351 Japan

It was found that methylviologen (MV^{2+}) adsorbed on cellulose can be
photoreduced effectively in the solid phase.   Visible and ESR spectra
showed the photochemical formation of the cation radical ($MV^{\ddot{+}}$).   The
optical density (O.D.) is proportional to the spin concentration up to
O.D. = 1.2,   and 1 O.D. corresponds to 1.07×10^{-8} spins/cm^2.   Broadening and longer wave length shift of the absorption of the MV^{2+} adsorbed on cellulose enable the photoreaction with visible light up to
480 nm.   The cellulose molecule plays a role of reducing agent giving carbonyl groups after the photoreaction.   The presence of $Ru(bpy)_3$
complex alone has almost no effect on the $MV^{\ddot{+}}$ formation, while the
presence of both EDTA and $Ru(bpy)_3$ markedly enhances the photoreduction.
The action spectrum for MV^{+} formation in EDTA-$Ru(bpy)_3$-MV^{2+}/cellulose
system proved that the excitation of the Ru complex participates in the
sensitized photochemical reaction.   Since the emission from the Ru(
$bpy)_3$* at 604 nm was not quenched by EDTA, but was quenched by MV^{2+},
the primary photochemical reaction should be the electron transfer from
$Ru(bpy)_3$* to MV^{2+} giving $MV^{\ddot{+}}$.   The $Ru(bpy)_3^{3+}$ formed by the reaction is
then reduced to $Ru(bpy)_3^{2+}$ by EDTA.   The photochemical accumulation of
MV^{+} in the cellulose system is possible even in the presence of oxygen,
contrary to the known system in which trace oxygen inhibits its accumulation.   The quantum yield for the $MV^{\ddot{+}}$ accumulation in vacuo under
irradiation with 470 nm light was 5 %.

PHOTOCHEMISTRY OF $Ru(bpy)_3^{2+}$ IN CLAY MEMBRANES AT THE GAS-SOLID INTERFACE
S. Abdo, P. Canesson, M.I. Cruz, J.J. Fripiat and H. Van Damme, C.N.R.S.
Centre de Recherche sur les Solides à Organisation Cristalline Imparfai-
te, 1 b, Rue de la Férollerie, 45045 Orléans Cédex, France.

Considerable work has been done on the light induced electron transfer
chemistry of $Ru(bpy)_3^{2+}$ and its surfactant derivatives in homogeneous or
interfacial environment, but far less work has been done on the reactivity of the complex itself, particularly in interfacial conditions. We
have studied the photoreactivity of $Ru(bpy)_3^{2+}$ in artificial membranes
made of hectorite. The complex is adsorbed in interlayer position between the individual clay sheets, each membrane corresponding to about
10,000 superimposed sheets. Continuous photolysis at 70°C in the presence of oxygen leads to the formation of Ru(III) oxo-bridged dimers. The
primary reaction is a photoaquation. The aquated Ru(II) species then
react purely thermally to give the Ru(III) oxo-bridged dimer. A more
detailed investigation of the thermal oxidation-dimerization reaction
shows that it is reversible to some extent, suggesting that the Ru(III)
dimer is able to oxidize water on the surface. The acid-base equilibrium of the water molecules coordinated to the Ru(III) ions is a critical
factor in this reaction. The luminescence of $Ru(bpy)_3^{2+}$ is strongly
quenched by coadsorbed Ru(II) aquated species or Ru(III) oxo-bridged
dimers. The quenching efficiency is significantly enhanced in the presence of an excess of water.

The Commission of the European Communities is gratefully acknowledged
for financial support.

PHOTOINDUCED CHARGE SEPARATION AT SOLID-LIQUID INTERFACE UTILIZING
POLYMER PENDANT Ru(bpy)$_3$ COMPLEX. Masao Kaneko, Akira Yamada, The
Institute of Physical and Chemical Research, Saitama, Wako-shi, 351
Japan, Eishun Tsuchida, Waseda University, Ohkubo, Shinjuku-ku, Tokyo,
160 Japan, and Yoshimi Kurimura, Ibaraki University, Bunkyo, Mito-shi,
310 Japan

Polystyrene (PSt) pendant tris(2,2'-bipyridyl)Ru(II) complex (PSt-(
bpy)$_3$Ru) was prepared by reaction of bipyridylated PSt with cis-Ru(
bpy)$_2$Cl$_2$. PSt-(bpy)$_3$Ru was insoluble in H$_2$O or MeOH, while soluble
in DMF, benzene, chloroform or THF. The visible spectrum showed the
same absorption as Ru(bpy)$_3$. The relative intensity of the emission
from PSt-(bpy)$_3$Ru* at 610 nm in DMF was weaker than that from Ru(bpy)$_3$*.
The PSt-(bpy)$_3$Ru showed photocatalytic activity in DMF/H$_2$O homogeneous
solution for the photoreduction of methylviologen (MV^{2+}) using EDTA
or triethanolamine as reducing agent, although the activity is lower
than Ru(bpy)$_3$. When a mixture of PSt-(bpy)$_3$Ru and EDTA was deposited
in a MV^{2+}/MeOH solution and the solid-liquid interface was irradiated
with visible light, photoinduced MV$^+$ formation in the liquid phase was
observed. The presence of protons and platinum black in the liquid
phase gave hydrogen gas. Thus, photoinduced charge separation at
solid-liquid interface and the separation of the products were realized by utilizing polymer pendant Ru(bpy)$_3$ complex.

PHOTOSENSITIZED ELECTRON TRANSFER REACTIONS IN MICELLAR SYSTEMS. Sílvia
M.B.Costa*,Radiation Laboratory, University of Notre Dame,Notre Dame,
In 46556, U.S.A, J.K. Thomas, Chemistry Department of University of
Notre Dame, Notre Dame, In 46556, U.S.A.

The kinetics of the photosensitized reduction of anthraquinone dissulphonate (AQDS^{-2}) and of methyl viologen (MV^{++}) by zinc tetraphenylporphyrin (ZnTPP) has been studied in non ionic micelles of IGEPAL 630 and
in reverse micelles of benzyl hexadecyl dimethylammonium chloride
(BHDC) and sodium di(2-ethylhexyl) sulfosuccinate (AOT). The results
obtained indicate that whereas the electron transfer is efficient in
non ionic micelles for either acceptor, the charge at the interface in
the reverse micelles introduces a selectivity in these reactions which
is possibly of electrostatic nature. Thus, the interaction of ^{3}ZnTPP*
with (AQDS^{-2}) is very efficient in BHDC, but it is not observed in AOT,
whereas that with (MV^{++})is very efficient in AOT, but negligible in
BHDC. The occurence of an oxidative mechanism in these systems was
demonstrated using different donors such as sodium ascorbate,cysteine
and NADH. The quantum yield of reduction increases with the oxidation
potential of the donors.The radical ions of methyl viologen and anthraquinone dissulphonate were also detected by E.S.R. as permanent
products. The possibility of the utilization of these reactions as
model systems for solar energy conversion will be discussed.

*Present address: Centro de Química Estrutural,Complexo I, Instituto
Superior Técnico, Av. Rovisco Pais 1096 Lisboa Codex Portugal.

CONTRIBUTED PAPERS

Session VI

PHOTOCHEMICAL H$_2$ AND O$_2$ PRODUCTION IN METAL COMPLEX SYSTEMS

Plenary Speaker: Jean-Marie Lehn

Session Chairman: Norman Sutin

NEW COLLOIDAL OXIDATION CATALYSTS FOR THE PHOTOCHEMICAL CLEVAGE OF WA-
TER INTO HYDROGEN AND OXYGEN WITH VISIBLE LIGHT, Albert Henderikus
Alberts, K. Timmer, A. Mackor, J.G. Noltes, Institute for Organic
Chemistry TNO, P.O. Box 5009, Utrecht The Netherlands.

A series of new microheterogeneous (colloidal) oxidation catalysts was
prepared and used in combination with known homogeneous and microheterogeneous transition metal catalysts to the photochemical cleavage of
water into hydrogen and oxygen with visible light.

PHOTOCHEMICAL H_2-PRODUCTION FROM WATER CATALYSED BY METAL DITHIOLENES
AND RELATED COMPLEXES. Horst Kisch, Rosangela Battaglia and Rainer
Henning, Institut für Strahlenchemie im Max-Planck-Institut für Kohlen-
forschung, D-4330 Mülheim a.d. Ruhr

We found that metal dithiolenes and related metal sulfur chelates cata-
lyse the photochemical hydrogen production upon irradiation in tetrahydrofuran (THF) - water solution. THF or dihydrofurans (DHF) function
as solvents and as reducing agents. We have investigated a large
number of complexes of the type $\{[R(S)C=C(S)R]_nM\}^z$, n=2,3, z=0, -1,
-2, R=H, CH_3, Ph, CN, CS_3, C_2O_2, M=Ti, V, Cr, Mo, W, Mn, Re, Fe, Os,
Co, Ni, Pd, Pt, Cu, Au, Zn and we found that the zinc complex (R=CN
or CS_3, n=2, z=-2) exhibits the best catalytic properties. Appropriate
mixtures of ligands and metal salts may be also used instead of the
isolated complexes. If D_2O is used in these experiments, the average
composition of the gas evolved is 87% D_2, 10% HD and 3% H_2. Presence
of air has no significant influence on the rate of hydrogen evolution.
The catalysts are active during a period of four to five days and up
to five liters of hydrogen are obtained upon irradiation ($\lambda \geq 254$ nm).
The maximum turnover numbers are in the range of 2000 mmol H_2/mmol
catalyst. The hydrogen evolution is slower at $\lambda \geq 290$ nm. Metal dithio-
oxamides, metal dithiocarbamates, metaldithioacetylacetonates and related metal sulfur chelates do also exhibit some catalytic activity.

Acknowledgement: We are indebted to J. Bücheler, W. Schlamann, B.
Ulbrich and N. Zeug for helpful assistance.

HOMOGENEOUS CATALYSIS OF THE EVOLUTION OF MOLECULAR OXYGEN FROM WATER.
G. L. Elizarova, L. G. Matvienko, V. N. Parmon, and K. I. Zamaraev,
Institute of Catalysis, Novosibirsk 630090, U.S.S.R.

We have recently succeeded in developing efficient homogeneous catalysts for oxygen evolution from water by using one-electron oxidants
(Ox) of the $Ru(bipy)_3^{3+}$ type, which can be produced under light illumination simultaneously with a strong one-electron reducing agent
capable of evolving hydrogen from water. Our preliminary data testify
that a binuclear μ-dihydroxo structure is a possible catalytic intermediate for some of our systems. Oxygen evolution and the regeneration of the intermediate seems to occur through consecutive one-
electron oxidation and without evolution of the intermediate products
of water oxidation (OH$\cdot$, H_2O_2 or HO_2) in the free state. The
catalytic action of such catalysts based on Fe(III) may be interpreted
in terms of two alternative schemes, but the overall reaction in both
cases is: $4Ox + 2H_2O \xrightarrow{cat} 4Ox^- + O_2 + 4H^+$ (where M = Fe(III) and
Ox = $Ru(bipy)_3^{3+}$, $Fe(bipy)_3^{3+}$, or $Fe(phen)_3^{3+}$. The catalytic activity
of some compounds of Co(II), Cu(II), Ni(II), Mn(II), Ir(III), Ru(III),
and Sb(V) has also been determined.

COLLOIDAL RUTHENIUM OXIDE AS CATALYST IN OXYGEN PRODUCTION
FROM WATER. Edmondo Pramauro and Ezio Pelizzetti,
Istituto di Chimica Analitica, Università di Torino, Italy

A colloidal dispersion (mean particle radius 300 $\overset{o}{A}$) of
RuO_2 (obtained by mixing a neutral solution of RuO_4 in
water with an equivalent amount of aqueous solution of a
copolymer of styrene and maleic anhydride and by stirring
for 1 h at pH 8) (K.Kalyanasundaram, C.Mićić, E.Pramauro
and M.Grätzel, Helv.Chim.Acta, 1979, 62, 2432) has been
found to strongly accelerate the reaction

$$4\ Fe(bpy)_3^{3+} + 2\ H_2O \longrightarrow 4\ Fe(bpy)_3^{2+} + O_2 + 4\ H^+$$

The reaction rates were followed by means of a stopped-
flow spectrophotometric technique. In the range of pH between 8.7 and 9.6, the reaction rate is increased of ca.
100-200 times (RuO_2 3.5 mg/l) being the maximum increase
at pH 9.0 (where the maximum oxygen yield is located).
The increase in rate is proportional to the catalyst concentration. After 1 month, the colloidal catalyst starts
to reduce its efficiency. However, in the presence of NaOH
($1.3x10^{-2}$ - $6.3x10^{-2}$ M) the catalytic effect is decreased,
reaching only a factor of 3 at the highest NaOH concentration. (This work was supported by CNR (Rome))

PHOTOSENSITIZED DECOMPOSITION OF WATER USING A Mn-PORPHYRIN.
Itamar Willner, Howard Mettee, William E. Ford, John W. Otvos
and Melvin Calvin. Laboratory of Chemical Biodynamics, Lawrence
Berkeley Laboratory, University of California, Berkeley 94720

We propose a route of achieving the photodecomposition of water using
a Mn-complex. This is a cycle in which Mn(II) is photochemically
oxidized to Mn(IV) via two consecutive electron transfer processes to
an acceptor. The Mn(IV) thus formed may oxidize water to an active
oxygen atom, for example as a Mn-oxo intermediate. Transfer of the
active oxygen atom to a trap recycles the sensitizer while producing
an oxygenated trap. The reduced acceptor may be used for hydrogen
evolution and the appropriate oxygen atom trap may itself be a useful
chemical, or perhaps serve as a relay for oxygen evolution.

We have performed this cycle using a Mn-porphyrin sensitizer, methylviologen (MV^{2+}) as an acceptor and triphenylphosphine as the oxygen
trap. Upon illumination, methylviologen radical cation ($MV^{\cdot}$) was
formed with the concomitant production of triphenylphospine oxide.
The radical cation could be recycled in the presence of a Pt catalyst
to produce hydrogen.

DYE-PHOTOSENSITIZED REDOX REACTIONS THROUGH PHOSPHOLIPID VESICLE WALLS.
William E. Ford, John W. Otvos and Melvin Calvin. Laboratory of
Chemical Biodynamics, Lawrence Berkeley Laboratory, University of
California, Berkeley, California 94720

The work presented deals mainly with characterization of the model
system that we have developed for studying dye-photosensitized
electron transfer across vesicle walls. In our system, a long-chained
derivative of $[Ru(bipy)_3]^{2+}$ (abbreviated (Ru^{2+})) photosensitizes the
reduction of methylviologen (MV^{2+}) with EDTA^{3-} as the ultimate
electron donor, when the MV^{2+} and EDTA^{3-} are separated by the vesicle
wall. Based on a comparison of the visible and UV spectra of (Ru^{2+})
dissolved in vesicles with its spectra in other solvents, we infer that
(Ru^{2+}) is oriented in the bilayer membranes with the chromophore of
the complex exposed to water. The permeabilities of phospholipid
vesicle walls to both MV^{2+} and EDTA^{3-} were determined. By comparing
the permeabilities of MV^{2+} and EDTA^{3-} to the rate that MV^{2+} is
reduced photochemically, we conclude that electron transfer through
the membranes was being observed in our model system. From temperature dependence studies we estimate the activation energy of the trans-
membrane electron transport step is far less than the activation
energy of transbilayer diffusion of lipids. This result supports our
proposed mechanism of electron transport through the membrane by
electron exchange.

CONTROL OF PHOTOSENSITIZED ELECTRON-TRANSFER REACTIONS ACROSS A WATER-OIL INTERFACE. <u>Itamar Willner</u>, John W. Otvos and Melvin Calvin. Laboratory of Chemical Biodynamics, Lawrence Berkeley Laboratory, University of California, Berkeley, California 94720

The separation of redox species formed in a photoinduced process is required to prevent thermodynamically favored back reactions. We suggest a water-in-oil microemulsion system as a means of separating the active products. In this system the phase transfer of one of the redox products from the interface to the continuous oil phase is the important step in achieving the separation. To illustrate this principle we performed the photosensitized reduction of an azo dye, solubilized in the continuous oil phase, by EDTA dissolved in the aqueous phase. $Ru(bipy)_3^{2+}$ was used as sensitizer and benzylnicotinamide was the primary acceptor that mediates the reduction of the dye.

Similarly, the photosensitized reduction of MV^{2+}, using $Ru(bipy)_3^{2+}$ as sensitizer and thiophenol (RSH) as a donor at the interface, was accomplished. Under the conditions of the experiment, the reaction was energy storing by ca. 7 Kcal/mole. The net storage of energy is attributed to the separation of the oxidized donor, diphenyl disulfide, from the $MV^{\ddagger}$ dissolved in the aqueous phase. The same photoinduced process was performed using Zn-porphyrins as sensitizers (Zn-<u>meso</u>-tetraphenyl porphyrin sulfonate and Zn-tetramethylpyridinium porphyrin.) The relative quantum yields are influenced by electrostatic interactions between the charged interface and redox products.

WATER SWOLLEN PERFLUOROSULFONATE MEMBRANE AS A MEDIUM FOR PHOTOREDOX REACTIONS*. <u>Plato C. Lee</u> and Dan Meisel, Chemistry Division, Argonne National Laboratory, Argonne, Illinois 60439

The cluster network of a water swollen perfluorosulfonate membrane (Nafion) was utilized as a medium to study the kinetics of excited state quenching. When the excited state of $Ru(bpy)_3^{2+}$ is quenched by relatively inefficient quenchers, linear Stern-Volmer behavior is observed. When an efficient quencher is incorporated in the cluster network (and thus at small quencher per cluster concentrations) non-exponential quenching is observed. Kinetic analysis of the quenching reaction under such conditions provides detailed information on the morphology of the membrane (20 Å for the radius of the cluster). The biphasic nature of the swollen membrane is exemplified by incorporation of water insoluble compounds into the membrane. These are presumably deposited in the fluorocarbon backbone. Nevertheless, strong interaction with the aqueous phase is observed for these systems and is demonstrated by external heavy ion (Ag^+) enhancement of pyrene phosphorescence. The cluster network of the membrane also furnishes a convenient structure for stabilization of colloidal redox catalysts and the hydrogen production from thermally produced intermediates on such catalysts was studied.

*Work performed under the auspices of the Office of Basic Energy Sciences, U.S. Department of Energy.

TWO PHOTON ACTIVATED ELECTRON TRANSPORT ACROSS LIPID MEMBRANE AS
ENHANCED BY ELECTRON MEDIATORS.

Taku Matsuo, Kengo Itoh, Keisuke
Takuma, Kunitoshi Hashimoto, and Toshihiko Nagamura, Department of
Organic Synthesis, Faculty of Engineering, Kyushu University, Hakozaki,
Fukuoka 812, Japan

Photoinduced electron-transport across phospholipid vesicle walls have
been studied in order to construct a photosynthetic model system. Since
the reported rates of electron-transport are not rapid enough, an attempt has been made to improve the situation by the use of various
electron mediators incorporated into the vesicle wall. The electron
was delivered to the mediator by photoactivated amphipathic zinc porphi-
nato complex ($ZnC_{12}TPyP$) at the inner surface of vesicles. The photo-
oxidized $ZnC_{12}TPyP$ was reduced by EDTA in the inner aqueous phase of
vesicles. The rate of electron transport across the vesicle wall was
monitored by the reduction of disodium 9,10-anthraquinonedisulfonate
(AQDS) dissolved in the outer aqueous phase of vesicles. The following
materials were used as the electron mediator: vitamin K_1 (VK_1), 1,3-di-
butylalloxazine (DBA), and 1,3-didodecylalloxazine (DDA). Remarkable
enhancement effects were observed in the presence of alloxazine derivatives (10^{-5} M) as it is clear from the rate of reduction of AQDS: DBA,
4.25; DDA, 2.65; VK_1, 0.38; and without mediator, 0.18 (in the units of
10^{-5} M/min). Mechanistic studies indicated that photoexcitation of
$ZnC_{12}TPyP$ at the outer surface of vesicle is also required to reduce
AQDS by the transported electron.

This work was supported by a Grant-in-Aid for Scientific Research from
the Ministry of Education of Japan.

CONTRIBUTED PAPERS

Session VIIa

PHOTOGALVANIC CELLS AND EFFECTS

Plenary Speaker: Mary Archer

Session Chairman: Norman Lichtin

Posters VII-1 through VII-12

PHOTOGALVANIC EFFECTS IN AQUEOUS AND NON-AQUEOUS IODINE SOLUTIONS.
K.L. Stevenson and W.F. Erbelding, Department of Chemistry; Indiana-
Purdue University at Fort Wayne, Fort Wayne, Indiana 46805

When aqueous and non-aqueous solutions containing iodine, complexed
with iodide ion, are irradiated with visible light in a cell consisting of a transparent electrode and a platinum electrode, a current
is generated such that the illuminated, transparent electrode is pos-
itively-charged. An analysis of the open-circuit potentials in the
aqueous system suggests a model in which I_2^- is the active species
which accepts electrons from the transparent electrode. The potential
of the cell as a solar energy conversion device will be discussed in
terms of the thermodynamics. Although the potential efficiencies are
high, the most efficient cell studied is below that of a viable system
at present.

A number of factors appear to enhance the efficiency of the cell.
Among these are (1) the use of very concentrated (9M) I_2 solutions
in non-aqueous solvents such as acetonitrile, and (2) the use of
indium-tin oxide coating as the transparent electrode. The dependence
of the efficiency on such factors as electrode spacing, light intensity, and wavelength will also be discussed.

This work was supported by an Appropriate Technology Award from the
Department of Energy.

HYDROPHILIC N-SUBSTITUTED THIAZINE DYES FOR APPLICATION IN PHOTOGAL-
VANIC CELLS.David Creed, W. C. Burton, Newton C. Fawcett, and Mark T.
Williams, Department of Chemistry, University of Southern Mississippi,
Hattiesburg, MS 34901

In photogalvanic cells it is desirable to have as narrow a gap as possible between electrodes consistent with the requirement of high
optical density for light collection and minimal formation of photochemically inert aggregates. A number of thiazine dyes (eg. 2 and 3)
analogous to methylene blue and thionine but N-substituted with hydrophilic groups have been synthesized.

1 R_1 = R_2 = H, X = Cl

2 R_1 = R_2 = -CH_2CH_2OH, X = Br

3 R_1 = -CH_3, R_2 = -CH_2CH_2OH, X = Br

These dyes have improved water solubilities when compared with thionine
and methylene blue. For example 2 is totally miscible with water and
has ε_{max} = 89,000 at 662 nm. In contrast the maximum solubility of
thionine, 1, is about 10^{-3}M. The synthesis and solubilities in water
of 2 and 3 and several related dyes are reported together with a study
of the aggregation of 2 and 3 in water. Preliminary data on the photoreduction of 2 with Fe(II) and the performance of this system in a thin
layer photogalvanic cell is also reported.

This work was supported by grant number DE-FG02-79ER-10534 from the
U.S. Department of Energy.

A PHOTOGALVANIC CELL BASED ON THE OXONINE - Fe(II) PHOTOREDOX SYSTEM
Newton C. Fawcett, David Creed, David W. Presser, and Robert L. Thompson, Department of Chemistry, University of Southern Mississippi, Hattiesburg, MS 39401

Oxonine ($\underline{1}$, X=O) is a close analogue of thionine ($\underline{1}$, X=S). Oxonine, however has a relatively long-lived singlet state (τ = 2.6 ns in water) compared to thionine ($\tau \approx 0.3$ ns). From linear Stern Volmer plots for quenching of oxonine fluorescence by Fe(II) a quenching rate constant, of 2.9×10^9 M^{-1} s^{-1} was obtained for $\underline{1}$ (S_1) + Fe(II).

Flash photolysis shows that in water there is negligible isc to triplet oxonine. At [Fe(II)] = 2×10^{-3}M, where ~1.3% of $\underline{1}$ (S_1) is quenched, semioxonine absorbance appears immediately after the laser pulse on a time scale too short (~3µs) to be attributable to triplet quenching. This is strong evidence for singlet state electron transfer.

The monochromatic power efficiency of an oxonine/Fe(II) photogalvanic cell at 580 nm was 58% that of a thionine cell at the same wavelength and optical density. Photobleaching of oxonine by ferrous ion in the cell was observed. Evaluation of cell power output as a function of ferrous ion gave inconclusive results from which electron transfer quenching of the singlet could neither be confirmed nor ruled out.

This work was supported by grant number DE-FG02-79ER-10534 from the U.S. Department of Energy.

PHOTOVOLTAIC GENERATION IN A SOLID PHENOSAFRANINE DYE-EDTA SANDWICH CELL. K. K. Rohatgi-Mukherjee, Mandira Roy and Benoy B. Bhowmik, Physical Chemistry Laboratory, Jadavpur University, Calcutta-70032, India.

A sandwich cell was prepared using pellets of finely powdered and completely dried phenosafranine (3,7-diamino-5-phenylphenazinium chloride) and EDTA mixture in different molar ratios. The pellet was pressed between a semitransparent conducting glass (coated with SnO_2) and platinum foil by spring clips. The separation between the electrodes was maintained by a 4 mil thick teflon spacer. An open circuit photopotential developed when the cell was irradiated (under vacuum) by a 300 watt projector lamp, and it was found to vary with the compositions of the dye-EDTA mixtures. Upon varying the composition between 1 and 0 mole fraction of dye, we found that the value of the photopotential started at 90 mV in the pure dye, attained a maximum value of 320 mV at 0.7 mole fraction of the dye, and then decreased gradually to zero. This result suggests that the generation of a photovoltage in this sytem in the absence of EDTA is due to an exciton mechanism. However, in the presence of EDTA, the increase of the photovoltage may be due to a charge-transfer interaction between EDTA (acting as an electron donor or reducing agent) and the excited dye molecule (acting as an electron acceptor). [Note: this paper was accepted but not presented].

THE PHOTOREDUCTION OF SAFRANIN O BY ASCORBIC ACID AND EDTA. Herbert H.
Richtol, Charles E. Baumgartner and David A. Aikens, Department of Chemistry, Rensselaer Polytechnic Institute, Troy, New York 12181.

We have studied safranin O transients under various pH and photoreducing conditions using buffers with ascorbic acid (AA) and EDTA and report pK values and kinetic parameters associated with the process. Three protonated triplet forms are observed ($^3DH_2^{+2}$, $^3DH^+$ and ^3D) with pKa values of 7.5 and 9.2. The decay constants for these three triplets are 3.6 x 10^3 sec^{-1}, 2.2 x 10^4 sec^{-1} and 1.7 x 10^4 sec^{-1} respectively. In acid solution $^3DH_2^{+2}$ reacts with ground state dye to yield the semireduced radical. The addition of a 100 fold excess of AA results in complete triplet quenching and the formation of semireduced species with an acidity constant of 9.5. The decay constants for $\cdot DH_2^+$ is 1.1 x 10^9 M^{-1} sec^{-1} and for $\cdot DH$ is 2.5 x 10^9 M^{-1} sec^{-1}. The triplet state is the precursor to the semireduced species in all our experiments. The semireduced transient observed with EDTA is identical to that with AA, but the nature of the back reactions differ. The oxidized product of EDTA does not reoxidize the leucodye as does AA. Multiple specular reflection studies have monitored leucodye characteristics which exhibits reversible electrochemistry. Initially large photocurrents are rapidly reduced to several microamps in a Saf-O/EDTA photogalvanic cell.

MECHANISTIC ASPECTS OF THE QUENCHING OF TRIPLET METHYLENE BLUE BY COORDINATION COMPLEXES OF IRON AND COBALT AND BY ORGANIC QUENCHERS Takeshi Ohno, Prashant V. Kamat and Norman N. Lichtin, Department of Chemistry, Boston University, Boston, Massachusetts 02215

Laser flash photolysis-kinetic spectrophotometry has been used to investigate mechanism of quenching of triplet methylene blue ($^3MBH^{2+}$ and $^3MB^+$) by a number of coordination complexes of Fe(II), Fe(III), Co(II) and Co(III), and by several organic compounds including the ground state of the dye. Measured rate constants include k_q, for quenching, $k_{e.t.}$, for net forward electron-transfer in the quenching event, $k_{r.e.t.}$, for net reverse electron-transfer in bulk solution, and k_D, for diffusion-controlled encounter of the products of net forward electron-transfer. Solvent effects on the various rate constants were correlated with Kosower's polarity parameter Z. It has been concluded that, in those cases where the only significant mechanism of quenching of triplet involves reversible electron-transfer in the encounter complex, $F_1 + F_2 = 1$, where $F_1 = k_{e.t.}/k_q$ and $F_2 = k_{r.e.t.}/k_D$. A detailed model of this mechanism has been developed. Quenching by stable low-spin coordination complexes of Fe(II), by several organic quenchers which reduce $^3MB^+$ or $^3MBH^{2+}$ and by ground-state dye proceeds by this mechanism. Quenching by several high-spin complexes of Co(II) proceeds to a significant degree by at least two mechanisms, one of which involves reversible electron-transfer.

This work was supported by the U.S. Department of Energy under Contract EY-76-S-02-2889.

PHOTOGALVANIC STORAGE VIA AN IRON CONCENTRATION CELL WITH INTEGRAL
IRON-THIONINE PHOTOREDOX CHARGING <u>Robert K. Brenneman</u> and <u>Norman N.
Lichtin</u>, Department of Chemistry, Boston University, Boston, Massa-
chusetts 02215, USA

An Fe(III)/Fe(II) concentration cell has been repeatedly charged by
means of the iron-thionine photoredox system, which is incorporated as
an inherent part of the cell, and repeatedly discharged. The cell
consists of two half-cells which, in the discharged state, contain
essentially identical solutions except that one half-cell (DC) contains thionine while the other (DF) does not. The half-cells are
separated by an anion-exchange membrane and both are equipped with
identical electrodes which are reversible to the Fe(III)/Fe(II) couple.
DC is also equipped with a selective electrode which blocks the Fe(III)
/Fe(II) couple. Charging is accomplished by illuminating DC while the
selective electrode in DC is connected through the external circuit to
the electrode in DF. During discharge, the working electrode in DC is
one which is reversible to the Fe(III)/Fe(II) couple. Charging to a
storage level equivalent to many times the concentration of dye has
been accomplished. Dependence of sunlight engineering efficiency of
charging on electrode and membrane materials and on electrolyte composition has been examined. Major emphasis has been placed on investigation of the mechanism of generation and discharge of charge-carriers
during the charging process using a number of spectrometric and
electrochemical techniques.

This work was supported by the Boston University Community Technology
Foundation.

PHOTOELECTROLYSIS OF WATER BY PHOTOGALVANIC CELL WITH
ALKYLAMMONIUM ISOPOLYMOLYBDATES. <u>Toshihiro Yamase</u> and
Tsuneo Ikawa, Res. Lab. of Resources Utilization, Tokyo
Institute of Technology, 4259 Nagatsuta, Midori-ku, Yoko-
hama 227, Japan

This paper describes the application of the photogalvanic
effect based on the photoreduction of the aqueous alkylammonium molybdates solution to water decomposition ($H_2O \rightarrow
1/2\ H_2 + \cdot OH$). The photogalvanic cell is made of two test
chambers (for anode and cathode) connected by a KCl-agar
salt bridge. Anode solutions containing alkylammonium
molybdate are flushed with argon and irradiated with UV
light ($\lambda \leqslant 400$ nm). Cathode solution is 5N H_2SO_4. When
hexakis(isopropylammonium) heptamolybdate is used as the
molybdate, the splitting of water is given in terms of the
half-reactions:

Pt photoanode: $\quad Mo(VI)=O + H_2O \rightarrow Mo(V)-OH + \cdot OH \quad (1)$
$\qquad\qquad\qquad\quad Mo(V)-OH \rightarrow Mo(VI)=O + H^+ + e \quad (2)$
Pt cathode: $\qquad\quad H^+ + e \rightarrow 1/2\ H_2 \qquad\qquad\qquad\qquad (3)$

E.s.r. spectroscopy is employed to elucidate the mechanism
of photochemical reactions. The spectral sensitization of
the photoreduction to Mo(V) by riboflavin derivatives is
observed. Tungstate and vanadate can replace the molybdate
for H_2 production.

KINETICS AND QUANTUM YIELDS OF THE $Fe^{+2}/Fe_3^{+3}||I^-/I_3^-$ PHOTOCHROMIC
SYSTEM. <u>Tatiana Oncescu</u> and Savu G. Ionescu, Institute of Chemistry,
Department of Physical Chemistry and Electrochemical Technology,
Bucharest, Bd. Republicii nr. 13, 70031 Romania.

The kinetics of this system have been investigated at the following
wavelengths: 366, 405, 437, 546, 578, and 650 nm. First order plots
of $-\ln \beta$ vs. time were obtained, where $\beta = A(t)/A_0$ (A = absorbance)
in the early stages of the photochemical reaction. The rate constants
obtained were used to calculate the reaction quantum yields, which
show a maximum value of 0.73 at $\lambda = 436$ nm, whereas at 650 nm and
366 nm, the values are 0.01 and 0.04, respectively. Our data differ
significantly from these reported earlier [E. K. Riedel and E. G. Williams, J. Chem. Soc., 258 (1925); G. B. Kistiakowski, J. Am. Chem.
Soc. <u>49</u>, 976, (1927)] and are useful to appreciate the efficiency of
this photochromic system for photogalvanic conversion of solar energy
(H. R. Grüniger, J. Baumann, G. Calzaferri, Ext. Abstr. Second Intl.
Conf. Photochem. Conv. Storage Solar Energy, Cambridge, 1978, pp. 53-
54). we performed additional experiments in order to check out the
usefulness of this system as a photogalvanic solution, and we obtained
currents up to 60 μA. [Note: this paper was accepted but not
presented].

PHOTO-ELECTROCHEMISTRY OF TRIS-BIPYRIDYL RUTHENIUM(II) COVALENTLY
ATTACHED TO n-TYPE SnO_2. Pushpito Ghosh and <u>Thomas G. Spiro</u>, Department of Chemistry, Princeton University, Princeton, New Jersey 08544

$Ru(bipy)_3^{2+}$ (bipy = 2,2'-bipyridyl) has been covalently attached to
n-type SnO_2 via condensation of surface hydroxyl groups with Ru(4-
trichlorosilylethyl, 4'-methyl-2,2'-bipyridine)(2,2'-bipyridine)$_2$
$(PF_6)_2$. A thick coating (~1000 layers, based on the surface hydroxyl group concentration) was produced, presumably via oligomeriaz-
tion of hydrolyzed -$SiCl_3$ groups. The coating, which was stable to
organic solvents as well as to aqueous acids and bases, gave reversible cyclic voltamograms, with peak potentials shifted slightly from
those of aqueous $Ru(bipy)_3^{2+}$, but the number of electroactive molecules
corresponded only to a few layers.   The coated electrode gave a
photocurrent about twice that observed for SnO_2 in contact with aqueous
4 m$\underline{M}$ $Ru(bipy)_3^{2+}$, with a slightly red-shifted excitation spectrum. Only
a small fraction of the electroactive molecules appeared to participate
in excited state electron transfer, although a steady state current was
supported, presumably by slow electron transfer from the outer layers.
Prolonged illumination produced extensive hydrolysis of the outer layers of the coating, but a modest reduction of electroactivity, and only
a slight decrease in photocurrent. The photocurrent increased with
applied potential, then reached a plateau, and falls off again near the
reduction potential of $Ru(bipy)^{2+*}$; the fell-off is attributed to back-
electron transfer via tunnelling through the thin space charge layer.

THE PHOTOGALVANOVOLTAIC CELL: A NEW APPROACH TO THE USE OF SOLAR
ENERGY. H. Ti Tien, Department of Biophysics, Michigan State
University, East Lansing, Michigan 48824

As the name suggests, the photogalvanovoltaic (PGV) cell is a novel
type of photoelectrochemical cell based on a combined principle of
photogalvanic (PG) and photovoltaic (PV) effects. The principle element of the cell consists of a spray-coated tetraphenylporphrin semiconductor electrode (SnO_2) and a counter electrode (Pt or SnO_2) reversible to a photoactive dye in solution. The photo-emf generated
across the cell is equal to the sum of the voltages derived from the
photogalvanic and photovoltaic processes occurring at the respective
electrodes (J.M. Mountz and H. T. Tien, Solar Energy, 21, 291, 1978).
Typical results obtained for the second generation PGV cells are 450
$\pm$ 50 mV photovoltage and 350 $\pm$ 50 μA photocurrent, which were stable
for several hours. Of special interest is the mechanism that takes
place at the TPP-SnO_2 photocathode. The TPP layer is considered as a
p-type organic semiconductor. The interface between the TPP layer and
redox solution is likened to that of a Schottky barrier with the redox
solution playing the role of a metal. Upon illumination electrons and
holes are generated and separated in the high-field region at the interface. It is envisioned that the excited TPP most likely decays first
to the lowest singlet state and then crosses to the triplet state, at
which the triplet state exciton transfers an electron to an acceptor
(e.g., AQ-2S) in the redox solution. The vacant hole left in the TPP
layer is filled by an electron from the doped SnO_2 substrate (H. T.
Tien and J. Higgins, J. Electrochem. Soc., 127, 1474-1478, 1980).

A NEW COPPER PHOTOTHERMOGALVANIC CELL. Helena L. Chum, T.S. Jayadev
and R.F. Fahlsing, Solar Energy Research Institute, Golden, Colorado
80401.

Aqueous thermogalvanic cells, the solution analogs of solid-state
thermoelectric devices, are investigated for power generation.
Measurements on the copper copper formate copper system yield thermoelectric powers, $(\Delta E/\Delta T)_{I=0}$, of 1.25 - 1.9 mV/degree, which are
higher than those exhibited by other copper systems. In these solutions three copper formate complexes are present. Practical cells
were built and tested. The power output is largely limited by cell
resistance, though mass and charge transfer contribute to the
observed overvoltages. The coupling of this thermogalvanic system
with an electrochemical photovoltaic effect to produce a photothermogalvanic cell showed that the photovoltage and the thermovoltage were
additive. The photovoltage slowly but continuously decays and the
initial power is almost double that of the thermogalvanic cell alone
(2.1 μW/cm^2 vs. 1.2 μW/cm^2 $\Delta T = 30^{\circ}C$). The anode photoactivity can
be regenerated by using this electrode as the cathode in the hot
side, or if the temperature of both electrodes is the same, by
operating it as a cathode in the dark.

CONTRIBUTED PAPERS

Session VIIb

NEW INSTRUMENTAL TECHNIQUES

Posters VII-13 through VII-15

THE STEP EXCITATION METHOD FOR STUDYING REVERSIBLE EXCITED STATE
ELECTRON TRANSFER REACTIONS. D.G. Taylor and J.N. Demas, Department
of Chemistry, University of Virginia, Charlottesville, Virginia
22901, U.S.A.

Reversible excited state electron transfer reactions are being actively
pursued because of their potential use in solar energy conversion
schemes and their fundamental interest. A new approach to quantitative measurements on such systems will be described. This technique
uses a cw laser source with a step excitation of the sample. From a
single measurement lasting less than 1 second, both the probability
of detectable electron transfer per quenching encounter and the rate
constant for the thermal back electron transfer reaction can be
measured. A microcomputerized system for carrying out such measurements will be described. The system has an accuracy and precision
of better than 5%, and results are ready within seconds of com-
pleting the experiment. In principle, thermal reactions which proceed
at the diffusion controlled limit ($\sim 10^{10} M^{-1}-S^{-1}$) can be studied by
this approach. Use of the technique to study reversible excited
state electron transfer reactions of ruthenium(II) α-diimine complexes
will be presented.

This work was supported by the Air Force Office of Science Research
under grant 78-3590.

DIFFUSE REFLECTANCE FLASH PHOTOLYSIS. Frank Wilkinson and Rudolph W.
Kessler, Department of Chemistry, University of Technology,
Loughborough, Leics., LE11 3TU, England.

Direct observation of absorption by transient species formed following
pulsed irradiation at interfaces and within powder samples has great
potential for investigating heterogeneous photochemical reactions and
adsorbed species at semiconductor surfaces. We report here a new
technique which we have developed of diffuse reflectance flash photolysis which has enabled us to record triplet-triplet absorption spectra
of aromatic hydrocarbons adsorbed as monolayers on highly scattering
surfaces of catalytic interest and to obtain transient absorption
spectra from powders. In several cases the decay of observed phosphorescence and transient absorption were shown to correspond within
experimental error. This allowed the unambiguous assignment of the
absorption spectra which are strikingly different from triplet-triplet
spectra of these same hydrocarbons in homogeneous media. At high
transient reflectance values, i.e. as $R_t \rightarrow 1$, the Kubelka-Munk function
leads to the expectation that $(1-R_t)$ should be proportional to the
square root of the transient concentration. However, since the luminescence intensity and the transient absorption or more correctly
(R_b-R_t) where R_b equals the background reflectance, have almost
identical time dependences in many cases, it follows that $(R_b-R_t) \propto C_t$.
The reasons for this are explained using numerical solutions to the
differential equations for diffuse reflectance of light induced
inhomogeneous absorbing samples with some adsorbent absorption.

LASER EXAFS: FAST EXAFS SPECTROSCOPY WITH A SINGLE PULSE OF LASER-PRODUCED X-RAYS. P. J. Mallozzi, R. E. Schwerzel, H. M. Epstein, and B. E. Campbell, Battelle, Columbus Laboratories, 505 King Avenue, Columbus, OH 43201, USA.

The technique of Extended X-Ray Absorption Fine Structure (EXAFS) spectroscopy has become an increasingly important tool for the study of chemical structure in samples which lack long-range order. We have recently shown that it is possible to obtain well-resolved EXAFS spectra of light atoms (e.g., magnesium and aluminum) with a single pulse of soft (1-2 keV) x-rays produced by Battelle's neodymium-glass laser facility. In these experiments, a 100 J, 3-1/2 nanosecond laser pulse is focused onto a metal slab target. The metal target is chosen on the basis of its ability to produce predominantly continuum radiation in the vicinity of the X-ray absorption edge to be studied. Laser-EXAFS is well suited to the study of transient species having lifetimes as short as a few nanoseconds or less, and is a promising tool for determining the chemical structure of reactive intermediates, such as free radicals and triplet excited states, which play a role in photochemical or photobiological energy conversion processes.

CONTRIBUTED PAPERS

Session VIII

CHEMICAL ASPECTS OF PHOTOVOLTAIC CELLS

Plenary Speaker: Sigurd Wagner

Session Chairman: Helmut Tributsch

Posters VIII-1 through VIII-9

STABLE PHOTOANODES OF n-Si AND OTHER n-TYPE SEMICONDUCTORS COATED WITH ORGANIC MATERIALS. Y. Nakato, M. Shioji, S. Fujiwara, H. Osafune, <u>H. Tsubomura</u>, Department of Chemistry, Faculty of Engineering Science, Osaka University, Toyonaka, Osaka, 560 Japan.

Trials to stabilize photoanodes of small band gap n-type semiconductors, Si, GaP etc., by coating with organic materials will be reported. Firstly, n-Si coated with copper-phthalocyanin film, ca. 100 nm thick, was found to work as photoanode in the presence of some redox couples, $Fe(CN)_6^{3-}/Fe(CN)_6^{4-}$, Fe^{3+}/Fe^{2+} or I^-. The action spectra correspond roughly to the absorption spectrum of n-Si, but are depressed in the region of the absorption spectrum of the phthalocyanin.

Secondly, it was found that aniline or amines forms thin film on n-Si photoanodically in acetonitril electrolyte solutions. The n-Si electrodes covered with such films have been found to be somewhat more stable than naked Si in the presence of a proper reductant, e.g., hydroquinone. The stability was increased by addition of sodium dodecyl sulfate (SDS). The selective action of SDS to the electrodic processes was also studied on graphite and other electrode materials.

Very thin aniline film (possibly, monomolecular) was formed on the n-Si surface by the above method and gold was deposited on it. This MIS electrode was found to be more stable to the photoanodic corrosion in aqueous solutions, and gave higher photovoltage than the simple gold coated n-Si reported by us previously (Y. Nakato, et al., Chem. Lett. 883 (1975)).

ROLE OF ILLUMINATED AND DARK PORTIONS OF TiO_2 SUBSTRATUM IN THE PHOTO-CATALYTIC DEPOSITION OF PALLADIUM. <u>Hideo Tamura</u>, Hiroshi Yoneyama, and Narutoshi Nishimura, Department of Applied Chemistry, Faculty of Engineering, Osaka University, Yamadakami, Suita, Osaka 565, Japan

Heterogeneous reactions on semiconductor photocatalysts are interesting from points of view of solar energy conversion. However, the working mechanism of photocatalysts and effects of electrical properties of semiconductors on the rate of the reaction remain ambiguous in many respects. The present study was conducted to clarify the role of the illuminated and non-illuminated faces of a semiconductor photocatalyst in the occurrence of heterogeneous reactions. The model systems chosen were the photodeposition of palladium and platinum onto illuminated TiO_2 single crystals of 1 mm thickness having various donor concentration. Important results obtained are as follows; (1) When the front face was illuminated and the back was kept in the dark, being contacted with metal chloride solutions as the plating bath, the deposition occurred preferentially onto the dark face, except for a crystal of very high resistivity. (2) If the back face was covered with an insulating wax to block any electrical contact to the plating bath, the deposition occurred on the illuminated face with rates comparable to those obtained for the non-illuminated face in the case mentioned above. (3) The rate of the photodeposition was markedly affected by the doping level of the crystal, and the highest rate was achieved at a critical donor concentration of TiO_2 around which the maximum quantum yield for the photoassisted oxidation of water is obtainable.

PHOTOELECTROCHEMISTRY OF POLYPYRROLE ON DOPED TIN OXIDE AND PLATINUM
SURFACES. Arthur J. Frank, John A. Turner and Arthur J. Nozik, Solar
Energy Research Institute, Golden, Colorado 80401.

The photoelectrochemistry of polypyrrole-coated doped tin oxide and
platinum electrodes has been investigated. The polypyrrole films have
been synthesized by electrochemical oxidation of pyrrole on the surfaces of the electrodes (1). In the solid state, the conductivity of
the polymer ranges over five orders of magnitude from insulator to
metal. Current-voltage data indicate that the polypyrrole film (2 μm
thickness) in 0.5 M phosphate buffer at pH 6 is oxidized and reduced in
a reversible manner between 0.8 and -0.8 V vs NHE. At large positive
voltage excursions, about 1 V, oxidation of the film takes place.
Mott-Schottky plots ($1/C^2$ vs V) are consistent with the polypyrrole
possessing a depletion region and showing p-type conduction. While the
film is conducting in the dark, light in the near-infrared is effective
in generating a current. Spectrophotometric measurements reveal an
asymmetrical broad band with a maximum at ca. 1000 nm ($E_{max} \simeq 8 \times 10^4$
cm^{-1}) and a more symmetrical peak at ca. 400 nm ($E_{max} \simeq 5 \times 10^4$ cm^{-1}).
However, the positions and the extinction coefficients of the bands are
variable and are dependent on the conditions of synthesis and the history of the film.

1. A.F. Diaz and K. Keiji Kanasawa, J.C.S. Chem. Comm. 635-36 (1979).

This work was supported by the Division of Chemical Sciences, Office
of Basic Energy Sciences, U.S. Department of Energy.

PHOTOCHEMICAL ASPECTS OF MEROCYANINE SOLID-STATE SOLAR CELLS.
Geoffrey A. Chamberlain, Shell Research Limited, Thornton Research
Centre, P.O. Box 1, Chester CH1 3SH, U.K.

Some solid-state photovoltaic cells based on merocyanine semiconductor
dyes have been prepared. A typical cell consisted of thin layers of
aluminum, aluminum oxide, merocyanine dye and gold, sequentially
vacuum evaporated, so that active areas of about 5 cm^2 were formed.
Doping the dye layer with oxygen, water vapor or iodine by vapor diffusion increased the photovoltaic response, iodine being particularly
effective. The magnitude of the photoresponse has been related to
merocyanine molecular structure for a series of dyes, and a correlation between ionization potential and quantum yield of electron flow
is apparent. By judicious tailoring of electron-donating and elec-
tron-accepting fragments of the dye molecule, we have prepared cells
with an efficiency of 0.2% in 45 mW/cm^2 sunlight. The power-
conversion efficiency, based on photons absorbed by the dye at this
intensity of solar illumination, was calculated to be 2.1%, and
quantum efficiencies of about 20% have been observed.

QUANTUM EFFICIENCY OF THE SENSITIZED PHOTOCURRENT OF ZnO ELECTRODE
BY THE J-AGGREGATE OF CYANINE DYES. Hiroshi Hada, Yoshiro Yonezawa and
Hiroo Inaba, Department of Industrial Chemistry, Kyoto University,
Yoshida, Kyoto 606, Japan

Sensitized Photocurrent of the c-axis oriented ZnO film electrode is
measured for two dye aggregates adsorbed on the ZnO electrode. The
dyes employed are 1,1'-diethyl-2,2'-quinocyanine chloride (dye I) and
3,3'-dimethyl-9-phenyl-4,5, 4',5'-naphthothiacarbocyanine chloride
(dye II). The aggregates of these dyes have the well-known J-band at
the longer wavelength than monomer-, dimer- and H-bands. The ZnO film
is deposited on a Nesa glass (transparent electrode) by using a
magnetron sputtering technique. The ZnO film electrode is mounted as
a window of an electrochemical cell with the ZnO side facing the dye
solutions. Thus, light is irradiated from the outside of the cell to
avoid the absorption of incident photons by the free dye molecules and
aggregates in the bulk solution. Dye I shows a stronger anodic current
than cathodic current for the J-band of the dye and the intrinsic
quantum efficiency of the anodic current are in the range of 0.3 - 1.0
in repeated experiments at the electrode potential of 0.6 V. On the
contrary, Dye II shows a stronger cathodic current than anodic current
for the J-band and the intrinsic quantum efficiency of the cathodic
current are in the range of 0.5 - 1.0 in repeated experiments at the
electrode potential of 0.5 V. The high Quantum efficiency for the J-
aggregate of dye seems to be significant for applications to the electrochemical solar cell and to the design of the photosynthesis system.

MCD SPECTROSCOPIC MONITORING OF THE PERFORMANCE OF TRANSPARENT
PHOTOELECTRODES COATED WITH PORPHYRIN FILMS. Cooper H. Langford,
Bryan R. Hollebone and Daniel Nadezhdin, Departments of Chemistry,
Concordia University, Montreal, Quebec and Carleton University,
Ottawa, Ontario Canada.

There have been a number of studies on the use of porphyrin and
phthalocyanine films as chromophores for electrochemical photovoltaic
cells and photoelectrosynthetic cells. However, little is known of
the structural factors controlling photoelectrochemical behaviour.
The transparent n-SnO_2 coated glass electrode allows for parallel
spectroscopic and electrochemical study. A synthetic survey indicated
that Palladium tetraphenylporphyrin has favourable properties.
Photoanodic, photocathodic, and photoelectrosynthetic behaviour have
been examined as a function of alternative film preparation and
"doping" strategies. In each case, MCD spectra of the electrodes were
recorded. Photoelectrochemistry is seen to relate to alteration of
the bands associated with axial ligatation. Crystal phase is not an
important variable. Indeed, the clearly visible spectroscopic
manifestations of solid state interaction (e.g. Davydov splitting) do
not relate simply to photoelectrochemistry. Amorphous films which
have spectra indicating non-interacting monomers are effective. We
conclude a localized carrier hopping mechanism is important. It
requires formation of carrier sites at a large fraction of molecules
in the film.

THE RELATIONSHIP BETWEEN ADSORBED SPECIES AND THE PHOTO-VOLTAGE OF PORPHYRIN SURFACES. S. C. Dahlberg, Bell Laboratories, 600 Mountain Avenue, Murray Hill, New Jersey 07974.

This report deals with surface photovoltage studies of porphyrins and related organic semiconductors such as phthalocyanines. The goal of this work is a fundamental understanding of the effect of adsorbed species on the opto-electronic properties of these materials under well characterized surface conditions. Experiments were performed in ultra-high vacuum using the optically modulated retarding potential technique for probing fast (msec) photovoltaic changes in the work function of these thin films following irradiation with a laser pulse. Since many of the important reactions photocatalyzed by these materials are oxidations, this work has documented the presence of two forms of adsorbed oxygen which differ in their sticking coefficients. One form is neutral, and increases the photovoltage by an order of magnitude. The other form has a partial negative charge and forms a charge transfer species at the surface which greatly decreases the photovoltage. The nominally undoped nickel phthalocyanine is known to be a p-type semiconductor due to naturally occurring defects and this research demonstrates that the photovoltage was sensitive to the presence of additional electron acceptors such as o-chloranil.

PHOTOELECTROCHEMICAL SOLAR ENERGY CONVERSION BY POLYCRYSTALLINE FILMS OF PHTHALOCYANINE, R.O. Loutfy and L.F. McIntyre, Xerox Research Centre of Canada, 2480 Dunwin Drive, Mississauga, Ontario, L5L 1J9, Canada

A novel photoelectrochemical cell based on an organic paint has been developed. The principle element of the cell is a thin organic film of microcrystalline particles of x-metal free phthalocyanine (x-H_2Pc) dispressed in a polymer binder painted on a conductive substrate. These films functioned as photocathodes in electrochemical cells in contact with aqueous electrolyte containing Redox couples such as p-benzoquinone, $Fe(CN)_6^{3-/4-}$ or $MV^{2+/1+}$.

Under solar illumination the cells develop photocurrent in the order of few milliampares/cm^2 and photovoltage of 250mV. The mechanism of the photocurrent and voltage generation was investigated by measuring action spectra, capacitance-voltage dependence and current-voltage curves for various Redox systems. The action spectrum indicates that the cathodic photocurrent arises from excitons generated near the semiconductor (x-H_2Pc) electrolyte interface. p-Benzoquinone, oxygen, MV^{2+} or $Fe(CN)_6^{3-}$ act as electron acceptors, enhancing the photocurrent significantly. A power conversion efficiently of 1% for solar light has been obtained. The most striking features of these devices are, the linear dependence of the photocurrent on light intensity, stability and simplicity of fabrication.

SENSITIZATION OF TiO_2 AND $SrTiO_3$ ELECTRODES BY TRANSITION-METAL IONS
AND DYES IN THIN FILMS. <u>A. Mackor</u>, Institute for Organic Chemistry
TNO, P.O. Box 5009, 3502 JA Utrecht, G. Blasse, J. Schoonman and P. H.
M. de Korte, Solid State Chemistry Department, State University,
Utrecht, C. W. de Kreuk and J. van Turnhout, Physics Department, Divi-
sion of Technology for Society TNO, Delft, The Netherlands.

Three approaches have been followed to study the sensitization of
TiO_2 and $SrTiO_3$ electrodes towards visible light Application of a
surfactant ruthenium tris(bypyridyl) dye as a solid film onto the surface of single crystalline electrodes leads to sustained sensitized
photocurrents i_s in acid medium, following the relation log i_s = A -
B × pH. Transition metal ions (Cr, La, Nb) have been introduced into
the crystal lattice (bulk doping) or onto the surface, the latter by
intimate contact of powdered dopants with the crystal, followed by
firing. Photocurrents in the visible region have been obtained from
$SrTiO_3$:$LaCrO_3$, where the principle of charge-compensation of La^{3+} for
Cr^{3+} has been used. However, a homogeneously doped sample showed
anodic photocurrents at more positive bias and a lower quantum efficiency as compared to undoped $SrTiO_3$. The presence of Nb has been
found to exert no great influence. The stability of $SrTiO_3$ in acid
medium is a point of concern. Crystals of $SrSO_4$ are formed at the irradiated surface in 0.5 M H_2SO_4, and titanium ions are found in solution. The Sr/Ti ratio at the surface decreases with irradiation time.
[This work was supported by VEG-Gasinstituut NV and by the EC Solar
Energy R&D programme (Project D)].

CONTRIBUTED PAPERS

Session IX

ELECTROCHEMICAL PHOTOVOLTAIC CELLS

Plenary Speaker: Rüdiger Memming

Session Chairman: Adam Heller

Posters IX-1 through IX-20

SUB-BAND-GAP SPECTROSCOPY OF PHOTOELECTRODES,* <u>Michael A. Butler</u> and
David S. Ginley, Sandia National Laboratories,† Albuquerque, NM 87185

The sub-band-gap photoresponse for several semiconducting electrodes
will be discussed. In TiO_2 and $SrTiO_3$ such photocurrents are observed
in undoped and Cr-doped samples. This is shown to be a bulk effect and
not due to surface states. A qualitative model will be used to describe the photoexcitation process and to specifically explain the spectral dependence of the sub-band-gap photocurrents. From this analysis
comes some information about the distribution of defect related states
in the semiconductor gap. Sub-band-gap photocurrents will be reported
for p-GaP which depend on surface treatment and electrochemical aging.
Cathodic aging produces a sub-band-gap response but of different spectral shape than produced by surface polishing. Anodic aging removes
the photoresponse produced by both methods. Etching experiments show
that this effect arises within 1000 Å of the semiconductor surface.
The model developed for TiO_2 and $SrTiO_3$ will be applied to explain
the spectral dependence of the photoresponse in GaP.

*This work was supported by the Materials Sciences Program, Division
 of Basic Energy Sciences, U. S. Department of Energy, under Contract
 DE-AC04-76-DP00789.
†A U. S. Department of Energy facility.

ELECTROCHEMICALLY ION IMPLANTED PHOTOELECTRODES,* <u>David S. Ginley</u> and
Michael A. Butler, Sandia National Laboratories,† Albuquerque, NM 87185

Most potential electrode materials for photoelectrochemical cells have
crystal structures with open channels intersecting the normal electrode
face. These channels have diameters large enough to allow the migration, under appropriate conditions of bias, of small ions in both
directions across the semiconductor/electrolyte interface. We will
demonstrate how this migration of ions can substantially affect the
stability, photoresponse and flatband potential of various electrodes.
A particularly interesting aspect of these observations is the potential to intentionally modify doping profiles and surface compositions
by forcing the electromigration or electroinjection of selected chemical species into or out of a semiconductor. Results will be discussed
for experiments utilizing n-type TiO_2 and $SrTiO_3$ and p-type GaP photoelectrodes. The spectral response and flatband potential of the
transition metal oxides will be shown to be very sensitive to the
electromigration of interstitial dopant atoms out of the semiconductor
and hydrogen ions into the semiconductor. Small cations, when injected
into p-GaP, can give rise to luminescence as hot conduction band electrons relax into the newly generated empty electronic states and can
change electrode properties.
The technique of electrochemical ion injection shows considerable
promise for modifying photoelectrode characteristics.

*This work was supported by the Materials Sciences Program, Division
 of Basic Energy Sciences, U. S. Department of Energy under Contract
 DE-AC04-76-DP00789. †A U. S. Department of Energy facility.

PHOTOELECTROCHEMICAL PROPERTIES OF SILVER SULFIDE ELECTRODE. Kazuhito
Maeda, Keietsu Tamagake, Takashi Katsu, and Yuzaburo Fujita, Faculty of
Pharmaceutical Sciences, Okayama University, Okayama 700, Japan

The Ag_2S electrode had a narrow band gap of 1.2eV which was desirable
for solar energy utilization up to 1000nm and exhibited a large photocurrent in the presence of ferrocyanide ions in electrolyte. However,
the deactivation of the electrode occurred at high anodic polarization
accompanied by the increment of the dark current. It was found that
the improvement of the Ag_2S electrode was achieved by the heat treatment at $200^{\circ}C$ in a vacuum. We attempted to elucidate the mechanism of
deactivation by various methods; rotating-ring-disk electrode(RRDE),
atomic absorption spectrophotometry, and impedance measurement under
the injection of a step-wise potential pulse. The unheat-treated electrode had a much smaller leak resistance at the surface than that of
the heat-treated electrode, and showed a large dark current due to some
complex formation at the surface between silver ions of the electrode
and ferro- and ferricyanide ions. By the heat treatment, the leak
resistance increased significantly and the thickness of the space
charge layer increased to show characteristics of an n-type semiconductor electrode. The mechanism of the deactivation and improvement of
the electrode before and after the heat treatment was discussed with
these experimental results.

This work was supported by Grant-in-Aid for Scientific Research Nos.
254126 and 387023 from Japanese Ministry of Education.

ON THE APPLICABILITY OF SEMICONDUCTING LAYERED MATERIALS FOR
ELECTROCHEMICAL SOLAR ENERGY CONVERSION. Wolfgang Kautek, Jens
Gobrecht, and Heinz Gerischer, Fritz-Haber-Institut der Max-Planck-
Gesellschaft D-1000, Berlin 33, Western Germany.

Semiconducting transition metal dichalcogenides with layer structure
belong to the most stable electrode materials in electrochemical solar
cells. However, photoelectrochemical reactions are complicated by the
surface anisotropy. Results of photoelectrochemical cells with n-type
$MoSe_2$ and WSe_2 in contact with redox electrolytes demonstrate the importance of surface morphology for the energy conversion efficiency.

Photoelectrochemical hydrogen evolution has been investigated with p-
type WSe_2. It is shown that a large increase of the energy gain can be
reached by the catalytic action of platinum deposits on the surface of
the electrode. This is explained in terms of surface state catalysis.

The energy conversion yield of the layered semiconductors of such type
has turned out being lower than theoretically expected from their band
gaps. Based on an investigation of the nature of the band gaps and the
band structure, it can be shown that this is connected with the existence of an indirect band gap which restricts light absorption in a
considerable energy quantum range above the band gap. Besides this,
surface inhomogeneities reduce the quantum yield further more by surface recombination although the mean diffusion length of minority carriers is found being favourably large.

CONCENTRATION AND REDUCTION EFFECTS ON Nb-DOPED TiO_2 PHOTOANODES. J.F. Houlihan and R.F.Bonaquist, Dept. of Physics, Shenango Valley Campus, Penn State Univ., R. Dirstine, Corporate Applied Research Group, Globe-Union, Inc. and D.P. Madacsi, Univ. of Conn.

Nb-doped and undoped TiO_2 photoelectrodes have been fabricated using standard powder metallurgy techniques and reduced in a CO/CO_2 atmosphere over a range of pO_2 values (10^{-4} to 10^{-10}). Effects of pO_2 on optical conversion efficiency and spectral responses have been investigated. Optical conversion efficiencies for the production of hydrogen as high as 0.6% have been obtained for undoped TiO_2, while slightly lower values were obtained for Nb-doped electrodes. The efficiency is found to decrease as the Nb concentration increases (pO_2 held constant) --a result which is attributed to increased recombination occuring as a result of electrons being trapped at the Nb-impurity sites. Electron Spin Resonance studies which confirm the increased localization of electrons with increasing Nb concentration will be presented.

Spectral response measurements indicate that the TiO_2 bandgap is not significantly affected by Nb doping. In addition the existence of a low intensity, broad peak at ~ 2 eV in the spectral response ($\sim .55$ nanometers) results for both doped and undoped TiO_2 will be discussed in terms of the formation of suboxides of titanium at the surface of electrodes during the reduction process and subsequent optical transitions between the various d-electron states of the suboxides.

*This work was supported, in part by the National Science Foundation under Grant No. SPI-7907385.

STUDY OF Zr DOPED $\alpha-Fe_2O_3$ SINGLE CRYSTAL PHOTOANODES, G. Horowitz Commissariat A L'Energie Atomique, Demt/Service D'Etudes Energetiques Cen.Saclay -B.P No. 2 -91190 Gif-Sur-Yvette (France)

The photoelectrolysis of water with solar energy by using n-type semiconducting oxides such as TiO_2, $SrTiO_3$ or $KTaO_3$ have been demonstrated by many workers. Unfortunately, very low efficiencies are obtained with these photoanodes because of the mismatch between their bandgap, which is larger than 3 eV and corresponds to UV light, and the solar spectrum. Due to its lower bandgap (ca. 2 eV), Fe_2O_3 is a promising photoanode material for a photoelectrolysis cell. Several studies on this oxide have been published during the past few years. Almost all of them were performed with polycrystalline undoped samples. We present a study of Zr doped Fe_2O_3 single crystal electrodes. Doped samples present the advantage of having well defined and reproducible characteristics. Spectral response and differential capacitance measurements have been carried out on these samples. The results are presented as a function of the doping level.

* Work supported by DGRST contract No. 78-7-2148

THE INFLUENCE OF CYSTEINE SOLUTIONS ON THE STABILITY OF CdS PHOTO-
ANODES. A. Kirsch-De Mesmaeker, A.M. Decoster and J. Nasielski,
Department of Organic Chemistry, Université Libre de Bruxelles,
1050 Bruxelles, Belgium.

Although CdS is thermodynamically unstable against photocorrosion, it
has been shown by Manassen et al. and Wrighton et al. that polychal-
cogenide redox couples can prevent CdS photooxidation, however with
some superficial electrode modification indicating some CdS degradation.
We tested the influence of cysteine in aqueous alkaline solution, as a
reducing electrolyte on the photoanodic behavior and stability of
polycrystalline CdS films. Cysteine 1 mol.dm^{-3} enhances the photocurrents by factors of 3 or 4 compared to those obtained in NaOH
solutions at the same pH and shifts the zero-photocurrent potential
cathodically. At pH 11,5 at the beginning of the illumination, the
photocurrents slowly increase by factors up to 4, thanks to some
modification of the CdS surface, accompanied by some electrode photocorrosion.
Cysteine however is an efficient reducing agent since after a same
irradiation time, the optical density of the CdS film in contact with
the reductant has only slightly decreased whereas in the presence of
cystine, the oxidant, the CdS film is completely photooxidized and
dissolved.
At pH 9, where there is only minor photocorrosion, cysteine seems to
be a stabilizing agent almost as efficient as sulfide ions, with the
additional advantage that, unlike polysulfides, the cysteine-cystine
electrolyte is transparent in the CdS band gap region.

STUDIES ON ZnO AND WO$_3$ P.E.C. CELLS-EFFECT OF MODE OF PREPARATION OF
ELECTRODE AND NATURE OF ELECTROLYTE   K. Ravindranathan Thampi,
T. Varahala Reddy, V. Ramakrishnan and J.C. Kuriacose, Department of
Chemistry, Indian Institute of Technology, Madras 600036, India.

Photocatalytic properties of polycrystalline n-ZnO and n-WO$_3$ have been
investigated for use in a P.E.C. Cell. The emf in a cell consisting
of the polycrystalline semiconductor deposited on Pt as the irradiated
electrode, a Pt foil as the counter electrode and KCl as the supporting electrolyte is modified on the addition of alcohols, amines, and
amides in the photo anode compartment. Reproducible results are obtained with the following types of electrode preparation, (i) a Pt foil
painted with a paste of semiconductor in 0.1M KCl and partially dried,
(ii) a Pt foil coated from a colloidal solution and dried, (iii) a Pt
foil painted with a oxide-liquid crystal paste. Bubbling N$_2$ through
the electrolyte for ZnO and O$_2$ for WO$_3$ improved the cell functioning.
The cell functions only at 365 nm in the case of ZnO and at both 397
nm and 456 nm in the case of WO$_3$. With a liquid crystal binder the
spectral response is extended to 405 nm for ZnO. In the case of ZnO,
addition of benzyl alcohol gives the highest photo emf, whereas for
WO$_3$ formamide gives the highest value. Sintering of the oxide reduces
the effectiveness of these electrodes while doping with Al is bene-
ficial.

IN-SITU PHOTOETCHING OF SILICON PHOTOANODES IN NONAQUEOUS
ELECTROCHEMICAL PHOTOVALTAIC CELLS. H. J. Byker, A. E. Austin,
V. E. Wood, R. E. Schwerzel, and E. W. Brooman, Battelle, Columbus
Laboratories, 505 King Avenue, Columbus, OH 43201, USA.

The use of n-silicon photoanodes in electrochemical photovoltaic
cells, though attractive in terms of the small band gap and widespread abundance of silicon, has been hampered severely by the
sensitivity of the silicon surface to different methods of preparation and to traces of air or water. We have found that the
electrochemical behavior of n-silicon photoanodes in acetonitrile
solutions containing tetra-n-butylammonium tetrafluoroborate (or
chloride) can be altered markedly by an _in-situ_ photoetching
process, which appears to result from either the anodic breakdown of
the tetrafluoroborate ion or the anodic decomposition of the silicon
surface. Application of this technique can lead to the sequential
removal of several layers of surface from the silicon electrode;
this may provide a means for the removal of surface damage from
polishing. While the scope and mechanism of this process are not
yet fully known, the observation of _in-situ_ photoetching is sin-
nificant in that, unlike many photocorrosion processes, it may provide a possible means of altering the silicon surface during the
operation of an electrochemical photovoltaic cell. It also provides
new insight as to the location of the n-silicon anodic decomposition
potential in these nonaqueous media.

This research was supported by The Solar Energy Research Institute,
U. S. Department of Energy.

RAPID PHOTOCORROSION IN N-TYPE STRONTIUM TITANATE PHOTOANODES.
R. E. Schwerzel, E. W. Brooman, H. J. Byker, E. J. Drauglis, D. D.
Levy, L. E. Vaaler, and V. E. Wood, Battelle, Columbus Laboratories,
505 King Avenue, Columbus, OH 43201, USA

Strontium titanate (n-SrTiO$_3$) has found widespread use as a photoanode material for laboratory studies of photoelectrolysis, despite
its wide bandgap, because it has been reported to be both chemically
stable and capable of catalyzing the photoelectrolysis of water
without an applied bias potential. In acidic aqueous electrolytes
containing carboxylic acids such as acetic acid, however, n-SrTiO$_3$
photoanodes can suffer rapid photocorrosion under conditions where
the photo-Kolbe oxidation of the acid would normally be expected
to occur. This is manifested by the pitting and dissolution of the
photoanode in the region under illumination. While not all n-SrTiO$_3$
electrodes corrode, those that do appear to corrode most rapidly in
strongly acidic solutions (e.g., 1.0 M H$_2$SO$_4$ containing 0.5M HOAc).
Both SrSO$_4$ and Sr(OAc)$_2$ are formed under these conditions. This
behavior may result from the presence on some, but not all, n-SrTiO$_3$
photoanodes of surface states which serve to mediate effective hole
transfer to degradation products at low but not at high pH. Thus
these observations underscore the importance of understanding the
factors that govern the kinetic balance between competing, thermodynamically allowed reactions at the semiconductor-electrolyte interface, in this case oxygen evolution, the photo-Kolbe reaction, and
the unprecedented photocorrosion of SrTiO$_3$. This research was
supported by The Solar Energy Research Institute, U. S. Department
of Energy.

THE PHOTOELECTROCHEMICAL TRANSIENT BEHAVIOUR OF FLAME OXIDIZED IRON
C.F. Brammall and A.C.C. Tseung, Department of Chemistry, The City
University, Northampton Square, London, EC1V OHB, United Kingdom.

The transient photo currents of flame oxidized iron electrodes subjected to pulsed illumination have been studied. The characteristic anodic
and cathodic peaks can be explained in terms of formation of an oxy/
hydroxy intermediate on illumination and its subsequent reduction when
the illumination was removed.

This hypothesis was supported by cyclic voltammetry, high voltage bia-
sed transient measurement and X-ray crystallography results.

A simple model of photochemical evolution of oxygen on flame oxidized
iron will be presented and its implications for the choice of photoanodes discussed.

THE MEASUREMENT OF FAST TRANSIENTS IN PHOTOELECTROCHEMICAL CELLS: THE
EFFECTS OF SURFACE ETCHING, Z. Harzion*, N. Croitoru** and S.Gottesfeld*
Department of Chemistry* and School of Engineering**, University of
Tel Aviv, Ramat Aviv, Israel

The response of photoelectrochemical cell to short light pulses may
provide information on the kinetics at the illuminated semiconductor-
electrolyte interface. Light pulses of 10nS width were directed onto
a single-crystal CdSe electrode immersed in an alkaline $S^=/S^o$ solution.
The photoelectrochemical circuit was completed by a load resistor (R_L)
and a reversible counter electrode of a large area. A pronounced effect
of surface etching on the measured decay time was observed. To allow
quantitative analysis of the photocurrent decay curves, the light intensity was attenuated creating quasipotentiostatic conditions under which
simple exponential decay was observed. A plot of τ vs R_L was used to
analyse the decay rate determining factors. At small values of R_L the
slope of this plot gives the photocapacity C, while the intercept gives
$R_{series} \cdot C$ where R_{series} is a combination of ohmic elements through which
the photocapacitor discharges. Etching was shown to result in the lower
ing of R_{series}. Interpretation in terms of a surface layer which contains traps and recombination centers is suggested.

LASER INDUCED PHOTOELECTROCHEMISTRY: NANOSECOND COULOSTATIC FLASH
STUDIES OF PHOTOOXIDATION AT CdX AND TiO_2 SEMICONDUCTOR
ELECTORDES.* J. H. Richardson, S. B. Deutscher, and L. L. Steinmetz,
Lawrence Livermore Laboratory, University of California, Livermore,
California 94550, U.S.A. and S. P. Perone, Department of Chemistry,
Purdue University, W. Lafayette, Indiana 47907, U.S.A.

Instrumentation developed for laser induced coulostatic-flash measurements of electron photoemission has been adapted to characterizing
photooxidation at n-type semiconductor electrodes. A nitrogen pumped
dye laser provides the excitation source; the temporal resolution is
therefore limited to 10 ns. The signal risetime with TiO_2 can be
interpreted as a function of solution resistance; the final magnitude
can be related to the net charge transferred. A long term decay is the
result of several different transient components. One component,
suggested by a $(time)^{-2/3}$ linear plot with both types of electrodes,
is space charge layer relaxation. A significant conclusion from the
TiO_2 work is that the electron transfer rate is proportional to the
excitation rate. CdX (X=S,Se) semiconductor electrodes behaved
significantly different in some respects. A two component risetime was
observed which depended on wavelength. The coulostatic response
dependent on the solution composition when uv irradiation was used.
These results demonstrate that the laser coulostatic technique can
yield new insights into the dynamics of charge separation both within
the semiconductor and at the solution interface.

*Work performed under the auspices of the U. S. Department of Energy
by the Lawrence Livermore Laboratory under Contract No.
W-7405-ENG-48 and No. DE-ACO2-77ER04263.A002.

KINETIC STUDY OF CHARGE TRANSFER REACTIONS IN PHOTOELECTROCHEMICAL
SOLAR CELLS. Frank Van Overmeire, Jean Vandermolen, Fernand Vanden
Kerchove, Walter P. Gomes, Laboratory of Physical Chemistry, Rijksuniversiteit Gent (Belgium) and Felix Cardon, Laboratory of Crystallography and Study of the Solid State, Rijksuniversiteit Gent (Belgium).

In order to obtain improved efficiencies in photoelectrochemical solar cells, it is important to know which factors govern the rates of
the photoelectrochemical reactions at the surface of semiconductor
electrodes. Although under normal circumstances the photocurrent depends on light intensity solely, the relative rates of different reactions, occurring simultaneously at the semiconductor-electrolyte interface, can be studied. For such measurements the rotating ring-disk
technique has proved to be a useful tool.
Different n-type semiconductors have been studied:
- III-V compounds such as GaP and InP. The competitive reactions are
the anodic dissolution of the electrode and the oxidation of added reducing electrolytes. The products, formed at the illuminated semiconductor disk, are detected at the metal ring and thus the stabilization
of the photoanode is determined. The light intensity appears to affect
the ratio of the competitive reaction rates in an unfavourable way: increasing light intensities cause a more than directly proportional enhancement of the photodissolution of the semiconductor.
- oxide semiconductors (TiO_2, $SrTiO_3$). With these relatively stable
materials there is competition between the photooxidation of water and
the photooxidation of other reducing agents. The influence of the
concentration of the competing reagents and of the light intensity on
the reaction rates has been studied.

PHOTOELECTROCHEMISTRY OF TRANSITION METAL DICHALCOGENIDE SEMICONDUCTOR ELECTRODES IN AQUEOUS HALIDE ELECTROLYTES. Margaret Levin, Department of Chemistry, University of Illinois, Urbana, Illinois, 61801, USA, Mark T. Spitler, Department of Chemistry, Mount Holyoke College, South Hadley, Massachusetts, 01075, USA.

It is known that halides, in particular I^-, greatly influence the power output of electrochemical solar cells which utilize the layer type transition metal dichalcogenides as photoelectrodes (Tributsch et al., Ber. Bunsenges, Phys. Chem., _83_, 655, 1979). In an attempt to understand this phenomenon, a systematic study has been undertaken to determine the nature of the interaction of the halides and trans-ition metal dichalcogenide electrodes. Conventional electrochemical techniques have been employed to analyze the effect of Cl^-, Br^-, and I^- in the electrolyte upon the photoelectrochemistry of MoS_2 and $MoSe_2$ single crystals. At concentrations as low as 5 mM, it has been found that Br^- and I^- cause anodic photocurrents negative of the flat band potential for MoS_2 and $MoSe_2$ electrodes, concomitant with reduction of the resultant halogen. These effects were found to be more pronounced with I^- than with Br^-, and with $MoSe_2$ than with MoS_2. It is proposed that these halides form surface states at the electrode surface which function as recombination centers for photogenerated holes and electrons: holes are captured by the halide which is then reduced by an electron from the electrode. A qualitative correlation was found between the quality of the electrode surface and the magnitude of these halide induced photocurrents.

This work was supported by the William and Flora Hewlett Foundation Grant of Research Corporation.

MEASUREMENT AND SIGNIFICANCE OF FLAT-BAND POTENTIAL DETERMINED FROM MOTT-SCHOTTKY PLOTS J. Cooper, J. Turner, and A.J. Nozik, Solar Energy Research Institute, Golden, Colorado 80401, USA

The flat-band potential is one of the most important parameters in pho-toelectrochemistry. It defines the energetics of the semiconductor-electrolyte interface by establishing the position of the semiconduc-tor electrode band edges with respect to the energy levels of the elec-trolyte. In the ideal case, the flat-band potential of a semiconductor electrode can be determined by the V-axis intercept of a plot of $1/C^2$ vs. V, where C is its capacitance, and V its potential; such graphs are called Mott-Schottky plots. However, in practice, Mott-Schottky plots are frequently non-linear and they also exhibit AC frequency dispersion effects. This non-ideal behavior makes the determination of the flat-band potential from Mott-Schottky plots difficult or sometimes impossible.

A technique has been developed to produce ideal Mott-Schottky plots for TiO_2 electrodes which show perfect linearity and little or no frequen-cy dispersion. It was found that the large scatter of flat-band poten-tial measurements of TiO_2 found in the literature can be explained by the impurity and defect variations in the different TiO_2 boules studied.

Another important fact that was uncovered was that the potential at which photocurrent begins to flow in TiO_2 electrodes is the same for electrodes showing ideal and those showing very anomalous Mott-Schottky plots. This means that the flat-band potential obtained from Mott-Schottky data cannot be used to predict photocurrent-voltage character-istics unless the data show ideal behavior.

NONAQUEOUS ELECTROCHEMICAL PHOTOVOLTAIC CELLS BASED ON n-GaAs AND n-Si.
Margaret E. Langmuir, Ronald H. Micheels and R. David Rauh, EIC Corporation, 55 Chapel Street, Newton, Massachusetts 02158

Factors influencing the output characteristics of electrochemical photovoltaic cells have been determined for n-GaAs and n-Si in several non-
aqueous solvents. In general, n-GaAs is the more ideal photoelectrode,
readily displaying hole-limited photocurrents. The saturation open
circuit photovoltages for cells of configuration n-GaAs|redox,solvent|Pt
generally vary linearly with E(redox). The photoelectrochemical behavior of n-Si, however, tends to be dominated by passivation and surface
states. Even in nonaqueous electrolytes, n-Si electrodes irreversibly
passivate on anodic polarization. Surfaces which are not carefully
polished and etched or chemically modified give rise to metallic behavior, accompanied by anodic dark currents. Clearly, the acceptance of
surface charge on Si allows the interfacial potentials of the bands to
vary with applied potential or according to the redox potential of the
species in solution. Positive effects of adsorbed Ru(III) and N hetero-
cycle polymers were noted for both electrodes, indicating a possible
route to the diminishing of surface state densities.

PHOTOCHEMICAL AND IMPEDANCE PROPERTIES OF SPUTTERED TITANIUM DIOXIDE.
a. Michael F. Weber, Lynn C. Schumacher, and M. J. Dignam, Department
of Chemistry, University of Toronto, Toronto, Ontario M5S 1A1, Canada.

In connection with achieving bias-free photoelectrolysis of water, we
are investigating the basic photo- and electrochemical properties of
intrinsic (undoped) oxides using sputtered thin films. As a first
study, titanium dioxide was chosen as a model system. By the use of
heated substrates (270°C) and a pure oxygen plasma, thin films of
titanium dioxide with photo responses almost identical to single crystals
were routinely obtained, with no pretreatment necessary. Photocurrent
onset occurs at potentials typical of a light to moderately doped single
crystal. Since photocurrent onset was found to shift ∿60 mV per decade
of light intensity, self doping due to the photoexcited electrons is
postulated. In addition, a photoshift of 55 mV was observed for hydrogen evolution. The shift was in the direction of lower hydrogen overpotentials and is directly proportional to the light intensity. Thin
films sputtered under other conditions required electrochemical doping
with hydrogen before high photocurrents were obtained. To study the
effects of hydrogen doping, the a.c. impedances of the films were
measured as functions of frequency and d.c. potential, pretreatment
history, and before and after illumination. Effects of surface coatings
were also studied. Approximately one monolayer (2 - 3 Å) of Pt was
sputtered onto oxide films. The photocurrent onset was shifted 60 mV
anodically but the photocurrent was otherwise unchanged. The overpotential for hydrogen evolution was reduced almost to that of pure
platinum.

Acknowledgement: This work was supported by the Natural Sciences and
                  Engineering Research Council of Canada.

ELECTROCHEMICAL PHOTOCELL OF $CdSe_\alpha Te_{1-\alpha}$ MIXED CRYSTALLINE SEMICONDUCTOR ELECTRODE, <u>A. Fujishima</u>, T. Inoue and K. Honda, Department of Synthetic Chemistry, Faculty of Engineering, The University of Tokyo, Hongo, Bunkyo-ku, Tokyo 113, JAPAN

$CdSe_\alpha Te_{1-\alpha}$ mixed crystal was made by press of the mixed powder of CdSe and CdTe at ca. 1 ton/cm^2, packing into alumina, and then heating in a nitrogen atmosphere at 600 to 700 ^{o}C for several hours. We could obtain good photocurrent-potential curves. Action spectra of the photocurrent were changed with the mixing ratio α, but they were wide until about 1200 nm. Stabilization of $CdSe_\alpha Te_{1-\alpha}$ electrodes were tried by adding reducing agents (eg. Se^{2-}) into the solution.

Cell characteristics of an electrochemical photocell with the $CdSe_\alpha Te_{1-\alpha}$ mixed crystalline semiconductor electrode/redox solution junction were investigated from the viewpoint of solar energy conversion.

THIN-FILM PHOTOELECTROCHEMICAL CELL BASED ON A SPRAYED CdSe PHOTOANODE. <u>Chin-hsin J. Liu</u>, David R. Saunders, J. Olsen and Jui H. Wang, Bioenergetics Laboratory, Acheson Hall, State University of New York, Buffalo, New York 14214.

Photoelectrochemical cells have recently received considerable attention as an attractive alternative to all solid-state solar cells for solar energy conversion. We report here a photoelectrochemical cell based on thin film CdSe electrodes fabricated by spray pyrolysis on Ti substrates. The cell performance depends greatly on surface preparation: (i) Treatment of CdSe thin film with HCl or HBr improved the short-circuit current substantially, but H_2SO_4 and CF_3COOH have no effect. The possibility that halide ions incorporated into CdSe lattice may play a role in reducing surface recombination has been examined. (ii) Treatment of this cell under illumination and short-circuited condition with $Na_2S_2O_3$ solution improved both short-circuit current and open circuit voltage.

The energy conversion efficiency of this cell (Ti/CdSe/Na_2S-S-NaOH/Pt) thus measured was 9.6% at 680 nm monochromatic illumination and 7.4% under 71 mw/cm^2 white light.

Flat band potential was determined from current-voltage curve in the dark as -1.49 V vs. SCE.

This work was supported in part by a research grant from the National Science Foundation (PCM 775502).

CONTRIBUTED PAPERS

Session X

PHOTOELECTROSYNTHESIS AT
SEMICONDUCTOR ELECTRODES

Plenary Speaker: Arthur J. Nozik

Session Chairman: Kenichi Honda

Posters X-1 through X-13

UNPINNED BAND EDGES AT SEMICONDUCTOR-ELECTROLYTE INTERFACES. J. Turner, J. Cooper, and A.J. Nozik, Solar Energy Research Institute, Golden, Colorado 80401, USA.

A recent modification of the conventional model for photoelectrochemical reactions suggests that photo-generated minority carriers that have not reached thermal equilibrium within the semiconductor space charge layer (hot carriers) may, under special conditions, be injected into the electrolyte. In this way, more efficient conversion of optical energy into chemical energy may be possible; different mechanistic pathways for chemical reactions in the electrolyte also becomes available in the new model. Experiments to test for hot carrier injection were conducted by probing photoinduced redox reactions with redox potentials apparently outside the semiconductor band gap; these experiments showed such reactions were possible. However, these results could be better explained by the unpinning of the semiconductor band edges at the semiconductor-electrolyte interface, rather than by hot carrier effects. This unpinning effect is also an unexplored and very important phenomenon in semiconductor photoelectrochemistry. With p-type Si electrodes, the unpinning effect arises from the creation of an inversion layer at the semiconductor surface. The effect is enhanced by a surface oxide, making the photoelectrochemical cell behave very similarly to a metal-oxide-semiconductor (MOS) device. Capacitance measurements as a function of frequency and light intensity confirm the presence of an inversion layer and unpinned band edges. Recent work by Bard and Wrighton also proposes band-edge unpinning effects arising from high densities of surface states, and is attributed to "Fermi-level pinning". Work is continuing to probe for hot carrier effects in the absence of unpinned band-edges.

FORWARD CURRENTS ON ILLUMINATED n-TiO$_2$ ELECTRODES. a. R.Schumacher, H.-R.Sprünken and R.N.Schindler, Institut für Physikalische Chemie, Universität Kiel, 2300 Kiel, Olshausenstr.40/60, W.Germany.

Photoelectrochemical processes are repeatedly suggested to be utilized for the conversion of light into electrical and/or chemical energy. This study reports on photoreduction processes which can be obtained on cathodically polarized n-type rutile electrodes when reducible species are present in the electrolyte. It is shown that under specific experimental conditions the quantum efficiency η^c for that photoreduction process can reach values up to 1. This feature offers the interesting possibility of transferring light into storable chemical energy. To demonstrate this point reducible species were added to the electrolyte such as O$_2$,Fe^{3+},Ce^{4+},Fe(CN)$_6^{3-}$ and H$_2$O$_2$. The changes of photoreduction is investigated by varying the stirring speed and the amount of reducible species present in the electrolyte, as well as the wavelength and intensity of the incident light. An energy scheme is suggested to explain the mechanisms which are responsible for initiating the photoreduction. In this scheme a broad distribution of surface states is introduced which covers the forbidden gap of the semiconductor from a mid-gap position up to levels close to the lower edge of E$_{CB}$. It is assumed that this surface states mediate the transfer of electrons and holes from the illuminated electrode to the solution. The reaction channels are discussed.

COMPETITION REACTIONS AT α-Fe$_2$O$_3$ PHOTOANODES. <u>John H. Kennedy</u>, Ruth Shinar, and John Ziegler, Department of Chemistry, University of California, Santa Barbara, CA 93106

Oxidation of iodide and bromide ions at 1 a/o TiO$_2$-doped α-Fe$_2$O$_3$ (99.999% starting material) in competition with oxygen production was investigated in the pH range 9-14 in aqueous solution. The results were compared with those obtained at 0.1 a/o SiO$_2$-doped α-Fe$_2$O$_3$. At high pH the rate at which iodide ion reached the electrode surface was shown to be important because of a significant stirring effect. The fraction of photocurrent leading to iodine production approached 100% in 1F KI when the solution was stirred, but was only 30% for unstirred solutions. Over the range of iodide concentration (0-1.8F KI) the amount of oxygen collected decreased nearly linearly from <u>ca.</u> 40% to <5%. At pH 9 little difference between stirred and unstirred solutions was found. No bromine was produced at pH 13 but 40-60% of the photocurrent could be accounted for as bromine at pH 9. The amount of oxygen collected dropped rapidly to <5% even at bromide concentrations as low as 0.05F KBr. From these results it appears that adsorption and other surface effects play a significant role in the photochemical properties of α-Fe$_2$O$_3$. Photocurrents increased with time at pH 9 during iodide oxidation, and after this treatment, photocurrents in 0.1F NaOH containing no KI were up to three times higher (4.2 mA vs. 1.3 mA). The higher currents were stable until the electrode was heated to 130°C.

This work was supported by the Division of Chemical Sciences, Office of Basic Energy Sciences, U.S. Department of Energy.

PHOTOLYSIS OF WATER ADSORBED FROM THE GAS PHASE ONTO PLATINIZED SrTiO$_3$
<u>Roger G. Carr</u> and Gabor A. Somorjai, University of California and Lawrence Berkeley Laboratory, Berkeley, California 94720.

Light-assisted dissociation of water adsorbed from the gas phase on a metallized semiconductor surface has been observed. When a platinized single crystal of strontium titanate is irradiated with bandgap ultraviolet light in the presence of water vapor at two atmospheres and 120°C, hydrogen production is detected at a rate of as much as several hundred monolayers per hour, referred to the surface area of the crystal. In the range from 100°C to 150°C and at the saturation vapor pressure of water, production increases sharply, but appears to depend mostly on the increase of pressure. At a given temperature, but with less than saturation pressure, less production of H$_2$ is observed. Production of H$_2$ increases with increasing coverage of platinum; essentially no production is seen without the metal. This reaction appears analogous to the wet photoelectrochemical cell reaction, but we use no electrolyte, and run at temperatures and pressures heretofore unexplored. Moreover, we can make direct use of ultrahigh vacuum studies of the surface and molecules adsorbed on it, with and without illumination.

SOLAR ELECTROLYSIS BASED ON A GALLIUM ARSENIDE/CADMIUM SE-
LENIDE HETEROJUNCTION ANODE. Wm. Pinson, R. Howard, W. Mc
Curdy, Infrared Photo Ltd., Nepean, Ont. K2G 2W3, Canada

Theory indicates that a n-GaAs/CdSe anode, in contact
with a polysulfide electrolyte in a regenerative electrochemical solar cell illuminated with am2 radiation, is
capable of generating a photoemf$>$1.23V, the 20°C electrolysis threshold. A barrier of 0.9V is formed at the CdSe
(E_g=1.7eV) surface by contact with the electrolyte, so
that a photoemf $\cong$0.7V may be generated there. For the
heterojunction, the electron affinity difference between
GaAs (4.05eV) and CdSe (4.95eV) can give rise to a barrier =0.9V on the GaAs (E_g=1.4eV) side of the junction.
That portion of the am2 radiation reaching the GaAs barrier thru the CdSe film can cause a 0.7V emf at the heterojunction. Both voltages have like polarity and are additive, so that a total photoemf of 1.4V is possible.
Experimentally, the CdSe is deposited as a thin film on
the GaAs surface by a quasiequilibrium hot wall method of
evaporation. Anodes have been made having an emf of 1.0V,
the sum of 0.43V in the GaAs and of 0.57V in the CdSe, a
$J_{sc}\cong$5mA/cm^2, and a max efficiency $\cong$ 3%. These results
have been improved on recently.

Work supported by the Solar Energy Projects Office of the
National Research Council of Canada.

ON THE MECHANISM OF CHARGE TRANSFER AT THE n-TiO$_2$-ELEC
TROLYTE INTERPHASE IN THE PHOTOASSISTED OXIDATION OF
WATER. P. Salvador, Instituto de Catálisis y Petroleoquimica del C.
S. I. C., Serrano 119, Madrid-6, Spain.

The measurement of photocurrents vs. electrode potential at different
electrolyte pH shows the quantum efficiency for water photooxidation
to be pH dependent. This effect is interpreted in terms of the participation of surface states in the process of charge transfer between the
semiconductor and the electrolyte. Both surface structural hydroxyls
and physisorbed OH$^-$ ions are involved in this process, their surface
concentration depending on the pH. The following simplified tow-step
mechanism for hole capture and charge transfer at the s. c/electrolyte
interface is tentatively proposed. In the first step the participation of
surface hydroxyls (basic OH$^-$ groups) as traps for photogenerated holes, giving rise to the formation of OH^0 radicals, is invoked

$$Ti^{+4}\text{-}OH^-_{surf}+h^+ \longrightarrow Ti^{+4}\text{-}OH^0_{surf}$$

In a second step OH^0 radicals should act as intermediates for the tran
sfer of charge from the electrolyte bound states (OH$^-$ ions) to the s. c.

$$Ti^{+4}\text{-}OH^0_{surf}+OH^-_{ads} \longrightarrow Ti^{+4}\text{-}OH^-_{surf}+OH^0_{ads}$$

Finally, oxygen evolution might result from the reaction between OH^0
radicals

$$OH^0_{ads}+OH^0_{ads} \longrightarrow H_2O_{2ads} \longrightarrow 1/2\ O_2^\uparrow+H_2O$$

From the analysis of the experimental data a maximum density of surface centers for charge transfer of the order of 10^{14} cm^{-2} is obtained.

PHOTOELECTROLYTIC DECOMPOSITION OF WATER BY USE OF SrTiO₃ SINTER ELEC-
TRODES DOPED WITH VARIOUS METALS. M. Matsumura, M. Hiramoto, and H.
Tsubomura, Department of Chemistry, Faculty of Engineering Science,
Osaka University, Toyonaka, Osaka, 560 Japan.

Strontium titanate (SrTiO₃) is known to be one of the photoactive n-
type semiconductors with which water can be decomposed into oxygen and
hydrogen. However, its wide band gap allows us to utilize only the
ultraviolet skirt of the solar radiation. We have found that strontium
titanate sinter containing small quantities of dopants (metal oxides)
can work as photoanodes which decompose water photoelectrochemically by
light in the visible region.

The metals which showed such effects were Cr, V, Co, Ru, Rh, and Ce.
Those which showed very little or no such effects were Si, Ti, Mn, Fe,
Cu, Zn, Ga, Sr, Nb, Mo, Pd, Sn, La, Ta, W, Tl, Pb, and Bi. The photocurrent action spectra for the former group corresponded well with the
reflectance spectra of these materials. The photocurrents were of the
order of a few μA cm^{-2} under short-circuit condition for the light of
$\lambda > 400$ nm. Much larger photocurrents were observed by the same light
under anodic bias. For the case of Ru-doped sinter, photocurrent of 1
mA was observed for 10 hr with continued bubble formation. The gas,
after mass-spectrometric analysis, had the composition of $H_2 : O_2$ of 2 :
1, showing that water decomposition certainly occured. Bubbling was
also observed for Cr, Co, Ce, and V. The results suggest the formation
of dopant states in the band-gap region of the semiconductor, the excitation of which causes the photocurrents, yielding the decomposition of
water.

EXPERIMENTAL INVESTIGATION OF PHOTOELECTRODIALYSIS. George W. Murphy,
Solar Energy Research Institute, Golden, Colorado 80401 USA (Sabbati-
cal leave Sept. 1, 1979–Aug. 31, 1980 from the University of Oklahoma).

State-of-the-art electrochemical photovoltaic cells invented by Heller,
et al (Appl. Phys. Lett., 33, 521 (1978)) have been chosen for initial
testing of the photoelectrodialysis concept (G. W. Murphy, Solar Energy, 21, 521 (1978)). If C represents a cation- and A an anion-select-
ive membrane, the photoelectrodialysis cell

n-GaAs/0.18 M Na₂Se,(Na₂Se₂)-C-0.06 M NaCl(3)-A⎤

Pt/0.18 M Na₂Se,(Na₂Se₂)-C-0.06 M NaCl(1)-C-0.06 M NaCl(2)

upon irradiation of the n-GaAs leads to depletion of salt in Compartment 2 and enrichment in 3. (Na₂Se₂) means a small amount of diselen-
ide ion. The practical objective of this cell is solar driven desalination. Compartment 1 is a "buffer" compartment which keeps selenide
ions leaking through an imperfect membrane from reaching the product
water compartment. A six-compartment cell (not shown) containing a
bipolar membrane yields acid and base from salt. One objective of
this cell is solar energy conversion and storage, the latter strateg-
ically based on cheap raw materials salt and water. In the dark with
inert electrodes attached to an external circuit recombination of acid
and base generate electrical power.

Technical feasibility for both solar driven desalination and acid-base
production from salt has now been verified experimentally.

EFFECT OF BASIC MEDIA ON THE PHOTOFIXATION OF OXYGEN
ON POLYCRYSTALLINE TiO$_2$: THE ANODIC SYNTHESIS OF H$_2$O$_2$.
G. Munuera, J. A. Navio and V. Rives-Arnau, Dpto. Quimica Inorganica,
Facultad de Quimica, Universidad, Sevilla(Spain).

Photofixation of oxygen readily occurs on strongly hydrated polycrystalline TiO$_2$ samples giving H$_2$O$_2$, which slowly decomposes under uv-
-irradiation. These two electron demanding processes would decrease
the efficiency of TiO$_2$ anodes in photo-chemical cells. Basic OH groups
($\tilde{\nu}_{OH} \sim$ 3730 cm^{-1}) tightly attached at the TiO$_2$ surface were found to
be directly involved in this process. A study is now made on the con-
ditions to generate these hydroxyl groups, using water vapor (solid/
gas interface) or liquid water (solid/liquid interface) as reactants. I.r.
spectra and oxygen photofixation rates have been measured for the samples after different rehydroxylation treatments. Water vapor, even at
200C, was unable to restore hydroxyl groups on a thermally dehydro-
xylated TiO$_2$ surface, though chlorine "doping" of the surface is rather
effective to do so. Liquid water at room temperature and different pH
only produces rehydroxylation for pH > zpc of the TiO$_2$ (ca. 5.8). I.r.
spectra of these samples confirm that only in these conditions basic
OH groups are regenerated, restoring the oxygen photo-fixation capacity of the samples. Emphasis is made on the effect of these processes in the efficiency of TiO$_2$ anodes.

ENHANCED PHOTOCHEMICAL OXYGEN UPTAKE AND HYDROGEN PRODUCTION WITH
PLATINIZED CdS IN AQUEOUS MEDIUM. John R. Harbour, Robert Wolkow and
M.L.Hair, Xerox Research Centre of Canada, 2480 Dunwin Drive,
Mississauga, Ontario, Canada, L5L 1J9.

Photochemical redox reactions in pigment dispersions are of interest
for solar energy utilization. Cadmium sulfide is of particular interest as it absorbs a considerable amount of visible light and is known
to photochemically generate H$_2$O$_2$ in aqueous medium. Recently, we have
demonstrated that CdS can photochemically mediate the reduction of
methyl viologen to its cation radical (a precursor to hydrogen formation) at the expense of EDTA. Platinization of the CdS was accomplished and the effect of this surface modification on the photochemistry determined. It was found that an enhanced photochemical oxygen
uptake resulted with the platinized CdS and although this rate was 12
times greater, no H$_2$O$_2$ was observed. With methyl viologen present,
illumination resulted in hydrogen formation with platinized CdS demonstrating that the surface platinum is an active catalyst for the
production of hydrogen. Finally, a dark catalytic reaction with platinized CdS in the presence of formate was observed as monitored by
oxygen uptake.

ON THE MECHANISM OF THE PHOTOCATALYTIC REDUCTION OF CARBON DIOXIDE AND
FORMIC ACID ON TITANIUM DIOXIDE   Alain Monnier, Jan Augustynski and
Charles Stalder, Département de Chimie Minérale, Analytique et
Appliquée, Université de Genève, 1211 Genève 4, Switzerland

The photocatalytic reduction of carbon dioxide in the presence of
various n-type semiconductor powders has recently been reported by
Inoue et al. and independently by Halmann. An interesting aspect of
this reaction, leading essentially to the formation of formaldehyde
and methyl alcohol, is in the fact that it occurs equally at the semiconductors, such as TiO_2, WO_3 and Fe_2O_3, having the flat-band potential
more positive than the reversible potential of hydrogen electrode. In
the present work, the photocatalytic and electrochemical (in the dark)
reduction of both CO_2 and HCOOH was investigated using n-type polycrystalline TiO_2 electrodes. Cyclic voltammetric measurements performed in slightly acid solutions have shown that the reduction of both
CO_2 and HCOOH is definitely preferred to the hydrogen evolution and
takes place reversibly. The effect of such features as the carrier density (n-type) of TiO_2, the doping of TiO_2 with foreign elements and
the surface modification through selective metal deposition, on the
photocatalytic and cathodic reactions are discussed.

This work was supported by Swiss National Science Foundation

PHOTOCATALYTICALLY INDUCED FIXATION OF MOLECULAR NITROGEN BY NEAR U.V.
RADIATION ON RUTILE (TiO_2) Roger I. Bickley School of Chemistry,
University of Bradford, W. Yorkshire BD7 1DP U.K.   and Venkataraman
Viswanathan, Dept. of Chemistry, Lakehead University, Thunder Bay,
Ontario, Canada. P7B 5E1.

The interaction of U.V. radiation of energy at the optical absorption
edge of rutile (TiO_2) produces adsorbed hydrogen peroxide on rutile
surfaces containing adsorbed water. In vacuo the hydrogen peroxide
formation is accompanied by the formation of Ti^{3+} centres; whereas in
the presence of gaseous oxygen its formation is accompanied by the
photo-adsorption of oxygen. By outgassing the surfaces at progressively
increasing temperatures it is possible to demonstrate that the oxygen
photoadsorption activity is removed mainly through the disappearance of
the surface hydroxyl groups.

In photo experiments conducted in nitrogen or air, temperature pro-
grammed desorption spectra reveal the presence of a nitrogen containing
surface species which apparently desorbs from the surface as NO at
$\sim$500K. It is concluded that dinitrogen is oxidised by a very reactive
oxygen species which arises from the decomposition of hydrogen peroxide.

N_2 + O^*(ads) $\longrightarrow$ NO^*(ads) + N (ads)

N (ads) + O^*(ads) $\longrightarrow$ NO^*(ads)

X.P.S. measurements confirm the presence of a nitrogen containing
surface species with a N(1s) signal at 399.5 eV. This observation
together with the absence of e.s.r. signals suggests that the adsorbed
species is either NO^- (ads) or $N_2O_2^-$(ads)

HYDROCARBON PRODUCTION FROM BIOMASS BY PHOTOCATALYZED ELECTROLYSIS.
J. S. Cantrell, Chemistry Department, Miami University, Oxford, Ohio 45056, USA

The first step of this research is to ferment a suitable form of bio-mass (farm crop, municipal sewage, marine algae, etc.) to dilute organic acids in a nonsterile fixed packed bed fermenter. The organic acids are then removed from the fermenter and concentrated by extrac-tion. The concentrated acid solution is then electrolyzed to give aliphatic hydrocarbons via the Kolbe electrolysis or the photo-Kolbe electrolysis or a suitable combination of these two processes. The fermentation conditions and extraction procedure are selected based on the products desired. If liquid hydrocarbons are the desired product, the fermentation would be run to give primarily butyric, valeric, and caproic acids and the extraction would be run to remove these acids preferentially to the lower acids. This process has been studied using fresh water algae. The rate of conversion of algae to organic acids is very rapid. The rate decreases with increasing acid concen-tration, which is caused largely by pH inhibition. A kinetic study indicates that if the acid products could be removed as formed a 95% conversion of fermentable solids could be obtained in 6 days. With no removal of product acids the reaction was essentially complete in con-version to organic acids in 15 days. For the marine algal species the estimated maximum conversion is 40 - 44%, assuming each hexose molecule is converted to three acetic acid molecules. The semiconductor mate-rials studied are $ZnWO_4$, WO_3, and MoO_3 with platinum.

CONTRIBUTED PAPERS

Session XI

PHOTOCHEMICAL STORAGE

Plenary Speaker: James R. Bolton

Session Chairman: Wolfgang H. F. Sasse

Posters XI-1 through XI-11

LIMITS ON THE EFFICIENCY OF PHOTOCHEMICAL ENERGY
CONVERSION. <u>Robert T. Ross</u>, Department of Biochemistry,
The Ohio State University, 484 West 12th Avenue, Columbus,
Ohio, 43210 U.S.A.

The use of a single system to process photons of differing energy,
entropy limits, and an obligate irreversibility are known to limit
the efficiency of a single-threshold terrestrial flat-plate device to
33%. Even a modest rate of non-radiative decay within the absorber lowers the efficiency materially, and defines an optimal
absorption spectrum. Non-optimal absorbance lowers the efficiency further; maximal efficiencies (AM 1.5, 28^{0}C) for the absorption spectra of Si, GaAs, and chlorophyll (<u>in vivo</u>) are 26.4%,
28.8%, and 27.2%, assuming in each case that the rate of nonradiative decay is equal to the rate of radiative decay. Energy
conversion prior to thermalization within the absorber's electronic
states can complicate evaluation of efficiency limits, but it cannot increase the efficiency of an ideal device; plant photosystem-
I is an example of a non-thermal system. The need to stabilize
products against back reaction need not lower the efficiency materially, and one way to reduce such losses may be to have a
reaction which acts as a one-way valve, or rectifier. A catalyst
which responds cooperatively to reactant can accomplish this.
(Supported by DOE/SERI and NSF.)

PHYSICS CONSIDERATIONS OF SOLAR ENERGY CONVERSION. Alan F. Haught,
United Technologies Research Center, East Hartford, Connecticut 06108, USA

Radiant energy can be converted to useful work by either a thermal or a quantum
conversion process. Just as the Carnot cycle represents the limiting efficiency for
thermal energy conversion by a heat engine operating between any two
temperatures, thermodynamic considerations establish an upper limit to the efficiency with which a radiation conversion engine can derive useful work from radiant
energy of a given frequency spectrum and intensity. Ths paper develops a single formalism to analyze both the thermal and quantum conversion processes, displaying
their similarities and fundamental differences, and uses this formalism to determine
the limiting thermodynamic efficiencies of each. From the analysis the maximum
ideal thermodynamic efficiency of a single collector thermal converter with unconcentrated solar radiation and an ambient (reservoir) temperature of 300°K is 0.54;
for the same conditions the maximum ideal conversion efficiency of a single quantum system is 0.31. The analysis is extended to consider the effects of cascaded
operation, in which the reject heat of the quantum converter is used as the input to
a thermal bottoming cycle, and of multiple collectors and concentration on the solar
radiation on the conversion efficiency. The results obtained represent the ideal thermodynamic limits for radiant energy conversion by thermal and quantum processes,
and the caluclations with solar input serve as a reference against which to judge the
performance and capabilities of prospective solar energy conversion systems.

THEORETICAL EFFICIENCIES OF SOLAR PHOTOCONVERSION: DEPENDENCE ON TEM-
PERATURE, INTENSITY AND WAVELENGTH. R. V. Bilchak and J. S. Connolly,
Solar Energy Research Institute*, Golden, Colorado 80401, U.S.A. and
J. R. Bolton, Department of Chemistry, University of Western Ontario,
London, Ontario N6A 5B7, Canada.

We have applied the thermodynamic treatments of Ross and Hsiao (1) and
Bolton (2) to a variety of solar intensities and absorber temperatures
for air mass (AM) 1.2 solar flux. The solar intensity was varied from
1 to 10^4 suns at decadic intervals, and absorber temperatures ranged
from 300 to 500 K in 50 K steps. For these conditions, we calculated
the optimum thermodynamic power efficiencies (η_p) for single-photon
absorption and derived the optimum wavelength combinations for double
band-gap systems. The theoretical efficiencies are directly proportional to the log of the intensity at constant temperature and decrease linearly with increasing temperature at fixed intensity. We
have also extended this treatment to calculate the optimum absorber
wavelengths for multi-absorber systems ($3 \leqslant n \leqslant 8$) at fixed temperature and intensity. For $n > 5$, η_p approaches 60%. The results are
applicable to photovoltaic as well as photochemical and photobiological systems, and represent absolute (i.e., ideal) upper limits on conversion efficiencies analogous to Carnot efficiencies of heat engines.

1. R. T. Ross and T.-L. Hsiao, J. Appl. Phys. 48, 4783-4785 (1977).
2. J. R. Bolton, Science, 202, 705-711 (1978).

*A division of the Midwest Research Institute, operated for the U.S.
Department of Energy under Contract EG-77-C-01-4042.

PHOTOCHEMICAL STORAGE OF SOLAR ENERGY USING THE NORBORNADIENE-
QUADRICYCLENE INTERCONVERSION. Richard R. Hautala, R. Bruce King and
Charles Kutal, Department of Chemistry, University of Georgia,
Athens, Georgia 30602

A solar energy storage system based upon the interconversion of norbornadiene and quadricyclene possesses several attractive features,
including high specific energy storage capacity, long-term kinetic
stability of the energy-rich photoproduct at ambient temperatures, and
relatively inexpensive reactants. Two steps are required in this
cyclical system: (1) Energy storage through the sensitized photolysis of norbornadiene to quadricyclene; (2) Energy release through
the catalyzed reconversion of quadricyclene to norbornadiene. Introduction of the sensitizer and catalyst onto separate heterogeneous
supports facilitates the construction of an actual device in which
the energy storage and energy release steps are sequentially coupled.
An energy storage system based on these principles could result in
the use of solar energy for low grade (~100°C) heat applications such
as space conditioning and hot water production in buildings.

Recent results from studies involving structural variations of catalysts and developments regarding interfacial photoprocesses of heterogeneous photosensitizers will be discussed.

THE TEMPERATURE DEPENDENCE OF QUANTUM YIELDS OF SENSITIZED VALENCE
ISOMERIZATION.   THERMAL UPCONVERSION OF LOW ENERGY TRIPLET EXCITED
SPECIES. Guilford Jones, II, Xuan T. Phan, and Richard J. Butler,
Department of Chemistry, Boston University, Boston, MA 02215, U.S.A.

The isomerization of dimethyl 2,3-norbornadienedicarboxylate (NBD) to
its quadricyclene derivative sensitized by various energy transfer-
agents has been studied. Sensitizers having triplet excitation
energies >200 kJ/mol (>50 kcal/mol), such as benzophenone, 9-cyanophen-
anthrene, biacetyl, fluorenone, and camphorquinone, induced the
reaction with very high efficiency (limiting quantum yields = 0.6 -
0.7). The best long wavelength absorption edge for this group of
sensitizers was 550 nm. From the profile of rate constants for energy
transfer the lowest triplet state of NBD was located at approximately
220 kJ/mol (53 kcal/mol). A sensitizer, benzanthrone (E_T = 192 kJ/mol,
46 kcal/mol) requiring significantly endothermic triplet-triplet energy
transfer was employed successfully in the isomerization of NBD.
Quantum yields with benzanthrone sensitization showed a marked temperature dependence (for various NBD concentrations, two- to three-fold
increase between 5 and 92°) with limiting yields in the 0.2 - 0.3
range. The thermal "upconversion" of the 192 kJ/mol triplet of
benzanthrone was successful in raising the efficiency of a reaction
which stores 80 kJ/mol. The prospects for deploying the activated
processes of long-lived, low energy excited species at elevated temperatures are discussed.

This work was supported by the Office of Basic Energy Sciences, U.S.
Department of Energy.

PHOTOCHEMICAL STORAGE POTENTIAL OF AZOBENZENES   John Olmsted III and
Geary T. Yee, Department of Chemistry, California State University –
Fullerton, Fullerton, California, 92634

The trans-cis photoisomerization of azobenzene is endothermic, reversible, and clean; it therefore meets several of the criteria for photochemical solar energy storage. Major problems with the system are low
light absorption efficiency in the visible and a photostationary state
which strongly favors the lower energy trans form. Studies have been
carried out on a number of substituted azobenzenes and naphthalenes,
including o- and p-dimethoxyazobenzene, o-tetramethylazobenzene,
1,1'-azonaphthalene, and methyl orange (4-[4-dimethylamino]phenyl-
azobenzenesulfonic acid). A photocalorimetric technique has been used
in which the temperature rise accompanying illumination of the azobenzene solutions is monitored, compared with the temperature rise under
identical conditions for an inert absorber, and from these data the
storage efficiency can be computed. From this efficiency and tran-cis
photoisomerization quantum yields (also determined in this work along
with thermal reconversion rates) the overall endothermocities of the
isomerizations are calculated. The general findings are that (1) azo
compounds with a naphthyl group or an ortho substituent which is
H-bonding are not photoisomerizable and/or show rapid thermal reversion
rates; (2) the thermochemistry of the isomerization is somewhat sub-
stituent-sensitive and is approximately +100 kJ/mole; (3) photoisomerization quantum yields are of the order of 0.2 for most of these
compounds; (4) the photostationary state using visible light is
generally unfavorable. This work was supported by the U.S. Department
of Energy and in part by the Petroleum Research Fund of the A.C.S.

HIGHLY ENDOTHERMIC PHOTOCHEMICAL REACTIONS OF LANTHANIDE AND ACTINIDE
IONS IN SOLUTION. Terence Donohue, Laser Physics Branch, Naval
Research Laboratory, Washington, DC 20375.

We have succeeded in inducing the most endothermic photoreductions
ever observed in solution. These experiments have involved a number
of lanthanide and actinide ions in aqueous or alcoholic solution, with
various UV photolysis sources (excimer lasers, Hg lamps). The feature
most significant in these reactions is stabilization of the reduced
states with simple (poly)macrocyclic ligands, such as crown ethers and
cryptands. Most dramatic is the stabilization of Sm(II), in the most
endothermic reaction studied to date (nominal $E_0 = -1.55$ eV). In
methanol, the lifetime of this ion is less than 1 sec, but in the
presence of 18-crown-6 polyether or 2.2.2 cryptand, the lifetime is
extended to several hours. Other ions studied include Eu, U and Yb.
Quantum yields for photoreduction range from 60% (Eu) to about 10%
(Sm). Hydrogen is readily produced by the photochemical (and thermal)
oxidation of the reduced states at the photostationary state. In
alcohol solutions, H_2 and an aldehyde are produced photocatalytically;
in certain cases, oxygen is also produced in aqueous solutions. Complexation by macrocyclic ligands produces another benefit by shifting
the charge-transfer bands to longer wavelengths. For example, the
photolysis of Eu(III) (18-crown-6) at 351 nm is more effective than at
308 nm without the crown. Thus, macrocyclic ligands can increase
quantum yields and shift threshold wavelengths to lower energies,
features which are clearly useful in solar energy conversion schemes.

PHOTO-ENHANCED CATALYTIC DEHYDROGENATION OF 2-PROPANOL WITH HOMOGENEOUS
RHODIUM COMPLEXES. CHEMICAL STORAGE OF SOLAR ENERGY AS THE HEAT OF
HYDROGENATION. Hiroshi Moriyama, Toshiya Aoki, Sumio Shinoda and
Yasukazu Saito, Institute of Industrial Science, University of Tokyo,
Minato-ku, Tokyo 106, Japan

The liquid-phase dehydrogenation of 2-propanol is an endothermic and
exoergic reaction, with hydrogen removed spontaneously. We have found
that the catalytic activity is tremendously enhanced by photo-irradiation. From the viewpoint of solar energy storage, the present reaction
is particularly interesting for the following reasons. 1)The heat of
hydrogenation (15.0 kcal/mol) is stored as the products of acetone and
hydrogen, which can be utilized as thermal energy at higher temperature
than that of solar heat by operating a chemical heat pump system. For
this purpose, it is necessary to remove acetone from 2-propanol by
fractional distillation, where heat is removed, with entropy decreased.
2)Higher quantum yield than unity has been attained ($\Phi = 2.2$ for a rhodium-tin complex catalyst : 254 nm, 82.5 °C) as a consequence of the exoergic reaction, compensating the defect of solar energy (low density of
incident photons). 3)Photo-enhancement of the catalytic activity at
the visible region is more pronounced for a binuclear rhodium complex,
$[Rh_2Cl_2(CO)_2(Ph_2PCH_2PPh_2)_2]$, than the rhodium-tin complex. 4)Low activation energies have been obtained under photo-irradiation (2.6 and
1.7 kcal/mol for the rhodium-tin complex and the binuclear rhodium complex, respectively). A mechanism is suggested for the interpretation
of high quantum yield and low activation energy in the photo-enhanced
catalytic dehydrogenation of 2-propanol with rhodium-tin complex.

PHOTOCHEMICAL ELECTRON TRANSFER REACTIONS OF PHENYLATED CAGE COMPOUNDS
Toshio Mukai, Keiji Okada, and Katsuhiro Sato, Photochemical Research
Laboratory, Faculty of Science, Tohoku University, Aramaki, Sendai
980, Japan

Electron transfer reactions between the ground state of phenylated
bis-homocubane derivatives and the excited state of colored sensitizers
(pyrylium and trityl salts) or semiconductors (ZnO and CdS) or metal
ions (Ag^+, Ce^{4+}) have been studied. In all cases, the cycloreversion
reactions took place giving the corresponding cyclic dienes in good
yields. Among them, the photoreactions using triphenyl pyrylium salt
involved the singlet state of the dye and proceeded via a chain reac-
tion ($k_q\tau$ = 1.14 x 10^2/M, ϕ = 0.7 ~ 55 for 10^{-4} ~ 10^{-2} M solution of
the cage molecules). For the semiconductor-catalyzed photocyclorever-
sion of the phenylated cage compounds, a mechanism involving perticipa-
tion of molecular oxygen has been proposed. It was also found that
triethylamine, 1,4-diazabicyclo[2.2.2]octane, and 1,2,4,5-tetramethoxy-
benzene quenched the ring opening reactions. In dark, cerium (IV)
ammonium nitrate reacted with the phenylated cage compounds and gave
the corresponding diene derivatives. Combined with the well known
photocyclization of the dienes, we established novel types of the valence isomerizations between the cage compounds and the dienes which
are induced by electron transfer process.

This work was supported by the Ministry of Education, Science and
Culture in Japan.

PHOTOELECTROCHEMICAL CELLS WITH REDOX STORAGE. Peter G.P. Ang
and Anthony F. Sammells, Institute of Gas Technology, IIT Center,
Chicago, Illinois 60616, USA

A photoelectrochemical cell consisting of an n-type photoanode and a
p-type photocathode having the capability of redox storage will be
discussed. Simultaneous illumination of these electrodes results in
the generation of oxidized and reduced species, respectively, at the
positive (oxidant) and negative (reductant) compartments of the cell.
Direct chemical reaction between the two redox species is minimized by
the use of a cell separator. Photoelectrochemical cells will also be
discussed where the photocathode is substituted for an inert counter-
electrode.

Performance characteristics of n-$MoSe_2$ and n-WSe_2 as photoanodes, and
p-GaP, p-GaAs, and p-Si as photocathodes will be presented. In particular, n-$MoSe_2$ in the redox electrolytes 1M HBr + 1M Br_2 and 1M KI +
0.1M I_2 typically gave photopotentials between -400 and -550 mV. Larger
photopotentials were obtained from selected single crystals of n-WSe_2.
Photopotentials between 50 and 500 mV were obtained from the candidate
photocathodes. The p-GaP and p-GaAs were able to generate large photocurrents (about 30 mA/cm^2), although the photopotentials were usually
small. Photocharge of some redox-storage systems, e.g. $Br^-/Br_2//I^-/I_2$
by n-$MoSe_2$ /Pt as well as by n-$MoSe_2$/p-GaP photosystems have been performed. This system has an open-circuit potential between 450 and 500
mV and has demonstrated good storage capability. The photocharge and
discharge characteristics of the photoelectrodes in selected redox
electrolytes will be compared.

ON THE GENERATION OF ELECTROCHEMICAL ENERGY BY SOLAR
EVAPORATION. Helmut Tributsch, CNRS Meudon,1 Place
Aristide Briand, France.

Nature is harvesting large amounts of solar energy by
evaporating water from thin capillaries which are extending from the roots to the leaves of plants. The negative
pressure which is thereby produced serves to pump water
more than 100 meters high (sequoia trees) to desalinate
seawater (mangroves), or to extract, with - 80 bars,
humidity from arid land (desert shrubs). Reasons why this
energy conversion mechanism has not found technological
application are (1) the difficulty to produce and maintain
thin capillaries, (2) the problems associated with
sustaining negative pressure (nucleation) and (3) the
limited use for hydrostatic energy. We propose a solar
technology which is modelled on this natural phenomen
but utilizes different materials and directly converts
solar energy into electrochemical energy. It is based on
the solar evaporation of intercalated atoms or molecules
from layer type polycristalline materials (e.g. graphite.
transition metal dichalcogenides) and their subsequent
electrochemical transformation and reintercalation on the
opposite electrolyte contact. These new solar devices
would be simple to prepare and could be made of cheap and
abundant materials. Suitable electrochemical coupling
mechanisms and possible overall efficiencies are
discussed. Considerable fondamental research will be
needed especially on the kinetics of evaporation.

Index

A

Acceptor catalyst
  in electron-transfer reaction, 4
    simulation, 5
Alcohol, from plant-carbohydrate fermentation, 2
Algae
  chlorophylls of
    ENDOR spectroscopy, 45
    EPR spectroscopy, 43
Amino acids, in CHl_{sp}, 50
ATP, in chloroplast, 32
Auger electron spectroscopy (AES), of silicon impurities, 237

B

Back reaction, in electron transfer, 2
Bacteria
  photosynthetic
    light quanta conversion by, 29
    photoreaction center chlorophyllin, 31
Band intensities, origin of, 90
Bacteriochlorophylls, 32
Bacteriochlorophyll a
  bacterial reaction centers in, 62
  covalently linked pairs of, 56
    absorption maximum, 58, 59
  fluorescence lifetime of, 75-76
  structure of, 35
  water adducts of, 39
Bacteriochlorophyll b, esterification with farnesol, 63
Bifunctional ligands, chlorophyll interactions with, 36-37
Bilipid membrane, electron-transfer reaction in, 4
Biomass
  examples of, 28
  hydrogen generation from, 175

Biomimetic processes
  examples of, 29-30
  for solar energy conversion, 27-78
Blackbody radiation, 79

C

Carbohydrates, of plants, fermentation to alcohol, 2
Carbon dioxide, photosynthetic reduction of, 32
Carbon-paste electrode, 226
Carrier, for artificial electron transfer, 7
Catalyst
  oxygen-generating, 5
  simulation of, for electron-transfer reactions, 5-6
Cellulose
  gasification of, 28
  in oxygen photogeneration, 175
Chemical potential model, for energy storage, 306-315
Chlorobium chlorophyll, 63
Chlorophyll(s)
  adducts of, 36
  antenna type, 31-32
    models for, 62-69
  dimers, optical properties, 78
  donor-acceptor coordination properties of, 34
  emission spectroscopy of, 51
  ENDOR spectroscopy of, 45
  fluorescence of, 66-67
    lifetimes, 76
  free radicals from 39
  interaction with bifunctional ligands, 36-37
  laser activity of, 76-77
  light-harvesting types, 62-64
  magnesium in, 34
    coordination number, 76
  manganese in, 13

437

nucleophilic functions in, 34-36
in nucleophilic solvent, 34
oligomer, as model for antenna, 64-67
optical properties of, 31
  variation in 75
photoreaction center type, 31
photosynthetic properties of, 33-39
in photosynthetic unit, 31
-protein complexes, 50
  structure studies, 62-63
relative nucleophylic strengths of, 36
role in photosynthesis, 30
as sensitizer, 5
in solvents, optical properties, 64-65
water adducts of, 37-38
  IR spectrum changes, 37-38
  structure, 38
Chlorophyll *a*
covalently linked pairs of, 56
diasteromer in, 57
monomeric form of, 31
structure of, 35
  oligomer, 65
from *Tribonema aequale*, 32
Chlorophyll *b*, structure of, 35
Chlorophyll P740
photoreactivity of, 39
structure of, 38, 39
Chlorophyll special pair(s) (Chl$_{sp}$)
antenna model, 67-69
biomimetic, 50-59
  chemically linked, 55-59
EPR spectroscopy of, 41-46, 58
models
  covalent linkage, 57
  non-conservation of oscillator strength
    in, 52
  electron transfer, 59-61
  requirements, 56
optical requirements for, 49
photoreaction centers of comparison, 60
protein participation in, 49
self-assembled systems of, 50-55
  ease of preparation, 55
  fluorescence properties, 55
  IR and visible spectra, 53
  macromolecular orientation, 54
  spin-sharing properties, 55
  water importance in, 52
structure proposed for, 48, 49
Chloroplast
electron-transfer in, 5
manganese in, 13
photosynthesis by

in lamellae, 2
simulation, 1, 2
synthetic diagrams, 20, 21
Chromium bipyridine complex, redox potential
  of, 105
Chromium *tris*-phenanthroline complex,
  properties of, 104
Chromium polypyridine complex
  as redox reactant, 121
  properties, 122
Coal, as fossil solar energy, 1
Cobalt (II), in dinuclear complex, 183
Cobalt oxide, as catalyst for electron transfer,
  10
Cobalt complexes
  as electron acceptors, 179
  hydrogen generation using, 168-169
  in photogeneration of oxygen, 199
  as photosensitizer, 173
  redox potentials of, 177
Colloidal assemblies, used in redox catalysis,
  134
Color, origin of, in condensed matter, 91
Concentration photogalvanic cell, diagram of,
  207
Condon principle, 108
Copper (I), in dianuclear complex, 183
Copper complex, as photosensitizer, 173
Coupled redox processes
  current voltage diagram of, 150
  cyclic water splitting by, 151
Cyclic decomposition, of water, 149-155
Cytochrome, in electron-transport chain, 32
Czochralski float-zone technique, silicon
  crystal growth by, 233

D

Diborane, in artificial vesicles, 7
Differential galvanic cell
  diagram of, 207
  characteristics, 208
Diffusion reaction, in electron transfer, 10
Dinuclear cryptates, as redox catalysts,
  167, 182
Dioxane, chlorophyll linkage with, 36-37
Donor system, in electron-transfer reaction, 4
Dyes, sensitization of, in photoelectrosynthesis,
  287-288, 292
Dye laser, Chl$_{sp}$ use in, 67

E

Electrochemical photovoltaic cells, 243-267
  conversion efficiency of, 245-247

electrode stability in, 248-250
energetics of, 244-245
Electrodes, thionine-coated, 228
Electronically excited states, as redox
  reactants, 98-106
Electrons, hydrated, 89
Electron ejection, in photosynthesis, 32
Electron-pool effect, in hydrogen formation,
  144
Electron-spin-echo spectroscopy, description
  of, 46
Electron Transfer
  across membranes, 25
  band, 84
  in Chl$_{sp}$ models, 59-61
  by classical compounds, 107
  description of, 4
  free-energy relationships in, 110-114
  induction of, 2
  "inverted," 83
  photochemical induction, 97-129
  in photogalvanic cells, 202
  photosensitized
    cofactor effects, 9
    dependence on light intensity, 10
  representation of, 3
  on ruthenium oxide, 149
  semiclassical treatment of, 110
Electron-transport chain, in photosynthesis, 32
ENDOR spectroscopy, of chlorophylls, 45
Energy, excited-state, storage of, 2
Energy transfer, solar quanta collection and,
  79-95, 90
Entropy, of reorganization, 108
EPR spectroscopy
  of chlorophyll special pair, 41-46
  of photoreaction centers, 39-41
  in photosynthesis studies, 34
Ethanol, from biomass, 28
Ethylenediamine tetraacetate (EDTA)
  photosensitized reduction of, 6
Euphorbiaceae, hydrocarbons from, 2
Europium ammonium complexes, redox
  reactions of, 118
Europium ion
  excited state oxidation and, 121
  properties, 122
Exchange reaction, in viologen photoreduction, 8
Excitation energy, in photosynthesis, 31
Excited states
  applications of, 101
  lifetimes of, 98

F

Farnesol, bacteriochlorophyll c esterifica-
  tion with, 63
Fermi's golden rule, 109
Ferrous polypyridine complex
  as redox reactant, 119, 121
  properties, 120
Fibers, hollow, artificial photosynthetic
  reactions in, 19, 20
Flash photolysis, in studies of electron
  transfer, 111-112
Fluorescent concentrators, 84-90
Franck-Condon principle, 81, 100, 111
Free-energy relationships, in electron
  transfer, 110-114
Fresnel lenses, properties of, 94
Fuels, from solar energy conversion,
  problems in, 131-132

G

Galactolipids, in photosynthetic membrane,
  65
Gels, artificial photosynthetic reactions in, 19
Gel filtration, in vesicle preparation, 6-7
Genetic engineering, 28
Glucose, in oxygen photogeneration, 175
Green plants
  photobleaching in, 39
  photosynthesis in, 28
  as quanta converting machines, 2

H

Heptyl viologen
  in artificial vesicle, 8, 25
  photoreduction of, kinetics, 11
Hevea rubber tree, hydrocarbons from 2
Hole traps, 268
  chemical nature of, 259, 260
  mobility of, 261
Hydrocarbons, from plants, 2
Hydrogen
  artificial generation from water, 18, 21
    iron-sulfide cluster catalysts, 23-24
  generation of, in electron transfer, 5
  production by water reduction, 139-147,
    161-200, 203
  components of, 169
  modified, 173-175
  quantum yields, 172-173
  redox catalysts, 174
  simultaneous with oxygen, 179-183

I

Infrared spectroscopy, in photosynthesis
  studies, 34
Intrinsic reorganizational parameter, 110
Iodide, optical transitions of, 81
Iridium chlorine complexes, as photocatalytic
  agent, 127
Iridium *tris*-phenanthroline, properties of, 104
Iron ammonium complexes, redox reactions
  of, 118
Iron catalysts, in photogeneration of oxygen,
  176-177
Iron-iodine system, as photogalvanic cell,
  220-221
Iron-sulfur clusters, as redox catalysts, 182
Iron-sulfur protein, as hydrogen-generating
  catalyst, 5
Iron-thionine half-cell, studies on, 214-219
Iron-thionine system, as photostationary state,
  209
Iron *tris*-phenanthroline complex, properties
  of, 104
Isoelectronic exchange reaction, electron
  transfer as, 10

L

Lamellae, of chloroplast, photosynthesis by,
  2-3
Langmuir behavior, of photogalvanic cells, 211
Lanthanide compounds, J-levels of, 80
Laser flash technique, in hydrogen production
  studies, 142
Layered compounds, for photoelectro-
  chemistry, 287
Leucothionine
  electroactivity of, 215
  photoreactions of, 210, 215
Lewis base, photochemical generation of
  oxygen and, 16
Ligand field effects, 80,82
Lipid bilayers, redox studies on, 135
Liposome, formation of, 5
Luminescent traps, 86

M

Madelung potential, 80
  of ligand field theory, 91
Magnesium, in chlorophyll, 34
Magnesium (II) tetraphenylporphin, in
  photogalvanic cell, 222
Manganese

binuclear compounds of, 13
  EXAFS, 13, 14
  in dinuclear complex, 183
  role in photosynthetic oxygen evolution,
    12-17
  in superoxide dismutase, 70
Manganese complexes, photoreduction of, 179
Manganese IV oxide, intermediate,
  "hot" oxygen atom in, 17
Manganese porphyrin, 13
  conversion to manganese oxide, 15-17
  oxidation of, 19
Manganese protein, in electron transfer, 13
Marcus "inverted region," 111, 113, 129
Marcus theory, 110-112
  limitations of, 128
Mass-diffusion reaction, in viologen photoreduction, 8-9
Membrane-separated system, for photochemical water splittign, 180
Metal clusters, as redox catalysts, 182
Metal dithiolenes, as hydrogen-generation
  catalysts, 127
Methyl viologen (MV)
  in artificle vesicles, 7, 25
  EDTA-induced reduction of, 6
  modification of, 145
  photoreduction of, 19
    kinetics, 8
  radical cation, 11, 142
    decay kinetics, 143
  water reduction by, 139
Methly pyrochlorophyllide *a*, optical
  properties of, 63
Methylene blue, photoreactions of, 210
Micellar systems, for photochemical water
  splitting, 182
Micelles
  electron-transfer schemes in, 4
  redox intermediate stabilization by, 135-139
  schematic structure of, 136
Microemulsions for photochemical water
  splitting, 182
  for synthetic chloroplasts, 21
  electron-transfer rates, 23
Monolayer assemblies, for photochemical
  water splitting, 182
Monolayers, redox studies on, 135
Mott-Schottky equation, 251, 261, 267

N

NADPH, formation in chloroplast, 32
Nernst potentials, for thionine/leucothionine
  couple, 213

Nickel (II), in dinuclear complex, 183
NMR spectroscopy, in photosynthesis studies, 34
Noble metal oxides, use in oxygen-generating reactions, 147

O

Olefins, phosphine replacement by, 18
Optical electronegativity, 83
Organic dyes, in fluorescent concentrators, 87
Organic wastes, hydrogen production from, 175
Osmium complex, as photosensitizer, 173
Osmium tris-phenanthroline complex, properties of, 104
Oxygen
  atom ("hot"), generation of, 16-18
  evolution, in photosynthesis, 69
  generation of
    manganese-sensitized, 16-16, 19, 21
    natural catalyst for, 11
    schematic diagram, 12
  production from water, 18, 147-149
    modified system, 179
    photoinduced, 161-200
    scheme for, 175-179
    simultaneous with oxygen, 179-183
    system components, 176-177

P

Paramagnetic species, in green plants, 39
Peroxide, generation of, by manganese catalyst, 25
Petroleum, necessity for replacement of, 2
Phase boundaries, electron transfer across, by chloroplast, 5
1, 10-Phenanthroline complexes
  in excited state electron transfer, 103
  properties of, 104
Pheophytins
  reinsertion of Mg into, 56
  structure of, 35
Phospholipids
  in bilipid vesicles, 4-5
  in electron-transfer reaction, 4
Photobleaching, in green plants, 39-40
Photochemical cell, for water-splitting, 162
Photochemical diodes, as semiconductor devices, 26
Photochemical energy storage, 297-339
  examples of, 326-328
  model comparison, 301-316
    summary, 315-216

nonidealities in real systems for, 322-326
thermodynamic limits of, 299-310
thermodynamics of, 316-322
Photochemical electron transfer reactions in homogeneous solutions, 97-129
Photoelectrochemical cells, stabilized regenerative type, properties of, 249
Photoelectrochemical devices, for solar energy conversion, 201
Photoelectrochemistry
  cells for, see Photoelectrosynthetic cells
  energy conversion by, 272-275
  particulate systems in, 285-286
Photoelectrolysis, efficiency of, 200
Photoelectrosynthesis
  dye sensitization in, 287-288
  layered compounds in, 287
  new developments in, 275-288
  at semiconductor electrodes, 271-295
Photoelectrosynthetic cells
  band-edge un-pinning in, 275-280
  classification of, 273
  energy-level diagrams for, 274
  potential advantages of, 274
Photogalvanic cells, 201-228
  comparisons among, 204
  criteria for, 204-209
  efficiency of, 215-216, 226
  electrode kinetics of, 205-207
  electrode selectivity in, 211-214
  electron-transfer reactions in, 202
  improvement of, 219-220
  iron-iodine system as, 220-221
  outlook for, 221-223
  output of, 216
  photocurrent potential curves at, 218
  photostationary state and, 209-210
  recent work on, 209-219
  schematic diagram of, 202
  storage capacity of, 203
  sunlight-engineering efficiency of, 219-220
  two-compartment type, 219
  types of, 207
Photogalvanovoltaic cell, studies on, 222-223
Photogalvanic effect, definition of, 201
Photosensitizers, lifetime of excited states of, 7-8
Photostationary state, homogeneous processes and composition of, 209-210
Photosynthesis
  artificial systems for, 1-26, 28
    biomimetic, 27-28
  quantum efficiencies, 24
  efficiency of, 77-78

modeling of, 30
oxygen evolution in, 69-70
Photosynthetic bacteria
chlorophylls of
ENDOR spectroscopy, 45
EPR spectroscopy, 43
Photosynthetic organisms, light energy
conversion in, 30-33
Photosynthetic unit (PSU)
chlorophyll role in, 31, 33-39
concept of, 30-31
as electron pump, 32
mechanism of action of, 32
schematic representation of, 33
Photosystems I and II, 4
Photovoltaic cells
chemical aspects of, 229-241
electrochemical, 243-267
solar efficiencies of, 239
in concentrated sunlight, 240
solar energy use by, 84
Phthalic acid, as trap for peroxide, 25
Phytol, in bacteriochlorophyll a, 63
Plants
hydrocarbons from, 2
quantum conversion in, 3
Plastoquinone, in artifical vesicles, 7
Platinum
as catalyst
for electron transfer, 10
in phosphine oxide generation, 16, 19
colloidal, as redox catalyst, 139, 147, 203
ultrafine sols of, 142
Platinum electrodes, cyclic voltammetry curves
of, 214
Polyelectrolytes, redox studies on, 135
Polynuclear cryptates, as catalysts, 167
Polyoxometalates, as redox catalysts, 182
Polypyridine ligands, transition metal
complexes with, 103-106
Polyvinyl alcohol, as colloidal platinum
stabilizer, 140-141
Porphyrines
models of, electron tranfer in, 61
quantum efficiency of, 200
Porphyrin sensitizer, 138
Potassium tantalate, as photoelectrode, 184
Primary acceptor, of ejected electrons, 32
Proflavine
in galvanic cell, 203
as photosensitizer, 173
Proton, movement, in electron transfer, 24
Proton carrier, in artificial vesicles, 7
Pyrazine, chlorophyll linkage with, 36-37

Pyropheophorbide in Chl_{sp}-antenna models, 67
Pyrochlorophyll a
configuration of, 48
pair
optical properties, 58
proton resonances, 58
in special pair model, 76
structure of, 35

Q

Quantum
energy of, fraction converted by plants and
bacteria, 29
harvesting of, 79-95
simulation of conversion of, 1-26
yield, in hydrogen production, 141
Quantum chemistry, 80
Quaterpyridine, as redox catalyst, 183
Quinone(s)
as artificial electron carrier, 7
use in membranes, 24, 26
Quinone-hydroquinone redox system, in
artificial vesicles, 19, 26

R

Redox catalyst
bifunctional, water splitting by, 152
for hydrogen photogeneration, 174-175
Redox intermediates, stabilization by
micellar systems, 135-139
Redox reactants, electronically excited states
as, 98-106
Redox reactions, photoinduced, 131
Redox couples, effect on electron-transfer
process, 24
Reisfeld concentrator
advantages of, 88
theory of, 84-85
Rhodamine B, in photogalvanic system, 221
Rhodium cluster, as electron-transfer catalyst,
11
Rhodium complex
as photo sensitizer, 173
redox potentials of, 170
Rhodium tris-phenanthroline, properties of,
104
Russell-Saunders coupling, 80
Ruthenium (II)
in dinuclear complex, 183
oxidizing characteristics of, 88
Ruthenium ammonium complex, redox
reactions of, 118

Ruthenium bipyridine complex
  as photosensitizer, 6, 19, 24
    kinetics, 10
  redox potential of, 105
  as photosensitizer, 133
Ruthenium (II) complexes, 272
  cyclic decomposition of water by, 149-155
  in hydrogen generation, 171-173
    schematic representation, 171
  laser flash photolysis of, 146
  in photoelectrochemistry, 286, 288
  in photogeneration of oxygen, 177-179
  as photosensitizers, 140, 142, 152
    for hydrogen evolution, 168
  redox potentials of, 170, 177
Ruthenium dioxide, as redox catalyst, 176
Ruthenium-iron system, as photogalvanic cell,
    221
Ruthenium oxide, use in oxygen generation,
    147-148
Ruthenium polypyridine complex
  as electron-transfer photosensitizer, 103, 123
    wavelength limitations, 125
  as redox reactant, 119, 121
    properties, 120, 122
Ruthenium-sulfur complexes, catalytic
    properties of, 126
Ruthenium tris-phenanthroline complex,
    properties of, 104
Rydberg transitions, 81

                        S

Saturation of dangling bonds, in semi-
    conductors, 237
Schottky barrier diode, 232
Secondary ion mass spectroscopy (SIMS),
    of silicon impurities, 237
Semithionine, from photogalvanic cells,
    209-210
Semiconductor(s)
  doped, in photosynthetic processes, 189
  in electrochemical photovoltaic cells
    electrolyte interface, 245
    energetics of, 244-245
  electrodes
    derivatized, 282-285
    photoelectrochemical characterization, 251
    photoelectrosynthesis at, 271-295
  electronic properties of, 231
  impurities in grain boundaries of, 237
  in solar cells, 231
  n-type
    surface reactions of, 255-261

p-type
  hot-electron injection from, 280
  surface reactions, 262-265
  as photoactive supports, 183-187
  photochemical diodes as, 26
  redox properties of, 186
Sensitizer, in electron-tranfer reactions, 5
Siemens process, for silicon purification, 234
Silicon
  deep-level impurities in, 236
  sheets, growth of, 235-236
Silicon cells, practical use of, 84
Silicon photovoltaic cell, energy transfer by, 86
Silicon solar cell, 229
  cost of, 230-231
  description and types of, 230
  efficiency of, 230
  fabrication of, 233-237
  silicon preparation for, 233-235
Single-cell system (SCS), for photochemical
    water splitting, 180
Solar cells
  description of, 229-230
  fabrication of, 233-237
  ideal, 298
  operation of, 231-233
  theoretical conversion efficiency of, 245-247
Solar energy
  coal as fossil type of, 1
  in photovoltaic cells, 84
Solar energy conversion
  biomimetic systems for, 27-78
  fuel production by
    problems in, 131-132
  photoelectrochemical devices for, 201
  of photovoltaic cells, 239
    practical, requirements for, 271-272
  redox reactant for, 102
  selecting systems for, 100
  to stored chemical energy, 2
Solar furnaces, tracking equipment of, 88
Solar photochemistry, 88-90
Stokes' shift, 94, 95
Stokes threshold, 85, 87, 88
Storage, of photochemical energy, 297-339
Strontium titanate
  as photoelectrode, 184, 185, 188
  stability of, 198
Sunlight, as fuel source, 22
Superoxide dismutase, 69
  manganese in, 70
  in oxygen evolution in photosynthesis, 77
"Supersensitizers," experiments on, 126
Surface reactions

of n-type semiconductors, 255-261
of p-type semiconductors, 262-265
Surfactant
    as basis for photosensitizer, 6, 25
    tetrapyridyl porphyrin as, 13, 15

T

Tetrapyridyl porphyrin, synthesis and
    structure of, 13-15
Thermodynamic model, for energy storage,
    302-306
Thermodynamics, of photochemical energy
    storage, 316-322
Thionine, photoreactions of, 210-228
Titanium dioxide, in bifunctional redox
    catalyst, 151, 153, 159
Titanium trichloride, as electron acceptor, 179
Tobacco chloroplasts, optical properties of, 66
Transition-density coupling of chlorophyll *a*
    oligomer, 65
Transition metals
    electron-transfer spectra of, 83
    oxidation state of, 89
Transition metal complexes, 98, 99
    energy transfer by, kinetic aspects, 106-119
    with polypyridine ligands, 103-106
    as redox reactants, 102
Transition metal dichalcogenides,
    as semiconductors, 241
Transmission coefficient, in energy
    transfer, 108, 109, 115, 128
Tribonema aequale, chlorophyll from, 32
Triphenylmethane dyes, in photogalvanic
    systems, 221
Triphenylphosphine (TPP)
    formation of, 16, 17
    as possible oxygen acceptor, 16
Triphenylphosphine oxide, formation by
    oxygen transfer, 16
Tris (pyrochlorophyllide *a*) 1, 1, 1
    tris(hydroxymethyl) ethane ester,
    structure of, 68-69
Trivich and Flinn nonradiative model,
    for energy storage, 301-302

U, V

Uranyl atom, energy provided by, 79

Uranyl glass systems, 86
    practical use of, 87
Uranyl ion, in Reisfeld concentrator, 84-85
Vesicles
    bilipid type
        artificial, 4, 6-7
        preparation, 7-8
    cross section of, 8
    electron-transfer schemes in, 4
    kinetic experiments on, 8-11
    manganese porphyrin photo-reaction
        with, 15
Vibrational levels, involvement in energy
    transfer, 109
Viologen(s)
    chain length of, 23
    derivative, as electron relay, 136
    photoelectrochemical reduction by,
        262, 264

W

Water
    adducts of, with chlorophyll, 37-39
    artificial hydrogen generation from, 18
    cleavage by visible light, 155, 272
    cyclic decomposition of, 149-155
        schematic device for, 154
    as fuel source, 22
    hydrogen from, photoinduced, 161-200
    oxygen production from, 147-149
        photoinduced, 161-200
    photochemical oxidation of, 10
    redox reactions of, redox potentials for, 164
    reduction of, 139-147
    splitting of, photoinduction, 131-160
Water-splitting cell, photochemical, 162, 163

X, Z

Xenon, optical transitions of, 81
X-ray fluorescence spectroscopy (EXAFS)
    of cinuclear manganese compounds, 13, 14
Zinc, in vesicle preparation, 7
Zinc porphyrin derivative, as photosensitizer
    structure of, 133
Z-scheme, diagram of, 3